Measuring Roots

Stefano Mancuso
Editor

Measuring Roots

An Updated Approach

Springer

Editor
Prof. Dr. Stefano Mancuso
Dpt. Plant, Soil & Environment
University of Firenze
Viale delle idee 30,
Sesto Fiorentino (FI) – Italy
stefano.mancuso@unifi.it

ISBN 978-3-642-22066-1 e-ISBN 978-3-642-22067-8
DOI 10.1007/978-3-642-22067-8
Springer Heidelberg Dordrecht London New York

Library of Congress Control Number: 2011940210

Printed on acid-free paper

Springer is part of Springer Science+Business Media (www.springer.com)

Preface

The real complexity of and adult root system can be barely conceived if we think that one single plant of rye excavated by Dittmer (1937) consisted of 13,815,672 branches and had a length of 622 km, a surface area of 237 m^2 and root hairs for 11,000 km. Furthermore, this complex network of widespread roots and subtle rootlets is laid bare in the soil and can be recovered from it only with great struggle. Thus, it is easy to understand why determining the position, the area, the degree of branching and other root characteristics has been for centuries a peculiarly difficult problem.

Roots represent half of the plant body: possibly the most interesting. This invisible part of the plant spreads widely through the soil and adsorbs the water and nutrients that, together with the carbon dioxide taken from the air, represent the material out of which the world's food supply is manufactured by. They give anchorage to the plant, frequently accumulate reserve foods and in some cases also represent a reproductive organ. Furthermore, according to Charles Darwin (1880), roots are the "anterior" pole of the plant, characterized by "brain-like" characteristics in opposition with its posterior end bearing the organs of sexual reproduction.

Despite the obvious importance for the whole plant, our comprehension of the root apparatus has been for long time annoyingly limited, mostly due to inadequacy in the techniques available. This situation just recently changed thanks to the advancement in visualization and measurement of roots that resulted in a significant progress of our understanding of the architecture and behaviour of the plant's hidden half. However, this information is spread across many specialized journals and, consequently "out of the sight" for many more applied-oriented scientists. On the contrary, many agronomy-based papers and books dealing with various aspects of root methods have been missed by more "academically oriented" colleagues. This book represents an attempt to combine both academic and practical component of this topic in one volume, making this book a universal handbook for any researcher or extension person interested in aspects of root methods. The most updated innovations in root visualization and analysis and the most advanced techniques of observation, architecture and behaviour of the root are described in detail and discussed. *Measuring root* has been written for a rather broad audience, from professional academics to undergraduate students at tertiary institutions and extension people interested in practical aspects of growing crops.

The volume consists of 18 chapters grouped in two main parts, namely:

1. Lab methods
2. Field methods

which should answer the needs of a large audience.

In the end, the editor gratefully acknowledges the many contributors of the chapters here presented, the financial support given to the University of Florence by the *Fondazione Ente Cassa di Risparmio di Firenze* and the support of Dr. Andrea Schlitzberger, at Springer, for the invaluable guidance during the production of the book.

Stefano Mancuso

References

Darwin C (1880) The power of movement in plant. John Murray, London, UK

Dittmer HJ (1937) A quantitative study of the roots and root hairs of a winter rye plant (Secale cereale). Am J Bot 24:417–420

Contents

List of Contributors

Mariana Amato Department of Crop, Forest and Environmental Sciences, University of Basilicata, Via Ateneo Lucano, 10, Potenza 85100, Italy, mariana.amato@unibas.it

Elisa Azzarello Department of Plant, Soil and Environment, University of Florence, viale delle Idee 30, 50019 Sesto Fiorentino (FI), Italy

Craig Barton Forest Resources Research, NSW DPI, P.O. Box 100, Beecroft, NSW, 2119, Australia

Giovanni Bitella Department of Crop, Forest and Environmental Sciences, University of Basilicata, Via Ateneo Lucano, 10, Potenza 85100, Italy, ileven@tiscali.it

Stephan Blossfeld Institute of Bio- and Geosciences, IBG-2: Plant sciences, Forschungszentrum Jülich GmbH, 52425 Jülich, Germany, s.blossfeld@fz-juelich.de

John R. Butnor USDA Forest Service, Southern Institute of Forest Ecosystems Biology, Southern Research Station, 705 Spear Street, South Burlington, VT 05403, USA, jbutnor@fs.fed.us

Yang Cao School of Forest Sciences, University of Eastern Finland, P.O. Box 111, 80101 Joensuu, Finland, yang.cao@uef.fi

Jan Cermak Institute of Forest Botany, Dendrology and Geobiocenology, Mendel University, Zemedelska 3, 613 00 Brno, Czech Republic

Reinhart Ceulemans Department of Biology, University of Antwerpen, Campus Drie Eiken, Universiteitsplein 1, 2610 Wilrijk, Belgium

Randy Clark US Plant, Soil and Nutrition Laboratory, USDA-ARS, Cornell University, Ithaca, NY 14853, USA

Nuno Conceicao Instituto Superior de Agronomia, Technical University of Lisbon, Tapada da Ajuda, 1349-017 Lisboa, Portugal

Masako Dannoura Graduate School of Agriculture, Kyoto University, Kitashirakawa Oiwake-cho, Sakyo-ku, Kyoto 606-8502, Japan; Graduate School of Agricultural, Kyoto University, Kitashirakawa oiwake-cho, sakyo-ku, Kyoto 606-8502, Japan, dannoura@kais.kyoto-u.ac.jp

Teresa S. David Instituto Nacional de Recursos Biológicos, Av. da República, Quinta do Marquês, 2780-159 Oeiras, Portugal

Jorge S. David Instituto Superior de Agronomia, Technical University of Lisbon, Tapada da Ajuda, 1349-017 Lisboa, Portugal

Frank P. Day Department of Biological Sciences, Old Dominion University, Norfolk, VA, 23529, USA

Michal Dohnal Faculty of Civil Engineering, Czech Technical University in Prague, Thákurova 7, 166 29 Prague, Czech Republic

Nicola Everitt The Centre for Plant Integrative Biology, University of Nottingham, Sutton Bonington Campus, Loughborough LE12 5RD, UK; Materials, Mechanics and Structures Division, Faculty of Engineering, The University of Nottingham, University Park, Nottingham NG7 2RD, UK, nicola.everitt@nottingham.ac.uk

Suqin Fang Root Biology Center, South China Agricultural University, Guangzhou 510642, China

Maria Isabel Ferreira Instituto Superior de Agronomia, Technical University of Lisbon, Tapada da Ajuda, 1349-017 Lisboa, Portugal

Andrew French The Centre for Plant Integrative Biology, University of Nottingham, Sutton Bonington Campus, Loughborough LE12 5RD, UK, andrew.p.french@nottingham.ac.uk

Karibu Fukuzawa Field Science Center for Northern Biosphere, Hokkaido University, 250 Tokuda, Nayoro 096-0071, Japan, caribu@fsc.hokudai.ac.jp

Dirk Gansert Department of Ecology and Ecosystem Research, University of Göttingen, Untere Karspüle 2, 37073 Göttingen, Germany, dganser@gwdg.de

Karl Gartner Department of Forest Ecology and Soil, Research and Training Centre for Forests, Natural Hazards and Landscape, Seckendorff-Gudentweg 8, Vienna 1131, Austria

Roman Gebauer Institute of Forest Botany, Dendrology and Geobiocenology, Mendel University, Zemedelska 3, 613 00 Brno, Czech Republic

Giuliano Mosca Department of Environmental Agronomy and Crop Sciences, University of Padova, Viale dell'Università 16, 35020 Legnaro, Padova, Italy

Said Attia al Hagrey Department of Geophysics, University of Kiel, Otto-Hahn-Platz 1, 24118 Kiel, Germany, sattia@geophysik.uni-kiel.de

Kurt H. Johnsen USDA Forest Service, Southern Institute of Forest Ecosystems Biology, Southern Research Station, 3041 Cornwallis Rd, Research Triangle Park, NC 27713, USA

Yuji Kominami Kansai Research Center, Forestry and Forest Product Research Institute, 68, Nagaikyutaro, Momoyama, Fushimi, Kyoto 612-0855, Japan

Vincenzo Lapenna CNR-IMAA Contrada S. Loja - Zona industriale C.P. 27, 85050 Tito Scalo, Potenza, Italy, vincenzo.lapenna@imaa.cnr.it

Hong Liao Root Biology Center, South China Agricultural University, Guangzhou 510642, China, hliao@scau.edu.cn

Martin Lukac Department of Agriculture, Development and Policy, University of Reading, Whiteknights RG6 6AR, UK, m.lukac@reading.ac.uk

Naoki Makita Graduate School of Agricultural, Kyoto University, Kitashirakawa oiwake-cho, sakyo-ku, Kyoto 606-8502, Japan

Stefano Mancuso Dpt. Plant, Soil & Environment, University of Firenze, Viale delle idee 30, Sesto Fiorentino (FI) – Italy, stefano.mancuso@unifi.it

Marianna Bandiera Department of Environmental Agronomy and Crop Sciences, University of Padova, Viale dell'Università 16, 35020 Legnaro, Padova, Italy

Elisa Masi Department of Plant, Soil and Environment, University of Florence, viale delle Idee 30, 50019 Sesto Fiorentino (FI), Italy

Daniel G. Milchunas Long-Term Ecological Research program, Colorado State University, Ft. Collins, CO 80523-1499, USA, Daniel.Milchunas@Colostate.edu

Anthony N. Mucciardi Tree Radar, Inc., 512 Ashford Road, Silver Spring, MD, 20910, USA

Sergio Mugnai Department of Plant, Soil and Environment, University of Florence, viale delle Idee 30, 50019 Sesto Fiorentino (FI), Italy, sergio.mugnai@unifi.it

Valeriy Nadezhdin Institute of Forest Botany, Dendrology and Geobiocenology, Mendel University, Zemedelska 3, 613 00 Brno, Czech Republic

Nadezhda Nadezhdina Institute of Forest Botany, Dendrology and Geobiocenology, Mendel University, Zemedelska 3, 613 00 Brno, Czech Republic, nadezdan@mendelu.cz

Hiroyuki Oguma Center for Environmental Measurement and Analysis, National Institute for Environmental Studies, 16-2 Onogawa, Tsukuba, Ibaraki, 305-8506, Japan

Harry Ozier-Lafontaine INRA, UR 1321 Agrosystèmes Tropicaux Domaine Duclos, 97170 Petit-Bourg, Guadeloupe, France, harry.ozier-lafontaine@antilles.inra.fr

Camilla Pandolfi Department of Plant, Soil and Environment, University of Florence, viale delle Idee 30, 50019 Sesto Fiorentino (FI), Italy, camilla.pandolfi@unifi.it

Susanna Pollastri Department of Plant, Soil and Environment, University of Florence, viale delle Idee 30, 50019 Sesto Fiorentino (FI), Italy

Tony Pridmore The Centre for Plant Integrative Biology, University of Nottingham, Sutton Bonington Campus, Loughborough LE12 5RD, UK; School of Computer Science, University of Nottingham, Jubilee Campus, Wollaton Road, Nottingham NG8 1BB, UK, tony.pridmore@nottingham.ac.uk

Tapani Repo Joensuu Unit, The Finnish Forest Research Institute, P.O. Box 68, 80101 Joensuu, Finland, tapani.repo@metla.fi

Roberta Rossi Department of Crop, Forest and Environmental Sciences, University of Basilicata, Via Ateneo Lucano, 10, Potenza 85100, Italy, roberta.rossi_sky@alice.it

Rachel Schroeder Department of Biological Sciences, Old Dominion University, Norfolk, VA 23529, USA

Hideaki Shibata Field Science Center for Northern Biosphere, Hokkaido University, 250 Tokuda, Nayoro 096-0071, Japan

Raimo Silvennoinen Department of Physics and Mathematics, University of Eastern Finland, P.O. Box 111, 80101 Joensuu, Finland, raimo.silvennoinen@uef.fi

Alexia Stokes Mixed Research unit in Peant Architecture, Botany and Bioinformatics, INRA, CIRAD, PS2 TA/A51 2 Bld de la Lironde, 34398 Montpellier cedex 5, France

Daniel B. Stover Earthwatch Institute, North America Regional Climate Center, Edgewater, MD, 21037, USA

Lyudmila G. Tarshis University of Ekaterinburg, 620102, Belorechenskaj str., 9/4–27, Ekaterinburg, Russia, tarshis@etel.ru

Galina I. Tarshis University of Ekaterinburg, 620102, Belorechenskaj str., 9/4–27, Ekaterinburg, Russia

Miroslav Tesař Institute of Hydrodynamics of the ASCR, v.v.i., Pod Paťankou 30/5, 166 12, Prague, Czech Republic

Teofilo Vamerali Department of Environmental Sciences, University of Parma, Viale G.P. Usberti 11/A, 43100 Parma, Italy, teofilo.vamerali@unipd.it

Darren Wells The Centre for Plant Integrative Biology, University of Nottingham, Sutton Bonington Campus, Loughborough LE12 5RD, UK, darren.wells@cpib.ac.uk

Part I
Lab Methods

Chapter 1
Higher Plants: Structural Diversity of Roots

Lyudmila G. Tarshis and Galina I. Tarshis

Abstract At the present time, the necessity of accumulating information about structure diversity of roots and root systems of the species existing on Earth has occurred, due to development both of theoretical basis and methods of preservation of plant biodiversity. There are few publications on structural features of plant roots in botanical literature. Moreover, comparative anatomical studies of roots of higher plants are far behind the researches on structure of shoots. Due to this fact, so far there is an opinion among botanists about the structure uniformity of roots of higher plants, and root systems. We are not going to consider all the causes for lag of comparative anatomical studies on higher plant roots in botany. This was done as long ago as in 1960s by (Comparative plant anatomy, Chapter 7. Root. Holt, Rinehart and Winston, NY. pp 94–101), American anatomist. Let us remind just of the main causes:

- Weak knowledge about intraspecific variation of root structure
- Technical difficulties in collecting root samples, the same as shoot samples, etc.
- Deficiency of monographs on rhizology, with comparative anatomical studies of roots, similar to those of other plant organs

We had taken into account these causes in our long-standing rhizological researches from 1974 to 2010.

L.G. Tarshis (✉) • G.I. Tarshis
University of Ekaterinburg, 620102, Belorechenskaj str., 9/4 – 27, Ekaterinburg, Russia
e-mail: tarshis@etel.ru

S. Mancuso (ed.), *Measuring Roots*,
DOI 10.1007/978-3-642-22067-8_1,

1.1 Rhizological Research: Material, Methods, and General Principles

Base material for the analysis of structure diversity of roots of rhizophytes was collected in natural populations of species during our expeditions to the Baltics, Ukraine, Russia (Ural, Siberia, Altai, and Far East). Along with wild-growing plants, we examined root structural features of introducents, which grow in the Russian botanic gardens. We have analyzed a total of 1,200 species of higher plants, which belong to various taxa and biomorphs.

We extracted root systems from substrate using the method of dry excavation along the trench walls. Trench depth varied from 15 cm to 3 m, depending on the mechanical soil characteristics. We excavated roots of ten mature blossoming or sporificating plants of each species. The samples were described according to the biomorphological characteristics by Serebryakov's (1962). We washed the roots, sketched them or photographed them. After that we fixed them in 73% *ethanol* for further anatomical examination. Every new root sample was included into the collection of rhizomes.

In laboratory by microtome, or manually by razor, we made cross-sections of ten roots of each species through the basal zone, middle zone, and apical zone. We examined and prepared microslides under the optical microscope according to anatomy standards (Kivenheimo 1947; Voronin et al. 1972). We used the ocular micrometer to measure root microstructures. To evaluate microstructure variation we applied the variation coefficient CV%. This value helped to compare features with different characteristics. Along with that, we developed structure models of roots in a form of graphical schemes. We used map symbols to mark topographic zones, systems of tissues, and some specific root structures, which had been found during microscope examination of the cross-sections. To create schemes we used computational microscope and the "Paint" program. A brief description of anatomic features was provided for each scheme.

1.2 Root System of a Plant: Specifics of Root Variability Manifestation

All higher plants, rhizophytes, are characterized by structure variability of roots, though to a different degree. To a less degree structure variability manifests itself among primary homorhizic plants, which belong to the following divisions: Lycopodiophyta, Equisetophyta, and Pteridophyta. Conventionally, we named this form of intraspecific variability of root structures as endogenous variability. It means variation of structure features among the cognate roots of a specimen.

Among present-day spore-bearing plants the endogenous variability is most pronounced in the representatives of the division Equisetophyta. We have studied five species of horsetails from natural populations of various geographical areas of

Russia (*Equisetum arvense* L., *E. pratense* L., *E. sylvaticum* L., *E. hyemale* L., *E. fluviatile* L.), and have found a distinct dimorphism of adventitious roots. In all these species, the so-called extension roots and sucking roots can be distinguished in root systems of each specimen. These roots are conspicuous. Extension roots are thick, look longer, and have positive geotropism. They are few in number and vegetate one by one from plagiotropic rhizomes. *Sucking* roots are very thin and short; they vegetate in groups from rhizome nodes, and form small fibrils. Comparative anatomical analysis allowed to reveal at them characteristic structural features similar structure features, typical for these types of roots in all examined horsetails. Comparison of cross-sections of the extension roots showed that they are 3.5–3.7 times thicker than the sucking roots. This occurs due to the growth of wider multilayered primary cortex. It has external and internal zones with large aerenchyma cavities. For example, primary cortex in extension roots of *E. arvense* is 4.3–4.6 times wider than the sucking roots cortex (Fig. 1.1). The stele size is only 1.8–2 times bigger in the extension roots. That is why the number of xylem and phloem strands is 5–6 in steles of the extension roots, while it is not exceeding 4 in sucking roots. General amount of the tracheary elements reaches 12–14 in extension roots, and it is only 4–5 in sucking roots. Together with the distinctions between roots of two morphotypes, there are a number of similar identities among different species of horsetails. For example, all these roots develop the single-layered rhizodermis, whose cells have reddish membranes, and are divided into trichoblasts and atrichoblasts. All the horsetails roots have sparse, though long, root hairs. Their paucity seems to compensate by means of absorbing hairs, which appear on the epidermis cells of rhizomes. For example, in *E. arvense* a number of such tubular hairs varies from 15 to 22 (CV% = 11.7) along the perimeter of cross-section of the plagiotropic rhizome.

The representatives of Lycopodiophyta demonstrate root dimorphism less distinctly than horsetails. Root differentiation in a specimen root system has been found among species of the following genera: *Lycopodium* L. and *Diphasiastrum* Holub. (Tarshis 2007). The most specific anatomical features of roots of club moss are presented in the basal zone, where plagiotropic shoot produces the root. Here a root has a wide, multilayered cortex, differentiated into three zones, and

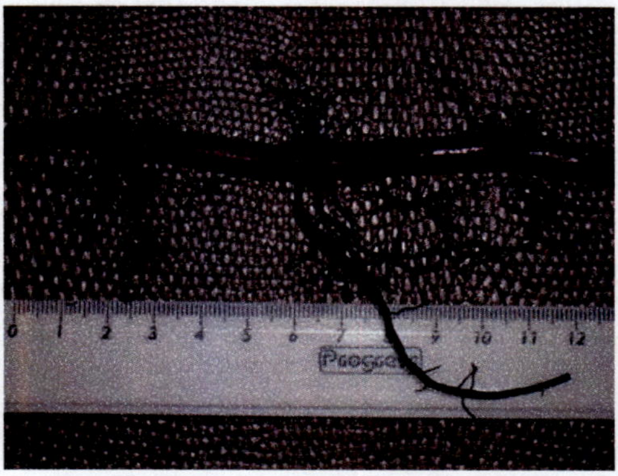

Fig. 1.1 Dimorphism roots in primarily homorhizophytic root system of a *Equisetum fluviatile* L

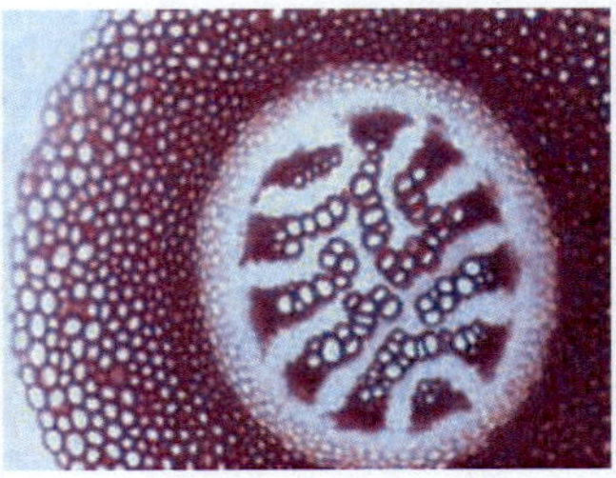

Fig. 1.2 A cross-section cut of an adventitious root of *Lycopodium clavatum* L

relatively small plectostele, which is similar in its structure to stele of shoot (Fig. 1.2).

We have registered two types of different by origin adventitious roots, in a root system of specimens, growing in the introduction conditions, which belong to the genus *Selaginella Beauv.: S. apoda* (L.) Fern., *S. emmeliniana* Van Geert., *S. kraussiana* (G.Kunze)A.Br., *S. vogelii* Spring. These two types are shoot-borne roots, which occur on the lower side of plagiotropic shoots, and rhizophore-borne roots, which occur on the apical tips of the orthotropous rhizophores. It must be emphasized that, notwithstanding the different origin, rhisophore- and shoot-borne roots are anatomically identical and have specific structures: protostele, tertiary structure of endodermis cells with wide Casparian strips, and long root hairs on the thin-walled cells of rhizodermis.

Among the representatives of the division Pteridophyta, in general, endogenous variability of roots manifests itself only in small variations of organs' thickness and length, and in dimensions of certain microstructures. Hardly ever can we find more significant differences in specimens of some species, which occur between thick aerial roots and thin roots, which grow in substrate. We can observe this among tropical epiphytic ferns, belonging to the genus *Platycerium* Desv.

The seed plants, allorhizophytes, as W. Troll called them (1949), show endogenous variability in the most distinct way. Representatives of the divisions Pinophyta and Magnoliophyta are known to undergo various underground root metamorphoses extremely frequently. Their roots greatly differ in their exterior and microstructure. As a rule, structure diversity of such roots is coming from their functions. For example, aerial roots of many representatives of the families Orchidaceae and Bromeliaceae are being developed in environment different than soil, and have quite specific structure. Contractile roots, which grow from the perennial bulb stem of *Lilium martagon* L., and draw it into substrate, differ as well. We will not recite various metamorphoses of roots of seed plants, which are resulting from their long adaptation to the certain environment. This problem was fully considered by many morphologists before (Serebryakov 1962; Tarshis 1975). But it must be emphasized that as well as root metamorphoses, seed plants undergo various shoot metamorphoses: rhizomes, stolons, tuber, and bulb. There are also metamorphoses of mixed shoot–root nature in seed plants, e.g., caudices. Many of these organs greatly resemble roots. That is why in the process of study of structure diversity of roots we identified the morphology of underground organs, which comprise the root system of each generative specimen, beforehand. For this purpose, we used method of anatomical diagnostics. We also took into account that roots may vary themselves within a species' group of plants. Due to such a careful approach, we established the root system structure of two types in species belonging to the division Magnoliophyta: isomorphic and heteromorphic (Tarshis and Tarshis 1998).

For example, species of the subfamily Pyroloideae are found to have isomorphic type of root system structure, while species of the various taxa of the division Magnoliophyta are found to have heteromorphic type of root system structure.

Isomorphic type manifests itself most distinctly in four species: *Pyrola rotundifolia* L., *Orthilia secunda* (L.) House, *Moneses uniflora* (L.)A.Gray, and *Chimaphila umbellata* (L.)W. Barton. Isomorphic type was found in all the specimens from 35 populations under study, located in various regions of Russia. We found that four species, which grow in wide range of environmental conditions, develop the uniform secondary homorhizic root system, which consists of numerous adventitious roots, growing from the plagiotropic stolon-like rhizomes (Fig. 1.3). Structure features of both roots and rhizomes in these Pyroloideae are alike and unique. Anatomical features of the underground organs show great stability and low level of variation. Roots and rhizomes of this taxon species do not change their specific anatomical features, although demonstrate the miniaturization of structures, even in the Far North, on the Yamal Peninsula, in extreme environmental conditions. Due to this characteristic, it is possible to describe features of these roots, using just one single structure model. The uniformity of the inner structure of the stolon-like rhizomes is represented on the second structure model.

Another type of structure of root systems, heteromorphic, was found in natural populations among many Magnoliophyta species. For example, great structural diversity of root systems was registered in generative specimens in the populations of *Lupinaster pentaphyllus* Moench and *Sanguisorba officinalis* L. (Fig. 1.4).

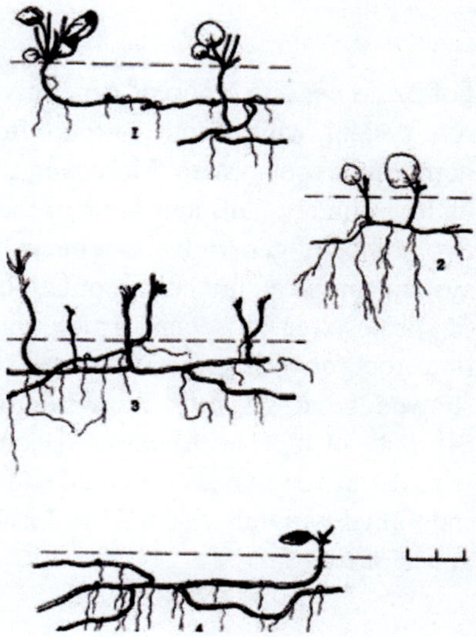

Fig. 1.3 Isomorphic type structural organization root system at species: (1) *Pyrola rotundifolia* L., (2) *Moneses uniflora* (L.) A.Gray, (3) *Orthilia secunda* (L.)House, (4) *Chimaphila umbellata* (L.) W. Barton

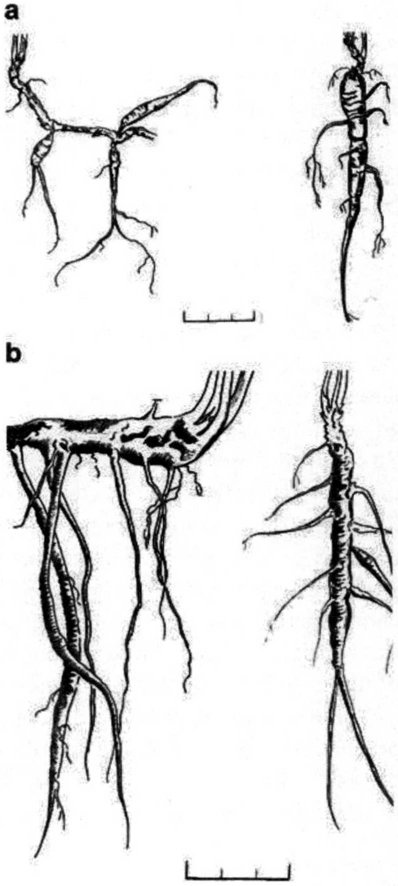

Fig. 1.4 Heteromorphic type of the structural organization of root systems at species: (**a**) *Lupinaster pentaphyllus* Moench, (**b**) *Sanguisorba officinalis* L

For example, one group of generative specimens *L. pentaphyllus* had an allorhizic root system, while another group, from the same cenopopulation, had a secondary homorhizic root system. Moreover, these specimens showed great polymorphism of underground organs, and some of the roots changed into tuber roots. There are also two groups of generative specimens with allorhizic and secondary homorhizic root systems in the nature cenopopulations of *S. officinalis,* distributed in different parts of species area. In second group, there were big rhizomes and numerous adventitious roots as well, turned into tuber roots. Microstructures of roots and root systems showed great stability, alongside with the clearly pronounced intrapopulation diversity of morphostructures. Detailed evaluation of variation of the anatomical features of roots in seed plants, showed a very narrow range of manifestations of individual variability (CV% = 2.2–13.2), registered in various phytogeographical zones of Russia.

1.3 Roots and Root Systems: Species-Level Study of Structure Diversity

Comparative assessment of intraspecific variability level of anatomical features of the different taxa roots showed their significant stability. This fact convinced us that root anatomical features, due to their uniformity, can be used to distinguish the structure features complexes, which reflect their species characteristics. Sporophytes have the structure features complexes with minimal quantity of anatomical features, whereas spermatophytes have complexes with maximal quantity of anatomical features. For example, there are features of rhizodermis, primary cortex, and stele, which are the same in various environmental conditions; this peculiarity has been distinguished among the species belonging to the genera *Lycopodium* L. and *Diphasiastrum* Holub. The rhizodermis structure features are manifested in the root hairs quantity, which appear from the trichoblasts one by one, by two, by three, and so on. Thus, *Lycopodium dubium* Zoega plants have root hairs appearing by one at a time, while *Lycopodium annotinum* L. have root hairs appearing by two at a time. Usually, the structure features of a species' primary cortex are coming from its differentiation into zones, occurrence of aerenchyma cavities and their sizes, disposal of mechanical tissues (collenchyma and sclerenchyma). For example, primary cortex of the *Lycopodium clavatum* L. roots has three zones: exterior, middle, and interior. Exterior zone has 4–5 layers, and it is formed by small parenchymal cells. Middle zone is formed by thin-walled cells, stretched in radial direction; breakage of their membranes results in formation of aerenchyma cavities and/or air space between zones. Interior cortex has many layers. It surrounds the polyarch stele in the root center, like heavy similarly to a casing from sclerenchyma. It is necessary to note that stele of the basal zone in adventitious roots of the species, belonging to the genera *Lycopodium* and *Diphasiastrum,* has a strong resemblance to stele of plagiotropic rhizomes, where these roots grow from. In this type of stele, there are primary xylem and phloem strands, arranged in a form of peculiar curved bands. We have borrowed this term to denote similar structure models of steles of shoots and roots among the species of the class Lycopodiopsida and named them "plectostelic" (Tarshis 2003).

We have developed the "haplostelic model" for the aerial adventitious roots of the tropic *Lycopodium carinatum* Desv. A one-layer rhizodermis with numerous and very long (up to 700 μg). There are also cells with thickened and hardened membranes in the exterior zone of the cortex. Interior zone of the cortex is formed by thin-walled cells. There are two rounded steles, divided by a narrow parenchymal diaphragm in the center of most aerial roots. Inside of each stele there is a single, bend-like, curved strand of primary xylem, surrounded by the primary phloem cells from all sides.

We have found that plagiotropic shoots, orthotropic rhizophores, and adventitious roots of some species of the class Isoëtopsida have the most primitive stele structure. Species *Selaginella apoda* (L.) Fern., *S. emmeliana* Van Geert, *S. kraussiana* (G.Kunze) A.Br., *S. vogelii* Spring have central strand of xylem tracheary elements, solidly encased by the primary phloem elements. All these species have a strong similarity in the inner structure of adventitious roots, shoots, and rhizophores. We have developed the "protostelic" structure model for them. There are vessels in metaxylem in roots with archaic protostele. Such a mixture of primitive and advanced root structures is considered to be a result of an early isolation of this group from other higher plants taxa during the evolutionary process.

Five wild species of the division Equisetophyta are found to have the minimal root structure diversity. Special feature of *Equisetum arvense* L., *E. pratense* L., *E. silvaticum* L., *E. hyemale* L., *E. fluviatile* L. is development of a thick, multilayered system of underground segmented shoots of two morphological types: plagiotropic and orthotropic rhizomes. It should be noted that all the species of horsetails are common to have poor developed adventitious roots of two morphotypes as well: thin sucking roots, and extension roots, which are thicker and longer. Roots of both types are incapable of secondary growth. All the roots are covered outside with one-layered rhizodermis, consisting of big cells, atrichoblasts, and short complanate trichoblasts. Cell membranes are brownish colored. All the species have long, but sparse root hairs, growing from trichoblasts. The primary cortex is built of parenchymal cells. Cell layers quantity varies from 4 to 10 in the root cortex of different horsetail species. Cells laying under rhizodermis (epidermis of root) have thick and lignification cellular walls other cortex cells layers are built of thin-walled parenchymal cells with light membranes. There are no mechanical tissue elements in cortex of all these species, but aerenchyma cavities occur. The endodermis is one-layered and consists of 10–11 rectangular cells with the Casparian strips. The stele is round, small, and inlaid with a narrow layer of pericycle cells. Primary xylem is exarch, and represented by a diarch ray or single tracheary element in thin roots. In thick roots a triarch, tetrarch, or pentarch xylem is developing in a form of rays, between which the phloem strands are arranged (Fig. 1.5). These root peculiarities of the species from the genus *Equisetum* L. are displayed in the "haplostelic" and the "actinostelic" models.

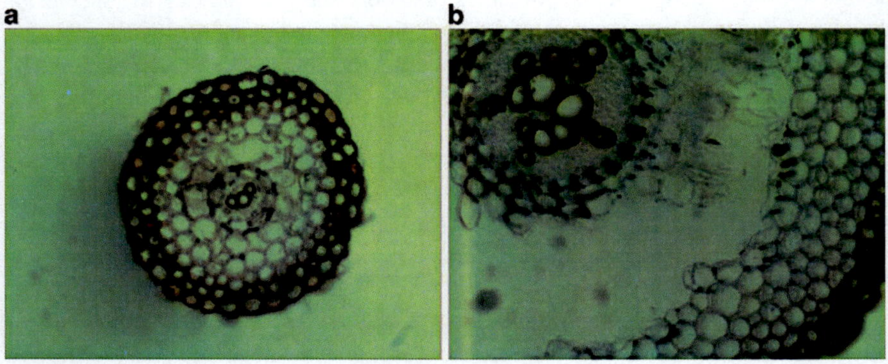

Fig. 1.5 Cross-section cuts of a roots horsetail (*Equisetum fluviatile* L.): (**a**) sucking root, (**b**) extension root

Analysis shows that the rhizome structure features, typical for the species, are more distinct than structure features of roots. Manner of the one/two endodermis layer arrangement in rhizomes is a very important diagnostic feature. For example, *E. silvaticum* L. has one sinuous endodermis layer, situated above the conducting bundles outside, while another one seems to wrap each of them inside. *E. fluviatile* L. has each conducting bundle, surrounded by own, or individual, endodermis. *E. arvense* L. and *E. pratense* L. have only one outer layer of endodermis in rhizomes. We have taken into account these particular structure features, when developing the three "artrostelic" model modifications for rhizomes of the various species of the genus *Equisetum* L.

We discovered that the division Pteridophyta species have the maximal diversity various microstructure features and feature complexes, which describe the specifics of the sporophytes' root structure. There were 80 species from 28 families of the classes Ophioglossopsida, Marattiopsida, and Polypodiopsida under study. Among them, samples of 37 species had been collected in natural populations, while samples of 43 species had been grown from spores. Fourteen structure models have been developed to describe the roots structure diversity of the division Pteridophyta species (Fig. 1.6). The following features have been included into the feature complexes:

- Presence or absence of root hairs and their characteristics
- Presence of mycorrhiza in the cortex cells
- Specifics of rhizodermis structure
- Specifics of primary cortex: differentiation manner, occurrence of aerenchyma and mechanical elements, quantity of layers, endodermis structure
- Specifics of stele: presence of pericycle and quantity of its layers, arrangement and quantity of the primary xylem and phloem strands

Roots of the species from the families Ophioglossaceae (R.Br.)Agardh and Botrychiaceae Nakai, belonging to the class Ophioglossopsida, are found to have

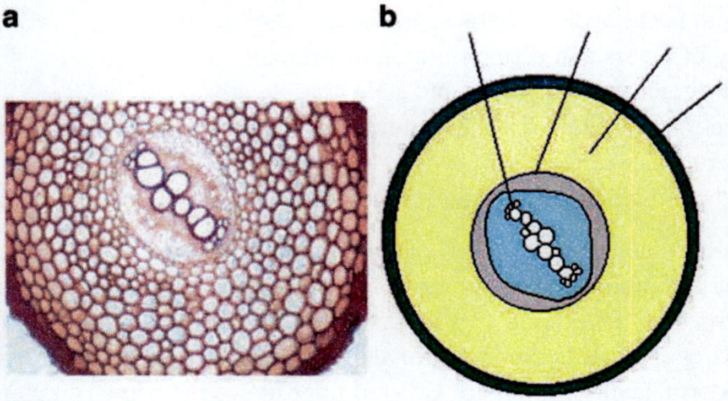

Fig. 1.6 A cross-section cut (**a**) and structural model (**b**) of a root fern *Matteuccia struthiopteris* (L.)Tod

a structure similarity, as well as great structure difference. Among the similarities there are the following:

- Absence of root hairs and presence of hyphae in the primary cortex cells
- Development of a wide, multilayered cortex, which is not divided into internal and external zones, etc.

Main differences are in the stele's structure features. For example, stele in the *Ophioglossum petiolatum* Hook. roots can be described as the simplest and the most primitive structure. In the stele center, which is called haplostele, there is an arcuated primary xylem strand, surrounded by the phloem continuous mantle. Other species, such as *Botrychium lanceolatum* (S.G.Gmel.)Ångstr., *B. lunaria* Sw., and *B. multifidum* (S.G.Gmel.)Rupr., have different structure, which is more specific; this structure is called actinostele, and is characterized by radial alternation of the primary exarch xylem and phloem. These differences became the basis for development of two structure models. A "haplostelic" model has been worked out for the roots belonging to species of the family Ophioglossaceae (R.Br.)Agardh, and "actinostelic" model has been worked out for the roots belonging to the species Botrychiaceae Nakai. Adventitious roots of the species *Angiopteris palmiformis* (Cav.)C.Chr., *A. polytheca* Tardien et C.Chr., and *A. crassipes* Wall., belonging to the class Marattiopsida, have a comparatively small stele and a very wide, multilayered cortex (32–45 layers). The stele has tubular structure, due to the pith's parenchyma cells in its central part, and primary xylem radial rays, whose quantity is up to 10–15. We have worked out a "siphonostelic" model for such a kind of root structure of the class Marattiopsida species.

The species of the class Polypodiopsida are found to have the greatest variety of root structures diversity. This class is known to have various life forms and ecological groups of plants, most of them grow at tropical and subtropical latitudes. That is why root structure peculiarities have been studied mostly in terms of species from the greenhouses of Russian botanical gardens. Twelve structure models have been developed during the interpretation of the results of comparative anatomical studies of roots. Root structure diversity of this taxon species manifests in as follows:

- Root hairs peculiarities, characteristics, and multitude
- Quantity of layers and differentiation of primary cortex
- Ratio of dimensions of cortex and stele
- Presence or absence of the coloration of cortex cells membranes with phlobaphenes
- Occurrence of aerenchyma cavities, mechanical tissues, and hyphae in cortex

Besides structure diversity, there are a number of similar features in the species of the class Polypodiopsida. Among these is presence of actinostele in the roots of all the species, belonging to this taxon. Most often a diarch actinostele was occurring (Fig. 1.7). It is common for these species to develop lateral roots from those endodermis cells, which are situated opposite to the primary xylem rays.

Among the representatives of this class, cortex without differentiation is being met in roots of the species *Osmunda vachellii* Hook., belonging to the order

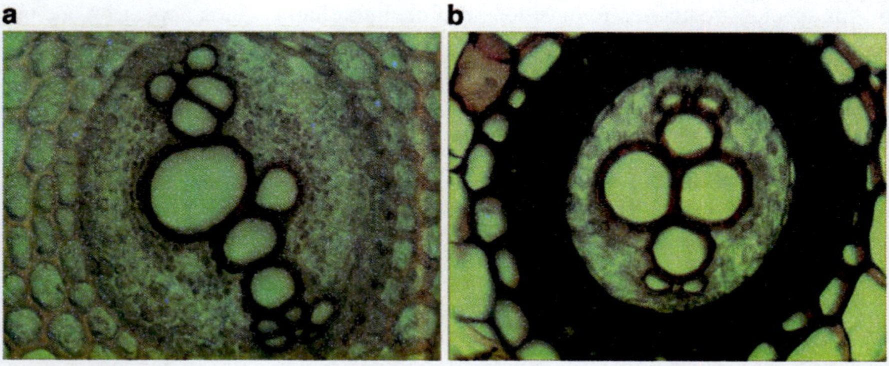

Fig. 1.7 Cross-section cuts stele of roots: (**a**) *Athyrium filix-femina* (L.)Roth, (**b**) *Pteridium aquilinum* (L.) Kuhn

Osmundales. On the contrary, in species of the order Schizaeales, root cortex divides into interior and exterior zones. Structure diversity is expressed most distinctly in structure features of water, morass, and epiphytic species of the order Schizaeales. For example, in roots of the genus *Ceratopteris* Brongn., there are very large aerenchyma cavities, located in cortex and separated by narrow, one-row diaphragms, consisting of parenchyma cells. Different in shape hexagon stele has been met in the species *Lygodium japonicum* (Thunb.) Swartz. Such a form of stele is conditioned by peculiar structure of endodermis cells, which engirdle the stele. Strong thickening of cell membranes in the exterior zone of primary cortex is a characteristic feature of roots of the epiphytic species *Vittaria flexuosa* Fee and *Pellaea viridis* (Forsk.)Prantl.

In the four species of the genus *Platycerium* of the order Polypodiales, just primary cortex interior zone shows process of significant hardening (lignification), while the seven to eight layered cortex exterior cells have thin and light membranes. It must be emphasized that in most species of this order, cortex inner cells have the thickened membranes, and are saturated with phlobaphenes (reddish-colored tannins oxidation products), rather than lignin. In many species belonging to the order Cyatheales, root cortex is divided into two zones, though thickening and saturation with phlobaphenes occurs only in inner zone. It is necessary to note that some species groups, belonging to this order, such as *Blechnum brasiliense* (Desv.) T. Moore, *Doodia dives* Kunze, etc., do not show differentiation of cortex at all. At the same time, in roots of *Onoclea sensibilis* L. there are cell membranes, which are thickening and browning in that cortex layer, which underlays the rhizodermis; it means that roots of this species may have exoderm.

Roots of amphibian and aquatic species from the family Marsileaceae are characterized by highest peculiarity. For instance, *Marsilea quadrifolia* L. entirely lacks root hairs, and there are numerous (up to 20) large (up to 140 μg in diameter) aerenchyma cavities. *Salvinia natans* (L.)All. do not have roots at all, which are reduced due to water environment adaptation. At the same time, mature specimens of this species are noticed to have long multicellular hairs, which appear by one on the rhizome epidermis cells (Fig. 1.8).

On study of root structure diversity of seed plants species, first, we analyzed structure features of their root systems. Development of heterogenous root system was found in all examined species of the classes Cycadopsida, Ginkgopsida, Pinopsida, Gnetopsida, belonging to the division Pinophyta, and of the class Magnoliopsida, belonging to the division Magnoliophyta. As a part of such root system there are primary root, lateral roots, and adventitious roots, functioning at the same time.

Structure diversity of roots, comprising the species root systems of the division Pinophyta, was studied both in wild and cultured species. For example, one primary root and two shoot-borne roots were found in the biennial specimens *Cycas revoluta* Thunb., raised from seed. Ten-year specimens had primary root, lateral roots, coralloid roots, and about 20 adventitious contractile roots. They differ both in external and internal structure. For example, primary root and first-order lateral roots demonstrated secondary growth (Fig. 1.9). Coralloids have blue-green algae and other symbionts, found in the layer of parenchyma cells, which are situated between exterior and interior zones of primary cortex. All the examined species of the division Pinophyta are noticed to have great structure diversity of roots, comprising root systems. For this reason, on development of structure features complex, we compared the even-aged specimens' roots of the same origin and position in root system. To discover specific root features of seed plants in

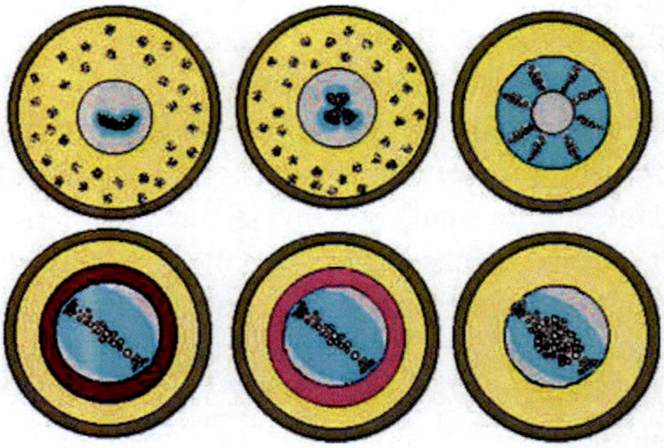

Fig. 1.8 Examples of schemes of structural models Polypodiophyta

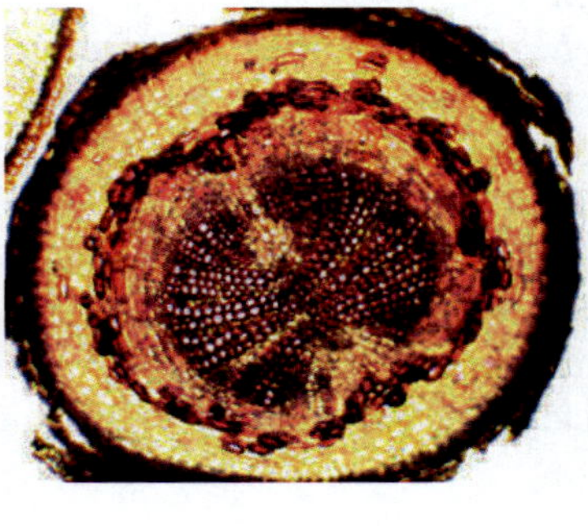

Fig. 1.9 A cross-section cut of a lateral root *Cycas revoluta* Thunb

comparison with spore plants, we extended a set of anatomical feature for structure model development up to the following:

- Characteristics of the cambium layer development in roots, its configuration, and activity
- Phellogen location and compound of tissues, formed by phellogen
- Way of primary cortex dying off and casting
- Structure features of secondary conducting tissues – phloem and xylem; characteristics of inner bark and wood elements; location of certain specific structures (resin canals, aerenchyma cavities, etc.)
- Ratio of thickness of cork (phellem), secondary phloem (inner bark), and secondary xylem (wood)

Comparative analysis of cross-sections of the lateral roots in four species *E. horridus* (Jack.)Lehm., *E. altensteinii* Lehm., *E. lehmannii* Lehm., *E. trispinosus* (Hook.)R.A.Dyer, of the genus *Encephalartos* Lehm., showed that they have the same inner structure due to secondary growth. This growth is common for roots of gymnosperms and dicotyledon angiosperms, and manifests itself in emerging of secondary conducting tissues – phloem and xylem from cambium, and secondary ground tissue from phellogen. Perennial roots of all the examined species of the division Pinophyta have a thin cambium layer with rounded contours; outside it is engirdled by a wide pale circle of bark elements; inside of cambium layer there is a heartwood, which is segmented by parenchyma rays. For example, in roots of species, which belong to the genus *Encephalartos* Lehm., heartwood is bisected; between these two sectors there is a diarch primary xylem ray. Outside there is a multilayered cork, covering the perennial roots. It replaces rhizodermis and primary cortex, which are being thrown in the process of root thickening.

The same structure features are found in roots of the species belonging to the class Cycadopsida. Minor variations in secondary growth of roots are noticed in *Stangeria eriopus* (G.Kunze)Nash. It has heartwood divided on six small sectors, rather than 2, and there are aerenchyma cavities in the inner bark. Though, in general, structure variations of perennial roots of the species from the class Cycadopsida manifest themselves only in the ratio of thickness between phellema, secondary phloem, and xylem.

Specific allorhizic root system develops in 5-year specimens, raised from seed of *Ginkgo biloba* L., the only representative of the class Ginkgopsida. Primary and lateral roots, which are capable of secondary growth, are noticed to have the same structure features. In wood of both roots there is distinct annual growth, and rounded fringed pores on the tracheid walls.

Significant diversity of root structures has been discovered in species, which belong to the largest class Pinopsida, comprising about 600 species. Many species develop mixed allohomorhizic root system, which is characterized by occurrence of primary root, lateral roots of several orders, and adventitious roots. For example, in 5-year specimens of *Picea obovata* Ledeb. and *Larix sibirica* Ledeb., primary and lateral roots have secondary structure, while adventitious roots have primary structure (Fig. 1.10). There are specific resin canals in the roots' wood of the

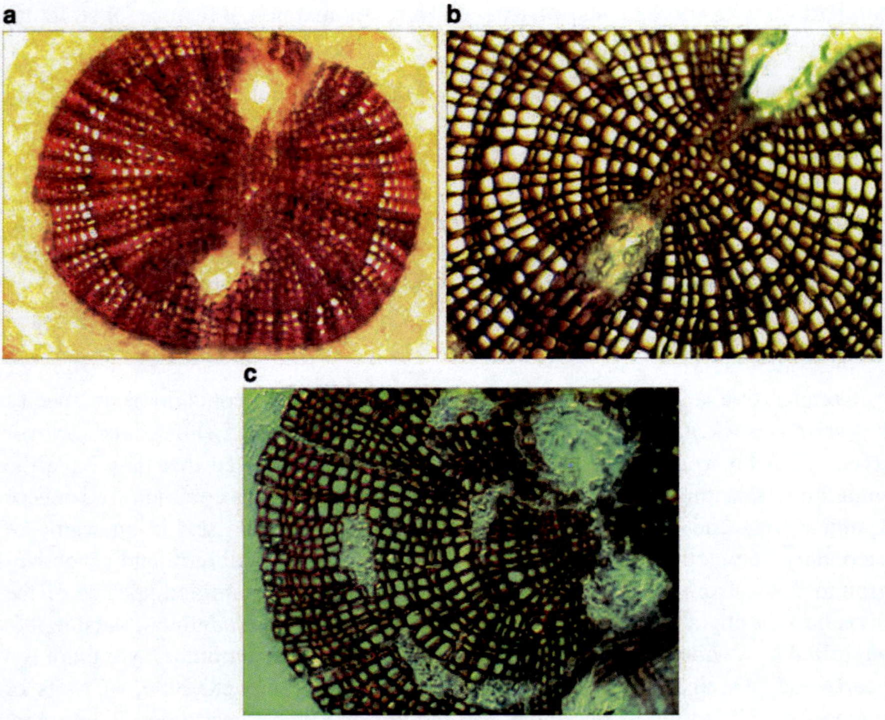

Fig. 1.10 Cross-section cuts of lateral roots of (**a**) *Picea obovata* Ledeb., (**b**) *Larix sibirica* Ledeb., (**c**) *Pinus sylvestris* L

species, belonging to the family Pinaceae Lindl. These specific structures are differently located in roots of species, belonging to different genera and families. Thus, peculiar vertical and horizontal resin reservoirs (cysts) are emerging in wood of roots of the species *Sequoia sempervirens* (D.Don)Endl., belonging to the family Taxodiaceae (Warm.)F.Neger. Perennial roots of the species of the genus *Metasequoia* Miki ex Hu et W.C. Cheng have large resin cysts, which are situated in wood opposite to the primary xylem rays. Such roots are noticed to have the high growth of wood, which is 1.8 times thicker than bark, and 6.4 times thicker than cork. Other more significant differences in root structure of most species of the family Taxodiaceae (Warm.)F. Neger have not been registered. The species *Sciadopitys verticillata* (Thunb.)Siebold et Zucc. does not have any resin canals or other resin reservoirs in wood of roots. Roots of *Callitris rhomboidea* R.Br. ex L.C.Rich. from the family Cupressaceae (A.Rich. ex Bartl.)F.Neger have bark fibers and sparse resin canals in bark and wood. Small ball-like resin reservoirs, which are situated in the phloem parenchyma, are discovered in roots of four species of the genus *Juniperus* L. It must be emphasized that lateral roots of these species show the same secondary anatomical structure, characterized by circular disposition of secondary ground and conducting tissues.

It is established that the species of the class Gnetopsida, comprising three orders Ephedrales, Gnetales, and Welwitschiales, differentiate by the greatest peculiarity of root structures. We studied roots structure features in the species, belonging to the orders Ephedrales and Gnetales. These features are occurrence of vessels in secondary wood and absence of resin canals. Wild growing in the Altai and Trans-Baikal species *Ephedra dahurica* Turcz. and *E. monosperma* S.G.Gmel. ex Willd., and raised from seed *E. americana* Humb. et Bonpl. ex Willd, have identical secondary structure of lateral roots. These roots are covered with multilayered cork (up to ten layers), then a narrow phellogen layer and strong secondary phloem band underlay. Wide parenchyma rays divide a solid wood into two triangular sectors. Tracheids are the main wood elements, but there are wide-bore vessels as well.

Roots of *Gnetum gnemon* L. of the order Gnetales are differentiated by the most peculiar structure. Roots of the grown in greenhouse specimens show secondary structure. Around the periphery they are covered with 5–7 layered cork, peeling from the surface. Bark has a form of wide band, engirdling the cambium layer. Sieve cells of bark are arranged in distinct radial rows. There is pith in the center of wood girdle, which is radially cut by parenchyma rays. Resulting from the comparative anatomical study of roots of the division Pinophita, the structure feature complexes have been created, and on their basis 12 structure models have been developed (Fig. 1.11). Among these models a so-called cyclic model is the most common, which displays circular arrangement of secondary tissues – ground tissue, conducting tissue, and meristem tissue.

Fig. 1.11 Examples of schemes of structural models of roots of species from department Gymnosperms

We discovered that species belonging to the division Magnoliophyta have the maximum structure diversity of roots. To interpret the data of comparative anatomical studies of the roots of flowering plants we used the Takhtajan system (1987).

Characteristic features have been discovered resulting from the study of the root structure in 22 species of the subclass Magnoliidae. Thus, the *Magnolia grandiflora* L. roots show secondary growth. There are parenchyma cells in pith of perennial roots, and on its periphery there are up to nine to ten primary xylem rays. The *Drimys winteri* J.R. Forst. et G.Forst. roots show only primary growth, and primary xylem is characterized by occurrence of tracheids and absence of vessels. It is established that primary growth is typical for roots of the families Piperaceae, Cabombaceae, and Nymphaeaceae. Numerous aerenchyma cavities and star-shaped sclereids are found in roots of the species, belonging to the family Nymphaeaceae.

It is found that 39 species of the subclass Ranunculidae have various types of root secondary growth. Thus, limited secondary growth is typical for the species *Ranunculus repens* L. and *R. acris* L., as well as for a number of species, belonging to the genus *Clematis* L., and preserving their primary anatomical structure for quite a long time. Species of the family Paeoniaceae are registered to have active secondary growth of roots. Moreover, peculiar metamorphoses occur in roots and caudices of a number of species in the subclass. There is active secondary growth in roots of the 40 species of the subclass Caryophyllidae as well. Rapid development of thick cork is typical for many of them. Thus, in species of the family Didieraceae root cork thickness is 24%, bark thickness is 24%, and wood thickness is 52%. It is registered that specimens of wild growing in the Altai semi-shrub *Krascheninnikovia ceratoides* (Z.)Gueldenst., of the family Chenopodiaceae, have polycambial roots.

Ten species of seven families of the subclass Hamamelididae have certain structure features, related to histological differences between bark and wood elements. In roots of *Simmondsia chinensis* (Link)C.K.Schneider, there is interxylary, or included, phloem and several growth layers, consisting of concentric xylem rings, segmented by narrow girdles of parenchyma cells.

There are numerous variations of secondary growth of roots in 77 species, belonging to the 26 families of the subclass Dilleniidae. Species from the families Ericaceae, Urticaceae, and Bombacaceae differentiate by the greatest structure diversity of taxon roots. For example, in *Chorisia speciosa* A.St.-Hil. and other species of the family Bombacaceae, secondary conducting tissues of roots are cut with wide rays of bark and wood; due to this they become similar to open, collateral conducting bundles.

There is great root structure diversity in 170 species belonging to 40 families of the subclass *Rosidae* (Fig. 1.12). It is displayed in 12 structure models. Numerous histological differences in secondary conducting tissues have been found in roots of most subclass species, excluding the family Droseraceae, whose roots have not demonstrated secondary growth. In the subclass Lamiidae, a group of species has been singled out from the 78 species belonging to 17 families, whose roots do not show secondary growth. Moreover, there are species that show slight secondary

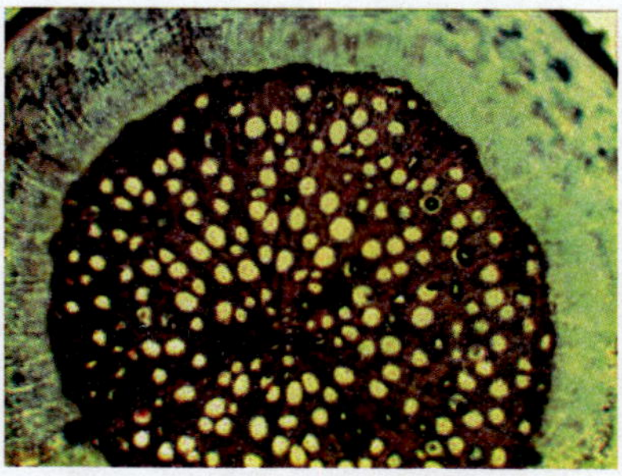

Fig. 1.12 A cross-section cut of a root *Malus domestica* Borkh

growth of roots, and large group of species showing rapid secondary growth of roots as well. Many species of this taxon have aerenchyma cavities in inner bark of roots, e.g. *Origanum vulgare* L. There are large sclereid groups in roots of a number of species of the family Apocynaceae.

In most of the 42 examined species, belonging to the subclass Asteridae, roots show specific secondary growth, though demonstrate numerous structure variations. Among these are the following:

- Wide rays of inner bark and wood, which are segmenting the ring-shaped layers of secondary tissues, similar to conducting bundles
- Segmented laticiferous vessels
- Numerous cells of parenchyma of storage parenchyma, etc.

The outstandingly peculiar species is *Barnadesia dianthiflora* Mart. ex Baker, whose roots do not have mechanical elements, show limited secondary growth, and at the same time have quite wide primary cortex with aerenchyma cavities.

Embryonic root of species of the class Liliopsida is known to die off quite soon, though numerous adventitious roots are being developed instead of it, and then form secondary homorhizic root system. Though roots of 269 examined species belonging to the subclasses Alismatidae, Liliidae, and Arecidae are not capable of secondary growth, they are found to have numerous structure features, which express inner variety (Fig. 1.13).

Among the most typical anatomical features of species of the subclass Alismatidae it is necessary to note the following:

- Occurrence of large aerenchyma cavities in the wide multilayered cortex
- Distinct radial and concentric arrangement of cells in inner zone of primary cortex
- Small steles with tetrarch and polyarch conducting bundles

Greater structure diversity has been found in roots of the subclass Liliidae. For example, species of the family Liliaceae are common to have:

- One-layered rhizodermis in roots with sparse root hairs of different length
- Wide primary cortex (up to 66%), which is not divided into internal and external zones
- Small stele (up to 34%) with triarch, tetrarch, or pentarch conducting bundle

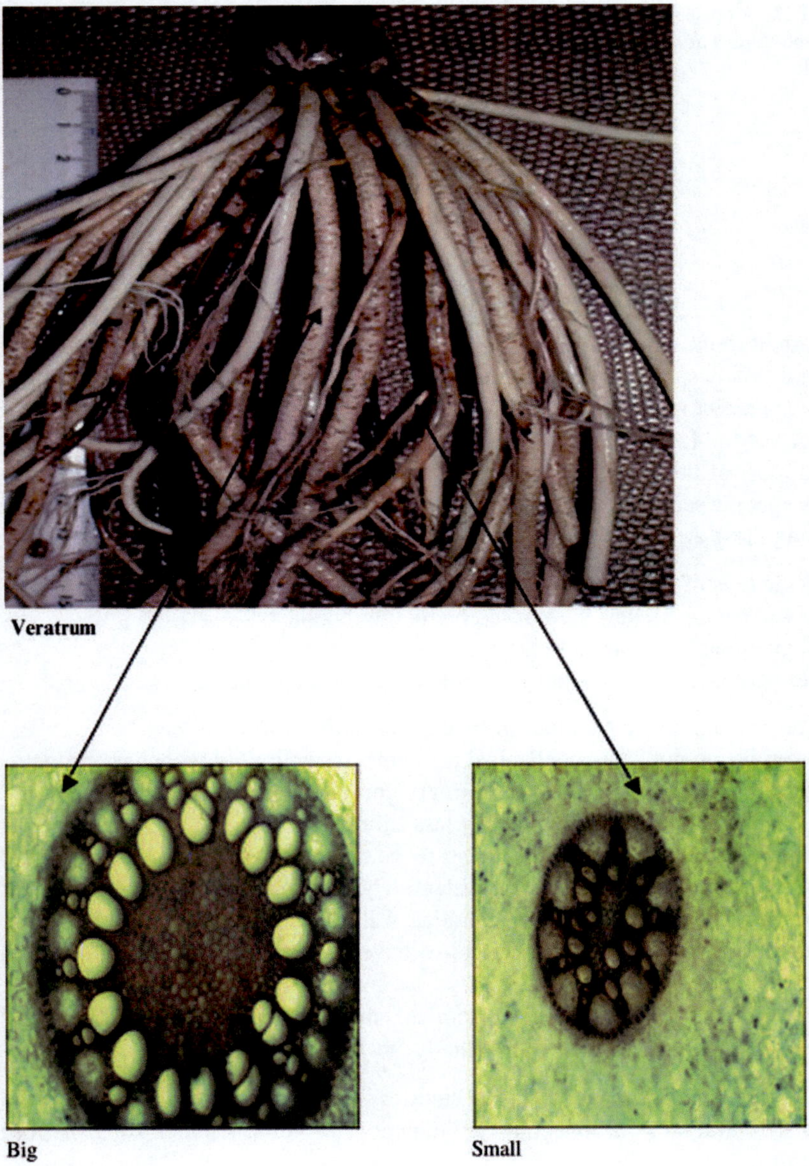

Fig. 1.13 Endogenous variability of root structures in secondary-homorhizophitic root system *Veratrum lobelianum* Bernh

Species in the families Agavaceae, Bromeliaceae, Dracaenaceae, etc., have several sclerenchyma layers developing in cortex, which is divided into two zones. There is peculiar sclerenchymal ring either in the internal or in the external zone. Structural features of ground tissue are the distinguishing anatomical features of roots of the family Orchidaceae. Thus, in wild growing Ural species *Platanthera*

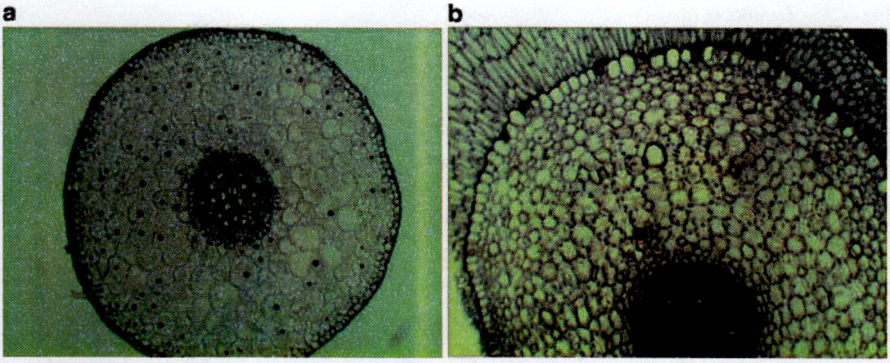

Fig. 1.14 A cross-section cut of a root *Cypripedium guttatum* Sw. (**a**) and *Cymbidium lowianum* Reichenb (**b**)

bifolia (L.)Rich., *Dactylorhiza incarnata* (L.)Soó, etc., there is a one-layered rhizodermis outside the roots, which has sparse root hairs or hyphae, which form a kind of mantle. Roots of tropical orchids are registered to develop velamen (Fig. 1.14). Interestingly, that velamen of ground roots and aerial roots of tropical species greatly varies by thickness (322–588 μm) and quantity of layers (2–9). Certain features of stele and cortex microstructures vary as well. Thus, significant variation in xylem strands quantity (from 7 to 27) have been registered between species.

Roots of the family Cyperaceae are found to have a one-layered rhizodermis with long root hairs. There is primary cortex divided into internal (with radial cells arrangement) and external zones. There is a narrow sclerenchyma ring in the external zone. Moreover, there are numerous aerenchyma cavities, arranged in seven to eight radial rows. Polyarch conducting bundle is being developed in root stele. Absence of mycorrhiza is typical for this taxon.

Complex characteristic features have been determined for 25 species of the family Poaceae. They are as follows:

- One-layered rhizodermis consisting of small cells with thickened walls
- Exoderm of cortex, consisting of cells, saturated with lignin
- Radial and concentric arrangement of cells in the inner cortex
- Occurrence of aerenchyma cavities
- Occurrence of polyarch conducting bundle with 6–24 xylem strands, and sclerenchyma elements in the stele center

Representatives of the family Arecaceae belonging to the subclass Arecidae, differentiate with the exceedingly great structure diversity of roots. We examined 20 species of palms growing in greenhouses. In root cortex we found suberinization, and sometimes lignification, of the exoderm cell membranes. There are large aerenchyma cavities and single sclereids in cortex. There is significant variation of xylem strands quantity (from 6 to 41) in stele. Root models of the species belonging to different families of the subclass Arecidae, alongside with features typical for all the species, show the certain properties of species. For example, a group of species,

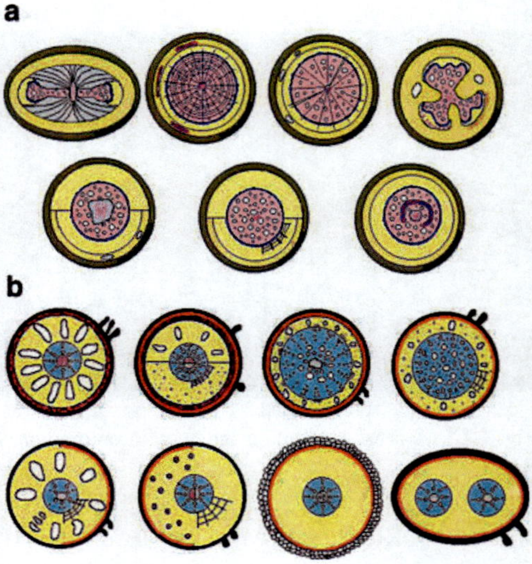

Fig. 1.15 Examples of schemes of structural models of roots of species Angiospermae. (**a**) Class Magnoliopsida, subclass Rosidae; (**b**) Class Liliopsida, subclass Arecidae

belonging to the genus *Philodendron* Schott, is found to have mucous canals, which are being developed in primary cortex of aerial roots; while another group of species has mucous canals in ground roots, as well as in aerial roots (Fig. 1.15).

1.4 Conclusion

As a result of comparative research of a structure of roots species of the higher plants, complexes of anatomic attributes, specific for representatives various taxons are revealed, have been determined, and graphical models have been developed to express their structure diversity. Anatomical characteristics of roots, due to their persistence, can serve as a reliable criterion to define the kin relationships between taxa (species, genera, families). But it must be borne in mind that variety of environmental conditions within the range of a species may cause certain changes in root tissues, e.g., development of mechanical elements and/or aerenchyma, etc. It is essential to continue gathering information about structure diversity of roots and root systems, and to use rhizological data to preserve the Earth's biological variety of flora.

References

Carlquist SH (1961) Comparative plant anatomy. Chapter 7. Root. Holt Rinehart and Winston, NY, pp 94–101

Kivenheimo VI (1947) Untersuchungen über die wurzeisysteme der Samenpflanzen in der Bodenvegetation der walder Finnlands. Ann Bot Soc Zool Bot Fenn. Vanamo, Helsinki T 22, No 2. p 180

Serebryakov IG (1962) Ecological morphology of plants. Vital forms angiospermous and coniferous. Moscow: the Higher school. p 378

Takhtajan AL (1987) System of Magnoliophyta. Leningrad, Nauka, p 439

Tarshis GI (1975) Underground organs of perennial grassy plants. Sverdlovsk: publishing house SGPI. p 135

Tarshis GI, Tarshis LG (1998) Morphological intraspecific diversity of below-ground organs in herbs and dwarf shrubs inhabiting the Urals/Developments in Plant and Soil Sciences, vol 82. Kluwer Academic Publishers, Dordrecht, The Netherlands. pp 155–163

Tarshis LG (2003) Structural variety of underground organs of the higher plants. Ekaterinburg: Izd. Ural Department of the Russian Academy of Science. p 196

Tarshis LG (2007) Anatomy of underground organs of the higher vascular plants. Ekaterinburg: Izd. Ural Department of the Russian Academy of Science. p 222

Troll W (1949) Űber die Grundbegriffe der Wurzelmorphologie. Wien: Springer Verlag. Bd 96, No 3–4, pp 444–452

Voronin NS (1972) Management to laboratory researches on anatomy and morphology of plants. Prosvechenje, Moscow, p 160

Chapter 2
Electrical Impedance Spectroscopy and Roots

Tapani Repo, Yang Cao, Raimo Silvennoinen, and Harry Ozier-Lafontaine

Abstract This chapter reviews studies dealing with the characterization of the plant root system by the electrical impedance method, i.e., with resistance and capacitance at a single frequency or at multifrequencies, according to the approach used in electrical impedance spectroscopy (EIS). Several studies have shown a correlation between electrical capacitance and resistance at a single low frequency with root biomass and morphology (e.g., surface area). It has not been possible to define clearly which part of the root system the electrical parameters represent. The circuitry in the stem–root–soil continuum has been analyzed in more detail by means of EIS. Using this approach, lumped and distributed models have been formulated that consider the role of roots in the context of other components in the circuitry. By means of EIS, a parameter referring to root capacitance was found to correlate positively with root biomass and root surface area. Several open questions remain with regard to the applications of the method. More studies are needed for the evaluation of longitudinal and radial electric field distribution between root interior and soil along the root system from root collar to root tips. In addition, further studies are needed under standardized measurement conditions

T. Repo (✉)
Joensuu Unit, The Finnish Forest Research Institute, P.O. Box 68, 80101 Joensuu, Finland
e-mail: tapani.repo@metla.fi

Y. Cao
School of Forest Sciences, University of Eastern Finland, P.O. Box 111, 80101 Joensuu, Finland

R. Silvennoinen
Department of Physics and Mathematics, University of Eastern Finland, P.O. Box 111, 80101 Joensuu, Finland

H. Ozier-Lafontaine
INRA, UR 1321 Agrosystèmes Tropicaux Domaine Duclos, 97170 Petit-Bourg, Guadeloupe, France

S. Mancuso (ed.), *Measuring Roots*,
DOI 10.1007/978-3-642-22067-8_2, © Springer-Verlag Berlin Heidelberg 2012

with soils of different ionic composition and texture as the growing substrate, and also by taking into account the role played by mycorrhizas.

2.1 Introduction

Studies concerned with the environmental responses of roots are scarce compared with those dealing with shoots. This is largely due to the technical problems involved in "seeing" roots hidden in the soil. In addition, the diverse structure of the root systems, along with the small diameter of fine roots, poses another challenge for their observation. In previous studies on roots, methods have been used that are destructive, cumbersome, or may not be fit for the study of fine root functionality, i.e., root excavation, the in-growth bag, microcosm, minirhizotron, or scanner set in the soil (Samson and Sinclair 1994; Majdi 1996; Costa et al. 2000; Smit et al. 2000; Hirano et al. 2009). Because of the soil, optical methods cannot be used without installing observation windows in the soil. In the radar technique, root observation is based on the penetration and reflection of electromagnetic radiation at a particular wavelength into the soil and on the detection of roots according to their effects on soil homogeneity. The radar method may detect coarse roots (diameter $>$19 mm), but currently the resolution does not permit detection of fine roots (diameter $<$2 mm) (Butnor et al. 2003; Hirano et al. 2009). In consequence, the challenge is to develop new nondestructive methods for studying plant roots in situ and especially for studying the functionality of fine roots.

As soil and roots contain electrolytes and water, they may conduct electric current. When the roots in soil are set in an external electric field of varying frequencies, current passes through the system, depending on the electrical properties of the different components in the circuit. This property has been used previously in developing the electrical method for characterizing plant root systems (Chloupek 1972; Dalton 1995; Ozier-Lafontaine and Bajazet 2005; Aubrecht et al. 2006; Cao et al. 2011). In general, when an electrical potential difference is gained between two electrodes, an external electric field appears between the electrodes, whether they are located in a vacuum or in real materials (as gas, liquid, solid, plasma). The appearance of potential difference generating an electric field will have an influence on the charge distributions in real material. Typically, electrons, positively charged electron holes, and dipoles formed by ions contribute charge distributions. Theoretically, when the spatial charge distribution is known, the electric field can be calculated by integrating the charge distribution within the space, and finally, when the electric field is solved, it is possible to find and define the electric potential distribution by integrating the electric field distribution within the space.

With reference to the electric potential difference between a metal with a crystal lattice (e.g., an electrode) and biomaterial, the majority of free charges (delivering the current) in a metal are typically electrons. In contrast, the structures of biomaterials are complicated, containing cells with cell membranes filled with electrolytes, which give the appearance of electric dipoles. In order to perform an electric impedance measurement to characterize the biomaterials, an alternating current is conducted through the tissue with the aid of metal electrodes. A polarization phenomenon will, however, be encountered in this simple procedure that will have an influence on the current delivery and also, at a later stage, on the impedance. The phenomenon is known as electrode polarization, which arises from the inverse population of charges at the interface between a metal and a biomaterial (e.g., Geddes and Baker 1989; Ragheb and Geddes 1990).

A general principle of the electrical impedance method for root studies is that the root system is exposed to an alternating electric field that would cause polarization and relaxation phenomena. In that case, the resistance and/or capacitance for the root system can be calculated either directly according to current and voltage relations or estimated by means of curve fitting, according to the method using electrical impedance spectroscopy (EIS). In the early studies in the field, resistance and capacitance were measured at a single low frequency. However, several different polarization and relaxation phenomena occur in the circuitry in the soil–root–stem–electrode continuum, depending on the frequency contributing to the resultant formation of several resistors and capacitors. In consequence, in recent studies a multifrequency approach has been used, and the system is characterized using electrical circuit models that suit EIS. The aim of this section is, then, to review the studies that deal with characterization of the plant root system according to its electrical properties. First, we deal with the theory of polarization and relaxation phenomena that form the basis for the impedance measurements, we provide a definition of electrical impedance, and we describe the principle of EIS. Second, we then review studies concerned with the characterization of a root system both at a single frequency and at multifrequencies using EIS.

2.2 Theory

2.2.1 *Polarization and Relaxation in Biological Materials*

In biomaterials an alternating electric field affects the occurrence of the polarization and relaxation phenomena that contribute to the formation of a current. The polarization of a particular component depends, e.g., on the frequency and the field strength applied (and accordingly also the current), ionic concentration, water content, and temperature. The phenomenon can be understood as equivalent to the conducting plates of a capacitor insulated by dielectric material. Capacitance is defined as:

$$C = \frac{A}{d}\varepsilon \tag{2.1}$$

where A and d are the area and distance of the plates, respectively, and $\varepsilon = \varepsilon_r \varepsilon_0$ is the permittivity, with ε_r indicating relative permittivity and ε_0 permittivity of a vacuum. In principle, the dielectric material has a high impedance for a constant electric field, and no current flows between the conducting plates. Capacitive impedance is defined as:

$$Z_c = \frac{1}{j\omega C} \tag{2.2}$$

where ω is the angular frequency (rads^{-1}) and $j^2 = -1$ is the complex unit that relates to the phase angle. An ideal capacitor does not dissipate energy. However, in real capacitances a very small amount of stray current can be measured, e.g., by a Coulomb meter. This current is due to the resistive component (R_s) in the capacitor (e.g., Guegan and Foulc 2009), and hence the capacitive impedance has the formula:

$$Z_c = R_s + \frac{1}{j\omega C} \tag{2.3}$$

Thus, the permittivity (ε) assumes a complex quantity that may be defined as:

$$\varepsilon = \varepsilon' - j\varepsilon'' \equiv (\varepsilon'_r - j\varepsilon''_r)\,\varepsilon_0 \tag{2.4}$$

where ε' and ε'' are real and imaginary parts of the permittivity (Fm^{-1}), and ε'_r and ε''_r are real and imaginary parts of the relative permittivity, respectively. In general, ε' relates to the fraction of the electric energy which is stored in the capacitances. If the voltage (v) and current (i) occur in the same phase, the imaginary part ε'' denotes the energy loss in the resistance (R), according to Joule's heat (P):

$$P = i \times v = i^2 \times R = \frac{v^2}{R} \tag{2.5}$$

Because the strength of the applied electric field in bioimpedance measurements is typically low, we may assume that the imaginary part of the dielectric permittivity describing the electric energy losses will also be low. Thus, the phenomenon related to the storage of electric energy in a biomaterial can be characterized by the real part of the dielectric permittivity.

The most important polarization mechanisms found in biological systems are interfacial polarization (Maxwell–Wagner effect), dipole polarization, and counter-ion polarization (also termed Faradaic or Warburg polarization), all of which have an effect at different frequency ranges (Ackmann and Seitz 1984; Foster and Schwan 1989; Schwan and Takashima 1991). The interfacial polarization appears at radio frequencies and it may exist in materials that have a heterogeneous microstructure, where regions of high conductivity exist within a matrix of low conductivity material (e.g., cell interiors surrounded by cell membranes). The dipole polarization process can occur in situations where dipoles exist and where dipoles are free to orient in an electric field. The relaxation frequency of dipole polarization is in the mega- to gigahertz range. Counterion polarization arises from the ionic diffusion in the electrical double layers adjacent to the charged surfaces. Their effects are typically prominent at frequencies below 20 kHz. In a simple system (e.g., isotropic material), the relaxation can be typically described by means of a first-order differential equation with one relaxation time. The relaxation time depends on the physical process involved: it can range from picoseconds in the case of the reorientation of molecular dipoles to seconds in the case of counterion effects (Foster and Schwan 1989). The change in the relaxation time or dielectric constant from one value to another as the frequency changes is termed dispersion.

In biological materials, three major distinct dispersion ranges, named α-, β-, and γ-dispersion, have been found in terms of dielectricity (Schwan 1957, 1963, 1988, Foster and Schwan 1989) (Fig. 2.1). They can be observed in the frequency response of the real part (ε_r') of relative permittivity or the real part (σ') of the complex conductivity (σ) that can be related according to the following formula (Grimnes and Martinsen 2008).

$$\sigma = \sigma' + j\sigma'' = \omega\varepsilon'' + j\omega\varepsilon' = j\omega(\varepsilon' - j\varepsilon'') = j\omega\varepsilon_o\ (\varepsilon_r' - j\varepsilon_r'') = j\omega\varepsilon \tag{2.6}$$

The different dispersion ranges overlap, but in general the α-dispersion extends from a few hertz to 10 kHz, the β-dispersion from 10 kHz to 100 MHz, and the γ-dispersion from 100 MHz to 100 GHz (Schwan and Takashima 1991).

Within each dispersion range, several polarization processes have been found where the relaxation cannot be described using a single time constant model. Moreover, in inorganic and organic (biological) materials several relaxation processes occur simultaneously, and the total electrical response is characterized by a distribution of relaxation times (Schanne and Ruiz-Ceretti 1978). Several distribution functions have been proposed to describe this phenomenon (Cole and Cole 1941;

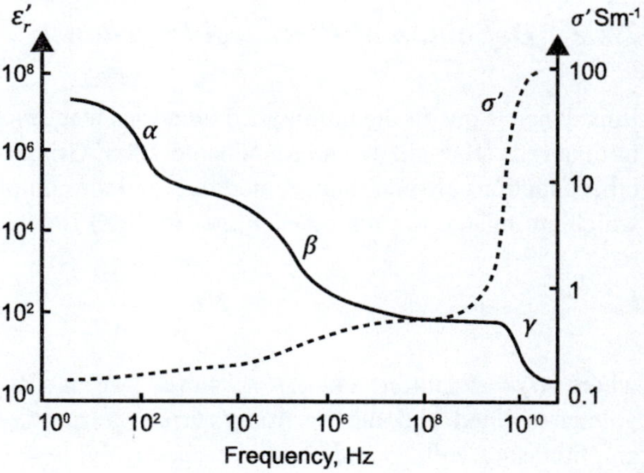

Fig. 2.1 Different dispersion ranges (indicated by α, β and γ) of biological materials, shown schematically. The real part of the relative dielectric constant (ε_r', *left axis*) and the real part of the conductivity (σ', *right axis*) as a function of frequency. (Reproduced from Schwan, Annals of Biomedical Engineering, by permission of Springer Verlag, 1988)

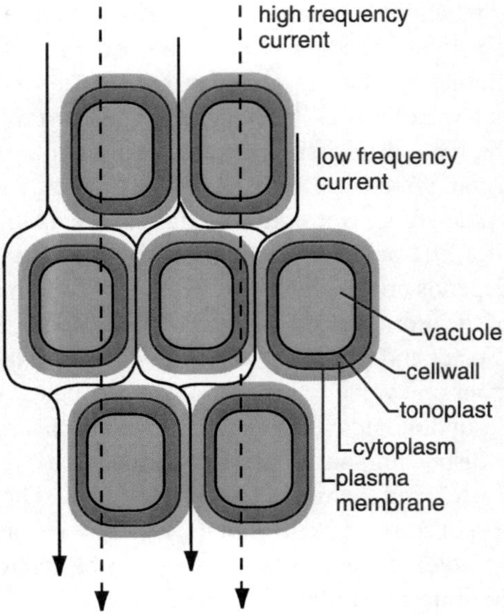

Fig. 2.2 Schematic diagram of the passage of an alternating current through plant tissue at low and high frequencies

Macdonald 1987; Foster and Schwan 1989). The distribution of the relaxation times appears as depressed center(s) in the impedance spectrum (see below).

One of the implications of the dispersion ranges is the passage of alternating current through plant tissue. Due to the lipid bilayer of cell membranes, they may not pass alternating current at low frequencies, but work like capacitors. In that situation, current will pass through apoplastic space and the total impedance is mostly due to the resistance of the extracellular space. The cell membranes become conductive with increasing frequency, and at sufficiently high frequencies (>100 kHz) the total impedance will be composed of a parallel combination of intra- and extracellular resistances (Fig. 2.2).

2.2.2 *Definition of Electrical Impedance*

Impedance is one of the immittance functions that describe the electrical properties of materials (Barsoukov and Macdonald 2005; Grimnes and Martinsen 2008). The other functions are admittance, modulus, and the complex dielectric constant, all of which are related to each other. Impedance (Z) is defined as:

$$Z(\omega) = \frac{v(t)}{i(t)} \tag{2.7}$$

where $v(t) = V\sin(\omega t)$ and $i(t) = I\sin(\omega t + \theta)$ are the monochromatic alternating voltage applied and the resulting current, respectively. Here, V and I are the amplitudes of voltage and current, respectively, $\omega = 2\pi f$ is the angular frequency

(f is frequency), and θ is the phase difference between the voltage and the current. In complex rectangular coordinates, the impedance vector has the formula:

$$Z(\omega) = Z_r(\omega) + jZ_i(\omega) \tag{2.8}$$

where j is the imaginary unit, and Z_r and Z_i are real and imaginary parts, respectively, which can be defined as:

$$Z_r(\omega) = |Z| \cos \varphi \tag{2.9}$$

and

$$Z_i(\omega) = |Z| \sin \varphi \tag{2.10}$$

The impedance phase angle is

$$\varphi(\omega) = \mathrm{a}\tan\left(\frac{Z_i}{Z_r}\right) \tag{2.11}$$

and the magnitude

$$|Z(\omega)| = \frac{V(\omega)}{I(\omega)} = \sqrt{Z_r^2 + Z_i^2} \tag{2.12}$$

The impedance vector defined earlier forms an impedance spectrum (IS) (Wessel diagram) in a complex plane with frequency as the parametric variable. According to the impedance spectrum, conclusions can be drawn about the equivalent circuit of the study object.

In a simple circuit composed of linear R–C circuit elements, the impedance arc is a semicircle with its center on the real axis. In the apex of the IS for such a circuit:

$$\omega_c \tau = 2\pi f_c \tau = 1 \tag{2.13}$$

where ω_c is the angular frequency, τ is the relaxation time for polarization and relaxation, and f_c is characteristic frequency (Fig. 2.3a, arc a). There are two equivalent circuits (model I and II) that may result in similar ISs (as shown in the arc a in Fig. 2.3a), but with different values for the circuit parameters (Fig. 2.3b). In addition, the relaxation time, direct current resistance, and resistance at a high frequency all acquire different formulas (Table 2.1).

The measurements of the impedance spectra for different biological systems have proved the existence of a depressed center or centers, i.e., that the center of the impedance arc(s) is below the real axis (at the apex $\omega_c\tau < 1$) (Fig. 2.3a, arc b). Such a property means that different polarization–relaxation processes overlap, which appears as a distribution of time constants. The systems can be modeled using several R–C circuits connected in series or parallel with overlapping arcs, i.e.,

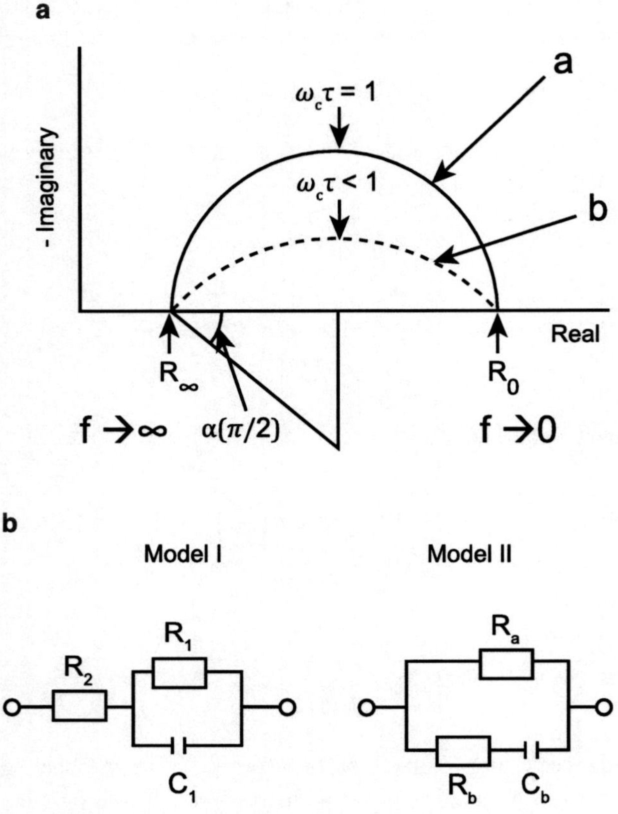

Fig. 2.3 (**a**) Two impedance spectra, denoted as arcs a and b, with their centers on and below the real axis (at the apex $\omega_c\tau = 1$ and $\omega_c\tau < 1$), respectively. (**b**) Arc a may be represented by two different lumped models, I and II, composed of resistors (R_1, R_2, R_a, R_b) and capacitors (C_1, C_b) (see Table 2.1). The intercept of arc a with the real axis at low and high frequency (*f*) is indicated by R_o and R_∞, respectively. ω_c is the angular frequency and τ is the relaxation time. The complex impedance for arc b, with its center below the real axis ($\omega_c\tau < 1$), is represented by a distributed model in (2.14)

Table 2.1 Definition of the relaxation time (τ), and the direct current resistance (R_o) and the resistance at high frequency (R_∞) for lumped models I and II (see Fig. 2.3b)

	Model I	Model II
τ	R_1C_1	$(R_a + R_b)C_b$
R_o	$R_1 + R_2$	R_a
R_∞	R_2	$R_aR_b/(R_a+R_b)$

closely related characteristic frequencies, or using a distributed model or models with corresponding characteristic frequencies. Several distributed circuit elements (DCE) have been developed for that purpose (Cole and Cole 1941; Macdonald 1987; Grimnes and Martinsen 2008). The complex impedance (*Z*) for one of the DCE models is given in (2.14). For the meaning of the model parameters, see Fig. 2.3a.

$$Z = R_\infty + \frac{(R_\infty - R_o)}{1 + (j\omega\tau)^{1-\alpha}} \tag{2.14}$$

2.2.3 *Electrical Impedance Spectroscopy*

In EIS the aim is that the system should be analyzed according to its equivalent circuit (macroscopic level) or at its physicochemical level (microscopic level) (Fig. 2.4) (Macdonald 1987). In the macroscopic approach, characterization of the system takes place using an equivalent circuit containing a complex nonlinear least squares (CNLS) curve fitting at both real and imaginary levels at the same time. That approach is used in the analysis of the root–soil circuitry.

Once the data is available, an estimation of the model parameters can be made using the CNLS program, e.g., LEVM 8.09 (J.R. Macdonald, University of North

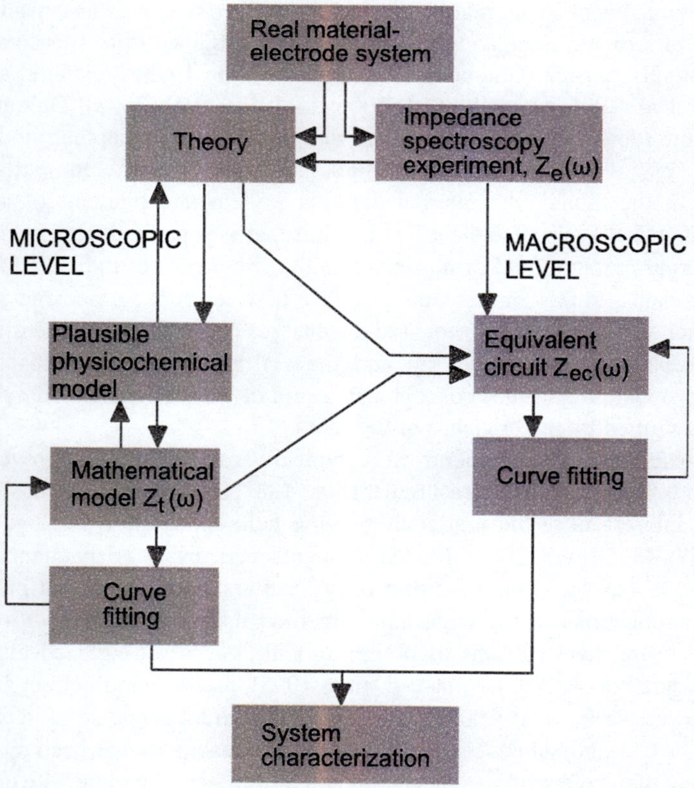

Fig. 2.4 Principle of electrical impedance spectroscopy. (Reproduced and modified from Macdonald, by permission of John Wiley and Sons, 1987)

Carolina, U.S.) (http://www.jrossmacdonald.com/levminfo.html). The calculation is based on the maximum likelihood method (Fisher 1922), which minimizes the sum:

$$S(P) = \sum_{j=1}^{M} w_j \left[Y_j - YC_j\,(P)\right]^2 \tag{2.15}$$

where M is the total number of data points, w_j is the weight associated with the jth point, Y_j is the jth data point value to be fitted, and $YC_j(P)$ is the value of the calculated fitting function involving the set of parameters P (Macdonald 1987).

2.3 Electrical Impedance for Assessing the Properties of Roots

2.3.1 *Single-Frequency Measurements of Roots*

A single frequency of an alternating current was applied to assess the capacitance or resistance of a root system, and the attributes were assumed to be measures of the active root surface area (Chloupek 1972, 1977; Dalton 1995; Preston et al. 2004; Aubrecht et al. 2006; Čermák et al. 2006; McBride et al. 2008). Dalton (1995) analyzed the root–soil–electrode network and proposed a conceptual model composed of several resistors and capacitors that represent different parts in the circuitry. In the model, the internal solution (xylem and phloem) of a plant's roots forms a low resistance electrical conduit that is separated from a low resistance external medium (soil or nutrient solution) by means of the isolation of the root membranes. In this analogy, the interface between root surface and substrate has a capacitance that is proportional to the charges accumulated on the membrane surfaces separating the xylem sap and the soil solution. Root capacitance – according to Dalton's (1995) concept – is a sum of parallel-connected cylindrical capacitors formed by the branches of the root.

The capacitance measurements were typically carried out at a frequency of 1 kHz and the resistance measurements at 128 Hz. Capacitance has aroused particular interest since the first studies in the field appeared in 1972 (Chloupek 1972, 1977; Rajkai et al. 2005). In measurements made using a capacitance bridge, one electrode was set at the base of a plant stem and another was set in the soil. Practical applications of the capacitance method have been tested with varying success for root investigations involving agricultural crops, e.g., red clover and alfalfa (Kendall et al. 1982), tomato (Dalton 1995), maize (van Beem et al. 1998), and sunflower (Rajkai et al. 2005), and even with different genotypes of a particular species. The correlation between root biomass and capacitance has been found to be independent of the electrode arrangement, i.e., results were the same with either one or several soil electrodes (Chloupek 1977). In hydroponic studies involving different tomato cultivars, a linear correlation between root capacitance and root mass was found (Dalton 1995). According to the same study using sandy soil as the growing substrate, the role played by soil water content in relation to root capacitance became evident. In contrast, however, studies involving woody species are few (Psarras and Merwin 2000; Preston et al. 2004). In young poplar trees, the capacitance of the root system was measured at 1 kHz using two electrodes, one in the soil and the other in the plant stem (Preston et al. 2004). The capacitance was found to correlate with the root dry mass (Fig. 2.5).

The earth impedance method using a single frequency (128 Hz) was introduced to facilitate estimation of the spatial distribution of the absorbing root surface area in the field (Aubrecht et al. 2006, Čermák et al. 2006). In contrast to the electrode configuration of the capacitance method, the earth impedance method applied four electrodes, two for the driving current and two for measuring the potential. By moving one potential electrode away from the stem, the mean distance of all of the absorbing roots from the tree could be determined by means of the potential characteristics. The estimated absorbing root surface area was found to be related to the basal area over a large range of stem diameters (Čermák et al. 2006). In a hydroponic study using willows, the electrical resistance of the root in a single-frequency current (128 Hz) correlated with the root surface area in contact with the solution (Cao et al. 2010). The resistance decreased with any increase in the root surface area (Fig. 2.6).

The capacitance and the resistance in the single-frequency measurements were found to depend on the properties of several different components between the measuring electrodes, i.e., the electrode–soil interface, soil, root, stem, and

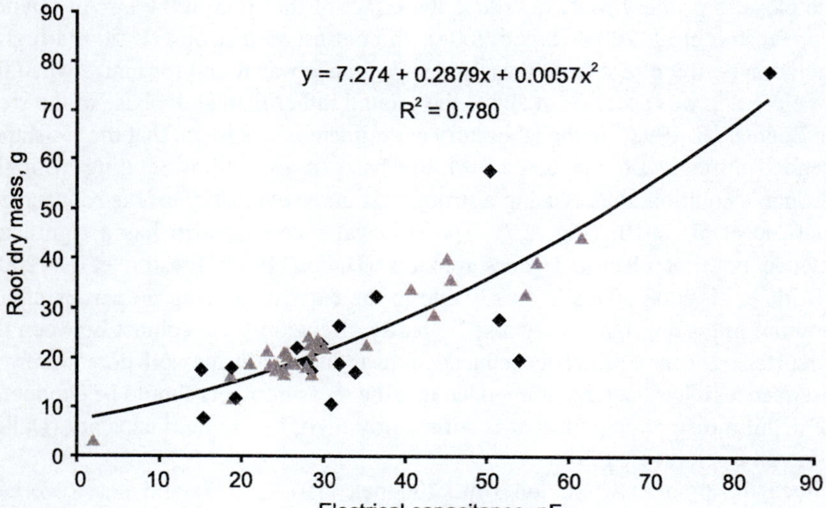

Fig. 2.5 Root dry mass versus root electrical capacitance for a pooled data of first- (*Filled triangle*) and third-year (*Black diamond*) hybrid poplars. The potting soil by weight was one-third a mixture of vermiculite and perlite and two-thirds a triple mix consisting of manure, peat, and loam. (Reproduced from Preston et al., Agroforestry Systems, by permission of Springer Verlag, 2004)

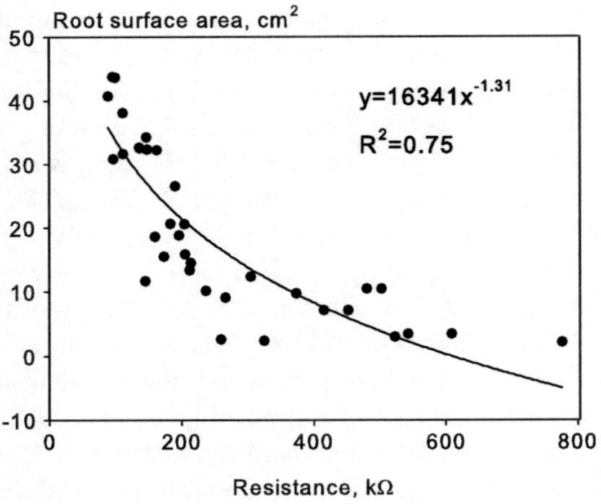

Fig. 2.6 Relation between root electrical resistance and root surface area for willows raised and measured in hydroponic solution. The contact of the stem with the solution was excluded. (Reproduced from Cao et al., *Journal of Experimental Botany*, by permission of Oxford Journals, 2010)

electrode–stem interface. All of these factors tended to confuse our interpretation of the results concerning the characteristics of roots. Dalton (1995) has presented his theory and assumptions based on the electrical measurement of root systems. His study demonstrated the important role played by the placement of the stem electrode close to primary roots to reduce the effect of the stem in the circuit (Dalton 1995; Preston et al. 2004, Rajkai 2005). In contrast to that observation, a higher correlation between root fresh mass and capacitance was found for maize when the stem electrode was set at 6 cm above the ground rather than at the base of the stem (van Beem et al. 1998). In the laboratory experiment it was found that the resistance depended strongly on the contact of the base of the willow cutting with the hydroponic solution, thus causing a strong bias in the evaluation of the root surface area (Cao et al. 2010) (Fig. 2.7). The soil water content also has a significant influence on root electrical measurements (Dalton 1995; Preston et al. 2004; McBride et al. 2008). This is mostly due to the current-carrying properties of soil containing moisture, but it may also be partly affected by the contact between the root surface and soil which is reduced by a reduction in the soil pore water. In consequence, it has been recommended that the measurements should be conducted with a soil moisture content that is sufficiently high, i.e., at field capacity (Dalton 1995; van Beem et al. 1998).

Since the pioneer work done by Chloupek (1972, 1977) and his coworkers (Chloupek et al. 1998), authors using single-frequency measurement of capacitance have pointed out the difficulty in using the method without specific calibration, which was shown to be the main restriction on any generalization of the method. Because the measurement includes the effects of the properties of both the soil and the roots, the method is prone to be sensitive to the soil electrical conductivity,

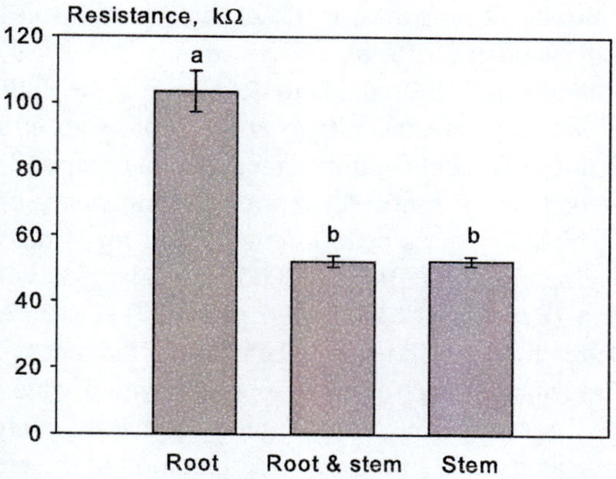

Fig. 2.7 The electrical resistance of intact willow root ("Root"), root, and a piece of stem ("Root and stem"), and the stem only with the root dissected ("Stem"), immersed in the nutrient solution. *Bars* indicate the standard error of the mean. The same *letter* indicates no significant difference at the level of 0.05 ($n = 16$). (Reproduced from Cao et al., *Journal of Experimental Botany*, by permission of Oxford Journals, 2010)

which is affected by the ionic status and water content. In addition, attention has to be paid not only to the plant species but also to the measuring frequency and voltage. It is also clear that the complexity of the soil–root system cannot be assimilated to a single ideal capacitor. There is a high probability of a wide spectrum of electrical interactions and ionic gradients at equilibrium between the components of the soil–root continuum (Hilhorst 1998). Hence, the single-frequency approach provides only a simplified view of the overall equivalent model of the system, while it does not consider multiple resistances and capacitances in circuitry. These conclusions have suggested the need for more extensive investigations based on the premises of the method so that a better electrical circuit representation and analysis of the soil–root continuum can be achieved, e.g., by using the EIS approach.

2.3.2 *Multifrequency Studies of Roots*

2.3.2.1 EIS in Plants

In the EIS approach, the characteristic behavior of the system in response to a range of frequencies of the electrical current can be represented by the components of the electrical circuits (resistors, capacitors). In the past, this method has been used to investigate the properties of plant, animal, and human tissues (Thomasset et al. 1973; Tiitta et al. 1999; Altmann et al. 2004; Bayford 2006; Grimnes and Martinsen 2008). In plants, it has been used to reveal the responses of detached, aboveground organs to cold acclimation, freeze-thaw and heat injury, and exposure to elevated ozone and carbon dioxide (Zhang et al. 1992; Zhang and Willison 1993; Repo et al. 1994, 2000, 2004; Ryyppö et al. 1998).

A number of attempts have been made to develop EIS for studying root system (Ozier-Lafontaine and Bajazet 2005; Repo et al. 2005; Cao et al. 2011). Using specific electric analogs to model the impedance spectra of the soil–root–electrodes continuum, the method can be useful for refining a noninvasive estimation of root traits, such as root fresh/dry mass or root length, both for tomato plants in a soil substrate (Ozier-Lafontaine and Bajazet 2005) and also for willow cuttings in hydroponic solution (Repo et al. 2005, Cao et al. 2011). Those studies proved that to achieve a better understanding of the behavior of the root system, it would be useful to use the impedance spectroscopic approach with a wide frequency range rather than a single frequency. They also pointed out that research in the field was still at an early stage as regards a thorough interpretation of the electrical behavior of the soil–root–electrode continuum, notably its ability to explain the impact of soil electrical conductivity on the overall response.

2.3.2.2 Measurement Setups

The basic equipment for running the IS measurements consists of (1) a impedance analyzer with a wide range of frequencies (from 10 Hz to 1 MHz or more), (2) a virtual pilot, such as Labview software, for driving the analyzer automatically using a computer equipped with a communication card and a specific driver, and (3) at least two electrodes (e.g., Ag/AgCl), one inserted into the plant stem and another into the soil, for driving current into the system and for measuring the voltage. The reason for preferring Ag/AgCl electrodes rather than purely metallic ones is to minimize electrode polarization. Because of their soft surface, Ag/AgCl electrodes are not, however, applicable for repeated measurements of hard tissues (e.g., wood). For standard measurements, the plant electrode may be inserted into the stem at 2 cm above the root collar and the soil electrode to a depth of 50 mm at a distance of 10 cm from the stem (Fig. 2.8). Once the IS data is available, an appropriate electric model can be formulated and the model parameters can be estimated by means of the CNLS curve fitting program [see (2.15)].

The recent EIS studies were all conducted under controlled conditions (photon flux density, temperature, soil moisture content, and electrical conductivity of growing medium) either in hydroponics or with soil substrate in containers. In the hydroponic studies, willow cuttings were cultivated in an aerated hydroponic culture in a growth chamber and measured for their IS in a container (Repo et al. 2005, Cao et al. 2011). The cuttings were mounted on floating pads in hydroponics with a given conductivity. The immersion depth of the plants in the containers was kept constant. The IS was measured either for the whole root system with the stem in contact with the solution and analyzed for a distributed model (Repo et al. 2005), or for different setups (with and without roots in contact with the solution) and analyzed for lumped models (Cao et al. 2011) (Fig. 2.9). The EIS model parameters were compared with either the root biomass or the root surface area.

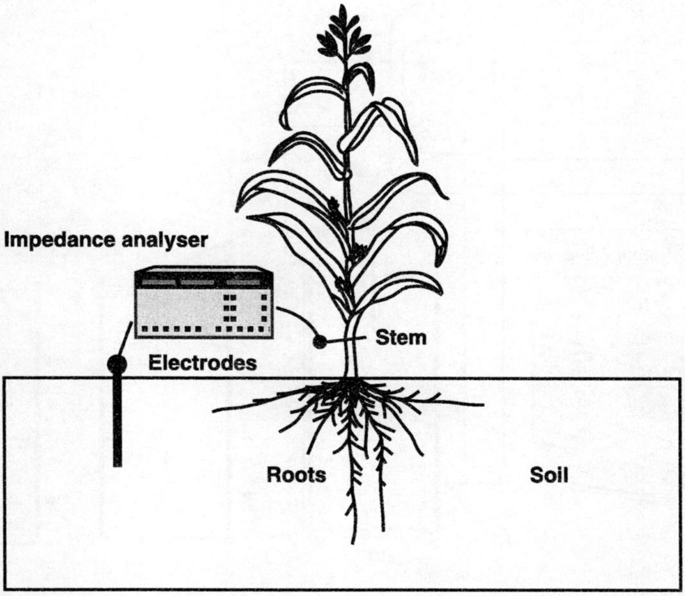

Fig. 2.8 The setup for measurement of the impedance spectrum of the soil/root system of a maize plant

The EIS experiment conducted with tomato was carried out under laboratory conditions with a constant temperature of 22°C (Ozier-Lafontaine and Bajazet 2005). The experiments were carried out in 50 dm^3 containers both in hydroponics at 930 μScm^{-1} and in pots with soil with different placement for electrodes (Fig. 2.10). The EIS results and root capacitance were correlated with the root growth (dry and fresh mass). The soil used was an oxisol (FAO-UNESCO, 34% sand, 33.4% silt, 32.6% clay) with a pH_{H2O} of 5.03 and a pH_{KCl} of 4.32, a *CEC* of 10.13 $mmol_c$/100 g, and a *C*/*N* of 12.5. The soil was packed with an average bulk density of 1.10 kgm^{-3}.

2.3.2.3 Components in the Circuit Affecting Current Passage and Role of Interfaces

Stem-Root Internal Composition

The interior of the stem at the root collar and in the roots is composed of tissues with apoplastic and symplastic solutions. Due to the cell membranes and their dielectric properties, the pathway between the extracellular (apoplast) and intracellular (symplast) spaces is characterized by biological capacitors that may not pass low frequency current, i.e., it bypasses the cells (Dvorak et al. 1981). At low frequencies, the impedance of that part of the circuit is mostly due to the apoplastic

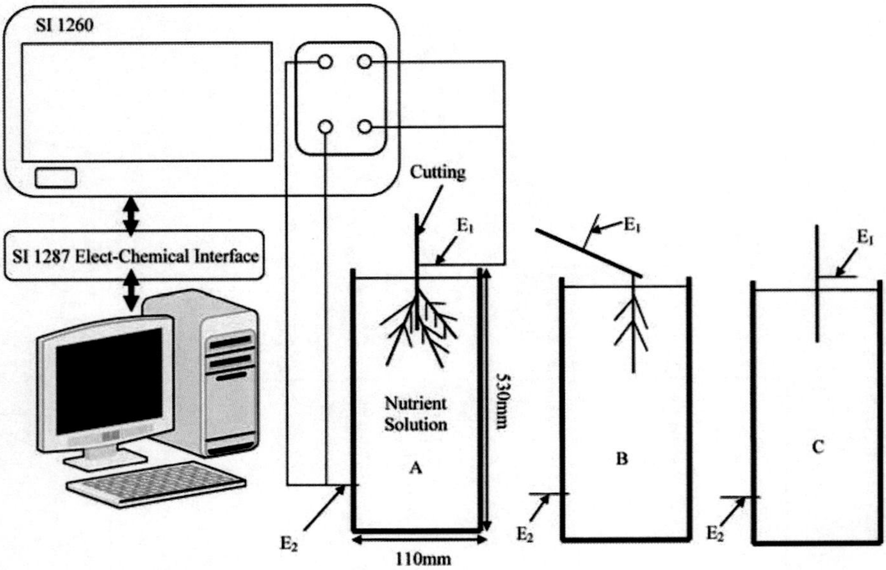

Fig. 2.9 The experimental design for EIS on willow cuttings in hydroponics. The impedance spectrum was measured for the whole root system together with the stem in contact with the solution (A), for a single lateral root in the solution (B), and for the stem only without its roots in contact with the solution (C). E_1 and E_2 refer to the Ag electrodes. (Reproduced from Cao et al. *Journal of Experimental Botany*, by permission of Oxford Journals, 2011)

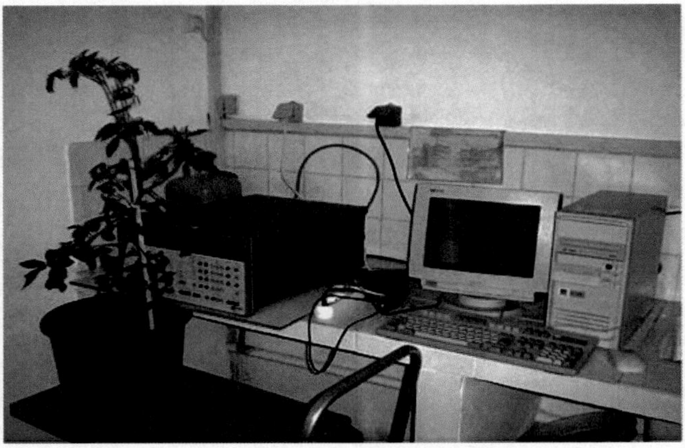

Fig. 2.10 Experimental setup for measuring the impedance spectrum of tomato roots

resistance, which is relatively high compared with the symplastic resistance (Figs. 2.2 and 2.11). At higher frequencies, the current starts to penetrate the cells, generating capacitive and resistive effects. When the frequency is sufficiently high, the capacitance due to cellular membranes is cancelled, and cells are completely crossed by the electrical current, the resistance being low. Accordingly, a simplified equivalent

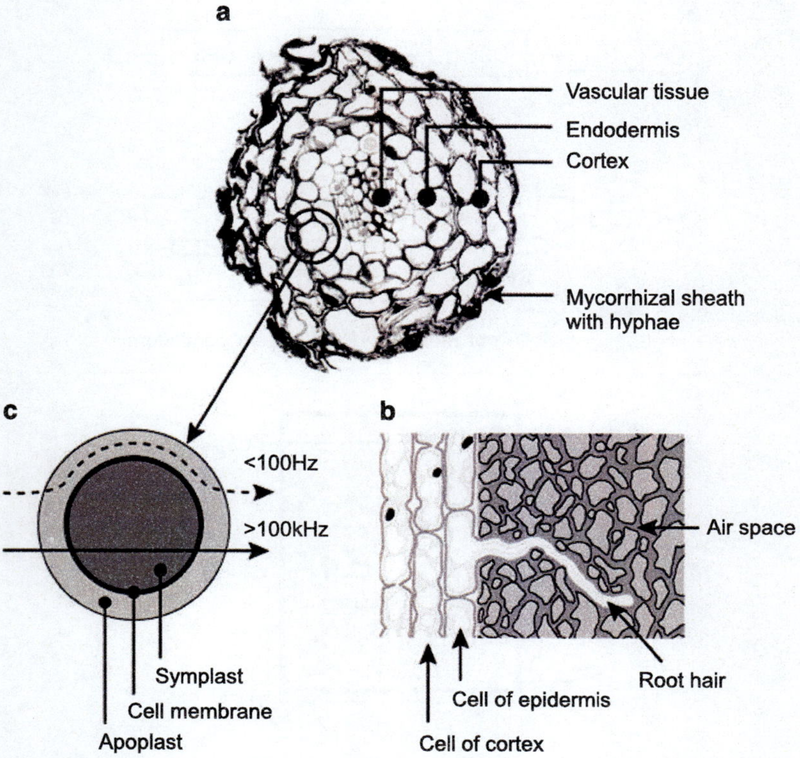

Fig. 2.11 Some typical features in a cross-section of a root (**a**) and at a root–soil interface (**b**) that affect passage of electric current between root and soil. The passage of low and high frequency current through a root cell, shown schematically (**c**)

circuit for the longitudinal pathway from root collar to root interior at different distances consists of two resistances and one capacitance (Fig. 2.12c).

Soil-Root Interface

The radial pathway between root interior and bulk soil includes several components, including apoplastic and symplastic spaces and interfaces, e.g., in rhizosphere (mycorrhizal hyphen, root hairs) and in cell membranes of root cells, which affect the passage of electric current (Fig. 2.11a, b) (Dalton 1995, Rajkai et al. 2005). The rhizosphere can be likened to a sheath that envelops the internal root system, and plays a dielectric role between the root's internal medium and the soil (Dalton 1995). As argued by Dalton (1995), in this part of the circuit the apoplast and symplast are mainly related to the root surface area, including the role of the mycorrhizas. Currently, there is no accurate technique that can discriminate

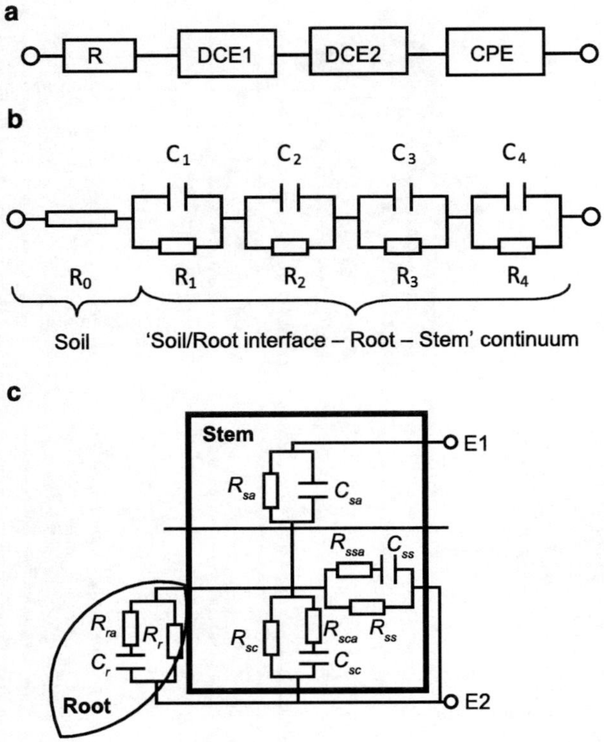

Fig. 2.12 The electric circuits used for the modeling of plant root systems. (**a**) Distributed model for willow cuttings grown in hydroponic culture: *R* is a resistor, DCE1 and DCE2 are distributed circuit elements, and CPE is a constant phase element (Repo et al. 2005). (**b**) Lumped model of tomato plants grown in a soil substrate: C_i are the capacitors and R_i the resistors along the soil–root–electrodes continuum. (Reproduced by kind permission of Springer Science + Business Media: Ozier-Lafontaine and Bajazet 2005, Plant and Soil 277, p. 308, Fig. 6b). (**c**) Lumped model of willow cuttings raised in hydroponics. Symbols: R_{sa} and $C_{sa,}$ the resistance and capacitance of the stem above the solution level indicated as a horizontal line in the middle; R_{ss} and $C_{ss,}$ the resistance and capacitance of the longitudinal interface of the stem with the solution; R_{sc} and $C_{sc,}$ the resistance and capacitance of the cross-sectional interface of the stem with the solution; R_r and C_r, the resistance and capacitance of the interface of the roots with the solution; R_{ssa}, R_{sca}, and R_{ra}, the auxiliary resistances for the bulk resistances properties of the surface of the stem, the cross-section of the stem, and the root immersed in the solution, respectively. E_1 and E_2 refer to the electrodes. (Reproduced from Cao et al., *Journal of Experimental Botany*, by permission of Oxford Journals, 2011)

between the proportion of the growth medium and soil–root interface in the resulting capacitance and resistance. There is probably an integrated effect that depends on the current frequency, the type of plant, its developmental stage, and the state of the soil environment.

Soil

In the soil itself, the current flows principally along the path delimited by soil pores. The resistance to its propagation is essentially conditioned by the soil texture and structure, the soil water content, and the electrical conductivity at the interfaces between the soil solution and the soil particles (Rhoades et al. 1976; Hilhorst 1998). Soil with charged colloid particles containing a liquid electrolyte has a relatively high dielectric constant at low (<1 kHz) frequency (Schwarz 1962). On the basis of these principles and the differences in the bulk dielectric properties of the soil constituents, Hilhorst (1998) proposed two equivalent circuit models that follow the Maxwell–Wagner dispersion: (1) the first model is a parallel circuit with equal layer thickness and different areas for water and solids, the water being conductive, while the solids are not; (2) the second model consists of two capacitors connected in series. One capacitor is formed by the water with a parallel conductor, resulting from ionic conduction of the water, while the other is formed by the solids.

Electrode Contacts with Plant and Soil

Electrical conduction in solutions and in biological substances, such as sap, is ionic because charges are carried by ions and are dependent on the dissociation of the solute. The purpose of the electrodes is to act as transducers between the ionic transport of the biological tissue or the solution of the substrate and the electron flow in metal wire. The junction between the electrode and the electrolyte permits the flow of the electric current, i.e., the flow of ions in the electrolyte gives rise to a flow of electrons (current) in the electrode, and vice versa. The phenomena at this interface are the source of polarization and, accordingly, of the electrode polarization impedance. If not eliminated, this usually results in a systematic error at low frequencies (<1 kHz). At high frequencies, stray capacitances may also affect the results. Some circuit analyzers include open and short circuit correction, which may eliminate the effects of cables and stray capacitances (Repo et al. 2000).

Electrode polarization is a manifestation of the reorganization of charged molecules, which occurs at the sample–electrode interface in the presence of water molecules and hydrated ions. In its simplest forms, the phenomenon is equivalent to a frequency-dependent capacitor in series. The high permittivity values observed at low frequencies are a manifestation of electrode polarization (Schwan 1992). Reversible electrodes are therefore recommended in order to minimize the influence of electrode polarization on the measurement.

2.3.2.4 EIS Models

Two types of equivalent electric circuit models have been used to characterize biological or physical materials: lumped and distributed. A lumped model consists of a limited number of ideal resistors and capacitors (Repo and Zhang 1993,

Zhang and Willison 1991). The distributed model is composed of DCE, and the complex impedance is represented by a mathematical expression (e.g., Repo and Pulli 1996, Repo et al. 2005). Following the conceptual model formulated by Dalton (1995), the soil–root–stem continuum of tomato was analyzed in terms of a lumped model using the EIS approach (Ozier-Lafontaine and Bajazet 2005). The circuitry was split into different components, which were represented by separate R–C circuits (Fig. 2.12a). If the use of lumped models is not appropriate, e.g., for highly inhomogeneous objects, the use of distributed models (DCE) (Fig. 2.12b) may provide good fitting results (Repo et al. 2005). In the DCE models, however, each component does not necessarily correspond to a biological structure. Their advantage is that the model usually takes the form of a highly compact mathematical expression that provides an excellent fit to the experimental data. A detailed analysis was recently made of the stem–root circuitry in terms of the resistance and capacitances for different components. A new lumped model was proposed (Fig. 2.12c) and tested with willows raised in hydroponics. The model was capable of distinguishing between the roles played by the roots and stem.

The circuit represented by the distributed model in Fig. 2.12a is composed of a piece of stem, the interfaces of roots and stem with cultivation solution, the cultivation solution, and the interfaces of the electrodes in the stem and the solution. The IS of this entity was modeled by a circuit consisting of two DCE elements (ZARC-Cole elements), one constant-phase element (CPE), and a resistor in series (for ZARC-Cole and CPE, see Macdonald 1987). The complex impedance of the circuit is represented by the equation:

$$Z = R + \frac{R_1}{1 + (j\tau_1\omega)^{\psi_1}} + \frac{R_2}{1 + (j\tau_2\omega)^{\psi_2}} + \frac{1}{\tau_3(j\omega)^{\psi_3}}, \tag{2.16}$$

where R is the value of the series resistor, R_i, τ_i, and ψ_i, $(i = 1, 2)$ are the resistance, relaxation time, and distribution coefficient of the relaxation time in the ZARC-Cole elements, respectively, while τ_3 and ψ_3 are the values of the parameters in the CPE.

The lumped model as shown in Fig. 2.12b and applied to tomatoes consists of a series resistor R_0 for the soil, and four parallel resistance–capacitance (R$\|$C) circuits for different parts in the continuum of the soil–root interface, root, and stem (Ozier-Lafontaine and Bajazet 2005). The complex impedance of the whole circuit is represented by the equation:

$$Z = R_0 + \sum_{i=1}^{4} \frac{R_i}{1 + (jR_iC_i\omega)} \tag{2.17}$$

where R_i, C_i $(i = 1–4)$ are the resistors and capacitors of the parallel circuits of the different soil–plant components and ω is the angular frequency.

2.3.2.5 Results of EIS for Roots

In the analysis and interpretation of the IS data, it is important to formulate a circuit model that provides a good representation of the electrical properties of the sample and a reasonable interpretation of the model parameters. In practice, it is usually possible to find more than one equivalent circuit that provides a good numerical fit for a given dataset, but only one of these is likely to provide a realistic representation of the electrical makeup of the sample. Three criteria are defined for the elaboration of the equivalent circuit (Srinivas et al. 2003) (1) intuition regarding the kinds of impedance that can be expected to be present in the sample and whether they are connected in series or in parallel; (2) examination of the experimental data to see whether the response is consistent with the proposed circuit; (3) inspection of the resistance and capacitance values obtained in order to check that they are realistic for the continuum under investigation.

When the impedance spectra of the tomato root system were measured at different stages of growth (see Fig. 2.10), the magnitude of the real and imaginary parts of the impedance decreased with the increasing root size (Ozier-Lafontaine and Bajazet 2005) (Fig. 2.13a). The general shape of the spectra remained approximately the same throughout the whole study period, and thus the same IS model could be used for the CNLS curve fitting all of the measurements (see Fig. 2.12b). According to the regression analysis for the different EIS parameters, the highest correlation with root dry mass was found to be in connection with C_3 ($r^2 = 0.963$)

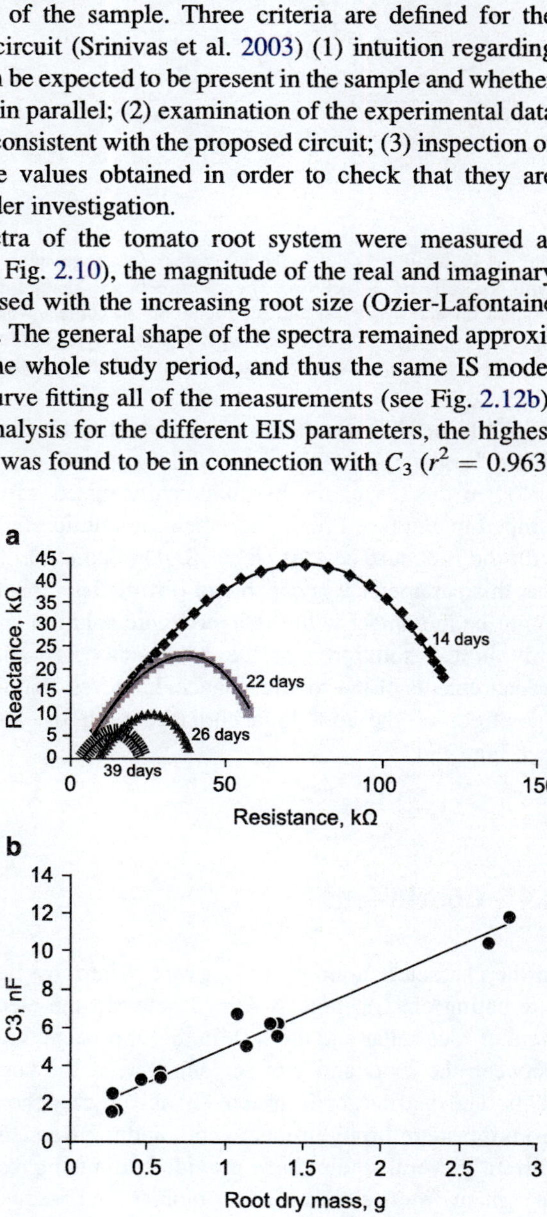

Fig. 2.13 (**a**) Electrical impedance spectra of tomato root systems at different times after planting, indicated by days. In each curve the frequency increases from right (10 Hz) to left (1 MHz). (**b**) The correlation between capacitance C_3 (see Fig. 2.12b) and the root dry mass. (Reproduced by kind permission of Springer Science + Business Media: Ozier-Lafontaine and Bajazet 2005, Plant and Soil 277 p 308, Fig. 6c and p 311, Fig. 8a)

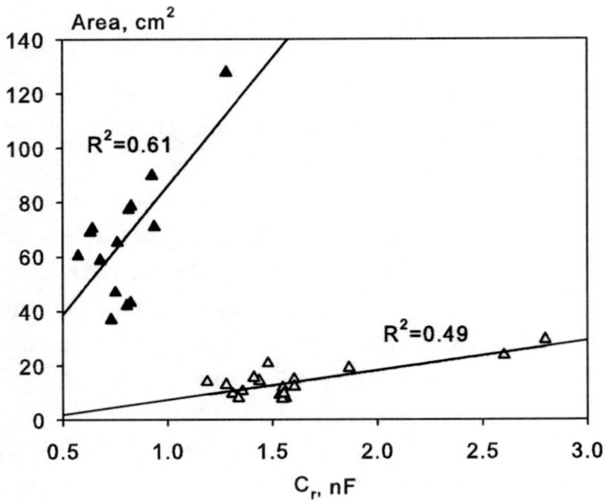

Fig. 2.14 Relation between the interfacial root capacitance (C_r) and the root surface area of hydroponically raised willows. The assessment was made for roots and stem in contact with the solution (*closed triangles*) and for roots only in contact with the solution (*open triangles*) (see Fig. 2.9a, b for measurement set-ups, respectively). (Reproduced from Cao et al. *Journal of Experimental Botany*, by permission of Oxford Journals, 2011)

(Fig. 2.13b). This correlation was higher than for the capacitance measurement at a single frequency of 1 kHz ($r^2 = 0.829$).

When the ISs of the hydroponically raised willows were analyzed using the lumped model (see Fig. 2.12c), the root-solution interfacial capacitance correlated with the root surface area ($R^2 = 0.61$) (Cao et al. 2011) (Fig. 2.14). It was found that this parameter was capable of distinguishing between the case where the stem might be in contact with the hydroponic solution and the case where the root was only in the solution (for the latter case $R^2 = 0.49$). It was found from the measurements of the root resistance, however, that the effect of the stem exceeded the effect of the root in accordance with the single-frequency measurements (see Fig. 2.7).

2.4 Conclusions

In the characterization of a plant root system by electrical impedance method, an alternating electric field is forced between the electrodes connected to the plant stem or root collar and the soil. In this approach, the distribution of an electric field between the roots and growing substrate is the key factor in the detection of the roots. The distribution is affected by the physicochemical properties in the pathway from the stem through roots to soil, and the frequency and strength of the applied current. Several studies have provided promising results for the characterization of the plant root system (e.g., biomass), based on an electrical capacitance measurement at a single low frequency. In addition, the electrical resistance measurements at a single frequency correlated with the root surface area of willow cuttings when only the root was in contact with the growing substrate, while the contact of the stem with the growing solution caused the estimate of the root surface area to be inaccurate. The single-frequency measurements gained a limited scope within the whole circuitry between the electrodes in the stem–root–soil continuum, and thus the different components in the circuitry could not be separated. A more detailed analysis could be carried out using a method based on EIS and equivalent circuit models that consider the role of the stem, root, and growing medium. As a result of the use of EIS, a parameter referring to root capacitance correlated positively with the root biomass and root surface area. Several questions remain open with regard to the applications of the method. More studies are needed to evaluate the longitudinal and radial electric field distribution between the root interior and the soil along the root system, from the root collar to the root tips. As far as the applications are concerned, further studies will be needed under standardized measurement conditions with soils of different ionic composition and texture as the growing substrate, and also by taking into account the role played by mycorrhizas.

Acknowledgements We should like to thank Dr Tarja Lehto and Dr Marja Roitto for their comments and Dr John A Stotesbury for the English revision of the manuscript.

References

Ackmann JJ, Seitz MA (1984) Methods of complex impedance measurements in biological tissue. CRC Crit Rev Biomed Eng 11:281–311

Altmann M, Pliquett U, Suess R, von Borell E (2004) Predication of lamb carcass composition by impedance spectroscopy. J Anim Sci 82:816–825

Aubrecht L, Stanek Z, Koller J (2006) Electrical measurement of the absorption surface of tree roots by the earth impedance method: 1. Theory. Tree Physiol 26:1105–1112

Barsoukov E, Macdonald JR (2005) Impedance spectroscopy theory, experiment, and applications, 2nd edn. Wiley, Hoboken, New Jersey

Bayford RH (2006) Bioimpedance tomography (electrical impedance tomography). Ann Rev Biomed Eng 8:63–91

Butnor JR, Doolittle JA, Johnsen KH, Samuelson L, Stokes T, Kress L (2003) Utility of ground-penetrating radar as a root biomass survey tool in forest systems. Soil Sci Soc Am J 67:1607–1615

Cao Y, Repo T, Silvennoinen R, Lehto T, Pelkonen P (2010) An appraisal of the electrical resistance method for assessing root surface area. J Exp Bot 61:2491–2497

Cao Y, Repo T, Silvennoinen R, Lehto T, Pelkonen P (2011) Analysis of willow root system by electrical impedance spectroscopy. J Exp Bot 62:351–358

Čermák J, Ulrich R, Stanek Z, Koller J, Aubrecht L (2006) Electrical measurement of the absorption surface of tree roots by the earth impedance method: 2. Verification based on allometric relationships and root severing experiments. Tree Physiol 26:1113–1121

Chloupek O (1972) The relationship between electric capacitance and some other parameters of plant roots. Biol Plant 14:227–230

Chloupek O (1977) Evaluation of the size of a plant's root system using its electrical capacitance. Plant Soil 48:525–532

Chloupek O, Skàcel M, Ehrenbergerova J (1998) Effect of divergent selection for root size in field-grown alfalfa. Can J Plant Sci 79:93–95

Cole KS, Cole RH (1941) Dispersion and absorption in diaelectrics. J Chem Phys 9:341–351

Costa C, Dwyer LM, Hamilton RI (2000) A sampling method for measurement of large root systems with scanner-based image analysis. Agr J 92:21–27

Dalton FN (1995) In-situ root extent measurement by electrical capacitance methods. Plant Soil 173:157–165

Dvoràk M, Cernohorska J, Janacek K (1981) Characteristics of current passage through plant tissue. Biol Plant 23:306–310

Fisher RA (1922) On the mathematical foundations of theoretical statistics. Phil Trans Roy Soc London Ser A222:309–368

Foster KR, Schwan HP (1989) Dielectric properties of tissues and biological materials: a critical review. In: Bourne JR (ed) Critical reviews in biomedical engineering. CRC, Boca Raton, FL, pp 25–104

Geddes LA, Baker LE (1989) Principles of applied biomedical instrumentation. Wiley, New York

Guegan Q, Foulc JN (2009) Electrical characterization of the solid phase (particles) of electrorheological fluids. J Physics: Conf Ser 149(012006):1–4

Grimnes S, Martinsen OG (2008) Bioimpedance and bioelectricity basics, 2nd edn. Academic, San Diego

Hilhorst MA (1998) Dielectric characterisation of soil. Doctorol thesis. Wageningen Agriculture University, Wageningen

Hirano Y, Dannoura M, Aono K, Igarashi T, Ishii M, Yamse K, Makita N, Kanazawa Y (2009) Limiting factors in the diction of tree roots using ground-penetrating radar. Plant Soil 319:15–24

Kendall WA, Pederson GA, Hill RR (1982) Root size estimates of red clover and alfalfa based on electrical capacitance and root diameter measurements. Grass For Sci 37:253–256

Macdonald JR (1987) Impedance spectroscopy: emphasizing solid materials and systems. Wiley, New York

Majdi H (1996) Root sampling methods – applications and limitations of the minirhizotron technique. Plant Soil 185:255–258

McBride R, Candido M, Ferguson J (2008) Estimating root mass in maize genotypes using the electrical capacitance method. Arch Agr Soil Sci 54:215–226

Ozier-Lafontaine H, Bajazet T (2005) Analysis of root growth by impedance spectroscopy. Plant Soil 277:299–313

Preston GM, McBride RA, Bryan J, Candido M (2004) Estimating root mass in young poplar trees using the electrical capacitance method. Agrofor Syst 60:305–309

Psarras G, Merwin IA (2000) Water stress affects rhizosphere respiration rates and root morphology of young Mutsu' apple trees on M.9 and MM.111 rootstocks. J Am Soc Hort Sci 125:588–595

Ragheb T, Geddes LA (1990) Electrical properties of metallic electrodes. Med Biol Eng Comp 28:182–186

Rajkai K, Vegh KR, Nacsa T (2005) Electrical capacitance of roots in relation to plant electrodes, measuring frequency and root media. Acta Agr Hung 53:197–210

Repo T, Zhang G, Ryyppö A, Rikala R (2000) The electrical impedance spectroscopy of Scots pine (*Pinus sylvestris* L.) shoots in relation to cold acclimation. J Exp Bot 51:2095–2107

Repo T, Zhang MIN (1993) Modelling woody plant tissues using a distributed electrical circuit. J Exp Bot 44:977–982

Repo T, Zhang MIN, Ryyppö A, Vapaavuori E, Sutinen S (1994) Effects of freeze-thaw injuring on parameters of distributed electrical circuits of stems and needles of Scots pine seedlings at different stages of acclimation. J Exp Bot 45:823–833

Repo T, Pulli S (1996) Application of impedance spectroscopy for selecting frost hardy varieties of English ryegrass. Ann Bot 78:605–609

Repo T, Oksanen E, Vapaavuori E (2004) Effects of elevated concentration of ozone and carbon dioxide on the electrical impedance of leaves of silver birch (*Betula pendula*) clones. Tree Physiol 24:833–843

Repo T, Laukkanen J, Silvennoinen R (2005) Measurement of tree root growth using electrical impedance spectroscopy. Silva Fenn 39:159–166

Rhoades JD, Raats PAC, Prather RJ (1976) Effects of liquid-phase electrical conductivity, water content, and surface conductivity on bulk soil electrical conductivity. Soil Sci Soc Am J 40:651–655

Ryyppö A, Repo T, Vapaavuori E (1998) Development of freezing tolerance in roots and shoots of Scots pine seedlings at non-freezing temperatures. Can J For Res 28:557–567

Samson BK, Sinclair TR (1994) Soil core and minirhizotron comparison for determination of root length density. Plant Soil 161:225–232

Schanne OF, Ruiz-Ceretti E (1978) Impedance measurements in biological cells. Wiley, New York

Schwan HP (1957) Electrical properties of tissue and cell suspensions. In: Lawrence JH, Tobias CA (eds) Advances in biological and medical physics, vol 5. Academic, New York, pp 147–209

Schwan HP (1963) Determination of biological impedances. In: Nastuk WL (ed) Physical techniques in biological research, vol 6. Academic, New York, pp 323–406

Schwan HP (1988) Biological effects of non-ionizing radiations: cellular properties and interactions. Ann Biomed Eng 16:245–263

Schwan HP (1992) Linear and nonlinear electrode polarisation and biological materials. Ann Biomed Eng 20:269–288

Schwan HP, Takashima S (1991) Dielectric behaviour of biological cells and membranes. Bull Inst Chem Res 69:459–475

Schwarz G (1962) A theory of the low frequency dielectric dispersion of colloidal particles in electrolytes solutions. J Phys Chem 66:26–36

Smit AL, Bengough AG, Engels C, van Noordwijk M, Pellerin S, van de Geijn SC (2000) Root methods: a handbook. Springer, New York

Srinivas K, Sarah P, Suryanarayana SV (2003) Impedance spectroscopy study of polycrystalline $Bi_6Fe_2Ti_3O_{18}$. Bull Mater Sci 26:247–253

Thomasset AL, Lenoir J, Jenin MP, Roudlet Cand Ducros MH (1973) Appréciation de la situation électrolytique tissulaire par le rapport des impédances corporelles du corps humain en basses et hautes fréquences. Rev Médec Aéronaut Spot 46:312–315

Tiitta M, Savolainen T, Olkkonen H, Kanko T (1999) Woody moisture gradient analysis by electrical impedance spectroscopy. Holzforsch 53:68–76

van Beem J, Smith ME, Zobel RW (1998) Estimating root mass in maize using a portable capacitance meter. Agr J 90:566–570

Zhang MIN, Stout DG, Willison JHM (1992) Plant tissue impedance and cold acclimation: a re-analysis. J Exp Bot 43:263–266

Zhang MIN, Willison JMH (1991) Electrical impedance analysis in plant tissues: a double shell model. J Exp Bot 42:1465–75

Zhang MIN, Willison JHM (1993) Electrical impedance analysis in plant tissues: impedance measurement in leaves. J Exp Bot 44:1369–1375

Chapter 3
Multi Electrode Arrays (MEAs) and the Electrical Network of the Roots

Elisa Masi, Eisa Azzarello, Camilla Pandolfi, Susanna Pollastri, Sergio Mugnai, and Stefano Mancuso

Abstract Electrically excitable cells are present in many multicellular organisms, especially in brains of animals, but also in lower animals such as sponges, which lack central nervous system and in animals having excitable epithelia, which can conduct signals via neuroid conduction. In plants, most cells are electrically excitable and active, releasing and propagating action potentials (APs), which may affect central physiological processes such as photosynthesis and respiration. The first report describing electrical signals in plants was published over 200 years ago on carnivorous plants. Since then, many researchers have made detailed analyses of the electrical activity of single cells by using microelectrodes for intracellular recordings. However, such techniques cannot address integrated issues of how large assembly of cells can combine information both spatially and temporally. This can be possible using a multi electrode array (MEA) approach, intensively used in neuroscience for any electrogenic animal tissues, and here presented as its first application also for plants tissues. The system allows noninvasive, long time and multisite recording and stimulation with high spatiotemporal resolution. After a short description of the MEA technique in terms of hardware and background, the chapter mainly focuses on the application of this technique in plant electrophysiology by showing some recent works concerning the study of both the intense spontaneous electrical activities and the stimulation-elicited bursts of locally propagating electrical signals generated by the root apex.

E. Masi • E. Azzarello • C. Pandolfi • S. Pollastri • S. Mugnai (✉)
Department of Plant, Soil and Environment, University of Florence, viale delle Idee 30, 50019 Sesto Fiorentino (FI), Italy
e-mail: sergio.mugnai@unifi.it

S. Mancuso
Dpt. Plant, Soil & Environment, University of Firenze, Viale delle idee 30, Sesto Fiorentino (FI) – Italy

S. Mancuso (ed.), *Measuring Roots*,
DOI 10.1007/978-3-642-22067-8_3, © Springer-Verlag Berlin Heidelberg 2012

Abbreviations

AP	Action Potential
DMSO	Dimethyl Sulphoxide
DNQX	6,7-Dinitroquinoxaline-2,4-dione
Gd^{3+}	gadolinium
L-glu	L-glutamate
MCS	Multi Channel Systems®
Mer	Meristem
MZ	Mature Zone
S/N	signal-to-noise
TiN	Titanium nitride
TZ	Transition Zone

3.1 Electrical Signals in Plants: A Brief Introduction

Electrically excitable cells are constitutive of many multicellular organisms, and are especially located in the brain of animals (Greenspan 2007). Lower animals such as sponges, even lacking central nervous system (Leys and Mackie 1997), can also possess electrically excitable cells, together with animals having excitable epithelia, which can conduct signals via neuroid conduction (Miljkovic-Licina et al. 2004; Miller 2009). Electrical events conducted in dedicated tissues may serve for the transduction of environmental information, obtained by sensory systems, into biological cues and processes.

In *plants*, most cells are electrically excitable and active. They can produce, release, and propagate action potentials (APs), which may affect such central physiological processes as photosynthesis and respiration (Fromm and Lautner 2007). Moreover, electrical signals play a central role in both intercellular and intracellular communication (Davies 2004; Felle and Zimmermann 2007) at all levels of evolution from algae (Beilby 2007) to bryophytes (Favre et al. 1999) and higher plants (Pickard 1973; Wildon et al. 1992; Mancuso 1999). This propagation of electrical events occurs in the membranes of nonnervous, nonmuscular cells, which is a general and basic characteristic of plants, and not only an oddity of carnivorous plants such as *Dionea* (Volkov et al. 2009). This electrical activity can be considered as equivalent to the neuroid conduction since it is propagated (1) in an *all-or-none* basis, (2) in nonnervous tissues, (3) from cell to cell via plasmodesmata, and (4) finally decline rapidly in amplitude and velocity due to the flow of the current in all directions.

Electrical signals in plants were described over 200 years ago (Bertholon 1783). After the early works of Burdon-Sanderson (1873) and Darwin (1875) on carnivorous plants, electricity became an important field of research in plants so that the analysis of the action potential in *Nitella* preceded that in the giant axon of the squid (Blinks 1930). Since then, many researchers have made detailed analyses of the electrical activity of single cells by using microelectrodes for intracellular recordings (Davies 2004). However, those techniques cannot obtain integrated issues of how large group of cells can combine information both spatially and temporally. In fact, it is technically difficult to record from and stimulate more than three cells concurrently using standard intracellular microelectrodes, and those cells usually die within minutes or, rarely, hours (Volkov 2006). Other techniques used in nonplant systems, such as optical techniques using voltage-sensitive dyes, although showing good spatial resolution, suffer from low signal-to-noise (S/N) ratio (Kerr and Denk 2008). A stable S/N ratio, necessary for the detection of signals in the lower range of 20–30 μV, can be achieved by the surface-mounted technology (SMT) of pre- and filter amplifiers where the complete electronic circuit and amplifier hardware are built into a single housing. This ensures optimal S/N ratio of the recording because no further cables are necessary other than a single SCSI-type cable connecting the amplifier to the data acquisition card. This results in an overall low noise level of the complete amplifier chain (×1,200, 12-bit resolution, 10–3 kHz) of ±3 μV, which is well within the ±5–10 μV noise level of a MEA TiN (titanium nitride) electrode (flat, round TiN electrode has low impedance, ranging from 20 to 400 kΩ).

3.2 The MEA Technology

The MEA technique was first developed for the detection of electrical activity produced in electrogenic cells of humans and animals, always accompanied by the flow of current through the extracellular fluid surrounding the cellular signal sources. In particular, it has been intensively used in neuroscience for spike recording in brain slices (Hampson et al. 1999), dissociated neuronal cultures (Madhavan et al. 2007), retinas (Torborg et al. 2004), cardiac myocytes (Friedman 2002), and generally in any electrogenic animal tissues. From the pioneering work of Masi and coworkers (2009), the MEA technique has been well standardized also for the analysis and detection of electrical activity in plant tissues, mainly root apices.

Thomas et al. (1972) introduced this technology almost 40 years ago. Since then, the MEA technology and the related culture methods for electrophysiological cell and tissue assay have been continually improved, with the development of high-quality arrays with novel electrodes, suitable for long-term monitoring with a sufficient S/N ratio, the establishment of advanced applications, and the analysis of data. The ability to record long-term electrophysiological information with extremely high spatio-temporal resolution and the correlation of this information with single cell and morphological data represents a paradigm shift from conventional techniques.

In principle, a MEA system is a two-dimensional arrangement of voltage probes designed for extracellular stimulation and monitoring of electrical activity of electrogenic cells. Its main advantage is linked to the possibility to carry out noninvasively multisite recording and stimulation. An extracellular voltage gradient, varying in time and space according to the time course of the temporal activity and spatial distribution of cells, is directly related to the current. Time resolution is limited only by the sampling rate of the data acquisition system: for example, 64-channel parallel, 50 kHz, and 16-bit sampling has already been achieved (Jimbo et al. 2003). When compared to optical recording, the latter offers finer spatial resolution, but MEA technique is able to offer a finer temporal resolution avoiding any damage to the recorded cell, together with direct access to cells for electrical stimulation. This permits the possibility of long-time recording.

The use of MEA is related to one (or a combination) of the following motivations:

1. To obtain information about the interactions between electrogenic cells at different locations in the same tissue. This can be used to monitor and analyze the spatial and temporal dynamics of the electrical activity
2. To permit the reduction of the time required for an experiment by a simultaneous recording at a multitude of sites. Thus, a comparison of tissue properties at different locations can be efficiently performed
3. To monitor changes of electrical activity in vitro, which cannot be performed by other conventional techniques, such as microelectrodes
4. To allow the recording of activity in excitable cells for a very long time (up to weeks), contrary to other conventional electrophysiological activity, such as patch clamping and single-electrode

Extracellular potentials deriving from cells are monitored and recorded at the 2-dimensional surface of a conductive sheet where a tissue slice with electrogenic cells (such as, for example, the transition zone – TZ – of the root apex) is put into contact with the planar substrate.

Basically, the system is composed by four main parts and components (Fig. 3.1):

1. The substrate with embedded microelectrodes
2. The interface between the tissue and the electrodes

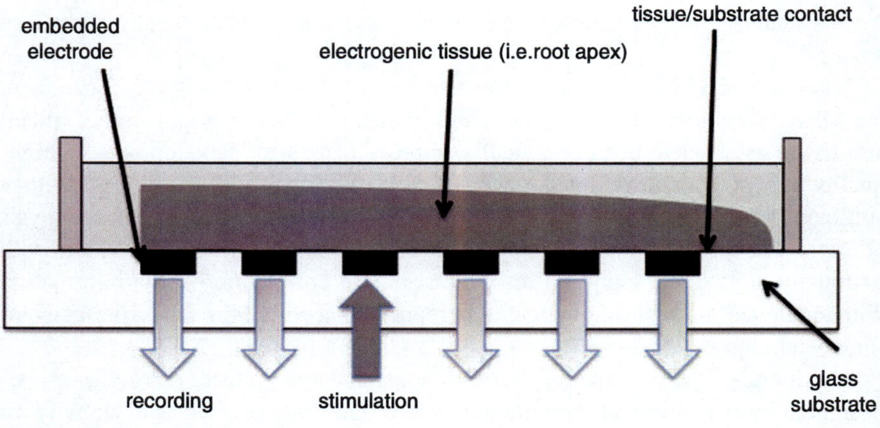

Fig. 3.1 Stimulation and recording of electrical activity in a root apex with MEA. The substrate-integrated planar electrodes can be use for both stimulation and recording

3. The cellular signal sources
4. The external hardware with stimulators, filter amplifiers, and recording system connected to the microelectrodes

The first component is a cell-culture dish (biochip) with an embedded microelectrodes array, multichannel extracellular recording setup, and an electrical stimulation system. The MEA substrate is fabricated using the technique of photolithography (Jimbo et al. 2003), or other techniques such as metal deposition, wet chemical etching, and back-end technologies (Heuschkel et al. 2006). The choice of the right material for the fabrication of the MEA biochip is very important as both the electrical and the experimental conditions can be deeply influenced by its quality. For example, all materials must not show any toxic effect on biological samples and should allow good adhesion of the biological preparation. Furthermore, the electrical characteristics of the electrodes should allow the measurement of small signal amplitudes with a good signal-to-noise ratio. Also, the MEA biochip should be as transparent as possible to permit the observation of samples through an inverted microscope.

The substrate can be made of glass (the most used material Heuschkel et al. 2006), polymers (Stoppini et al. 1997) and silicon (Offenhausser et al. 1997). The main advantages of glass substrates compared to the other materials are their transparency, good chemical and electrical isolating features, and low cost.

Many different metals can be effectively used for the construction of electrodes, and their choice mainly depends on the application and the specific requirements due to their different electrical characteristics. For example, to obtain very low noise levels for monitoring the electrical activity of biological samples with very small signal amplitudes, the use of porous materials such as platinum black and titanium nitride is advisable. For specific cell patterning, gold electrodes are recommended. On the contrary, indium-TiN oxide (ITO) electrodes are mainly used when optical and electrical recording must be combined, due to the optical transparency of the material, which allows the perfect transparency of the MEA biochips. However, the most commonly used electrode material is the platinum black because of its low noise level, even if it can present a weak adhesion on the underlying electrode that often leads to the removal of material during the cleaning process. A good compromise between good electrical characteristics and a cheaper fabrication process is the use of metallic platinum: in fact, the noise level obtained is in the range of 20 μV (Heuschkel et al. 2002).

Another important aspect is the electrode insulation, because it impacts the unavoidable parasitic capacitances between the electrode and the culture medium (normally a conductive saline solution). To reduce parasitic capacitances, the right material must be chosen on the basis of a low dielectric constant or have a larger thickness. The most widely used materials for insulation are silicon nitride and silicon dioxide (also combined together), deposited as thin-film layer ($<$1 mm) on the top of the MEA biochip. Other materials are polymers (polyimide and SU-8 epoxy) because of their lower cost and their larger thickness, which can better reduce the parasitic capacitance.

3.2.1 Pattern of the Electrodes

The main standard electrode patterns are composed by a 8 × 8 or 6 × 10 square grid of electrode terminals, with an interelectrode distance of 50–500 μm, depending on the type of sampled tissue (Jimbo et al. 2003; Shew et al. 2010; Table 3.1). The actual electrodes are formed by an opening through the insulator (5–50 μm of diameter), exposing the tip of the leads underneath (10–50 μm square). To resume the fabrication features of the system, we present here shortly (Table 3.2) the most widely used configuration and construction methods of MEA systems for electrical network analysis in roots (Masi et al. 2009). On a glass substrate, a thin indium-TiN-oxide layer is deposited to form the electrode-array pattern by etching. Electrode terminals are arranged in a 6 × 10 grid (Fig. 3.2), with interelectrode distance (IED) of 500 μm and electrode diameter of 30 μm (Multi Channel Systems (MCS), Reutlingen, Germany). The electrodes were coated with porous titanium nitride to minimize the impedance (about 100 kΩ at 1 kHz) and to allow recording of local field potentials at a high signal-to-noise ratio.

Table 3.1 The most common electrode sizes for different applications and types of tissues, according to the scientific literature

Electrode size (μm)	Type of tissue	Reference
10 μm diameter	Various	Stieglitz (2001)
100 μm × 1 mm	Hippocampal cultures	Tsay et al. (2005)
150 μm diameter	Spinal cord and retina	Rodger et al. (2006)
20 μm diameter	Cerebellar slices	Egert et al. (2002)
30 μm diameter	Maize root apex	Masi et al. (2009)
300–350 μm diameter	Retina	Maghribi et al. (2002)
500 μm × 1 mm	Peripheral nerves	Sahin et al. (1997)
500 μm diameter	Radial nerve	Schuettler and Stieglitz (2000)
600 μm square	Various	Schuettler et al. (2005)

Table 3.2 Common configuration and construction methods of MEA systems used for electrical network analysis in roots

MEA settings	
Amplifier gain	1,200
Input Voltage Range of Data Acquisition Board	±4,096 mV
Samplink frequency	1–50 kHz
ADC resolution	16 bit
Data format	16 bit
Total channels	64
Electrode raw data channels	60
Analog raw data Channels	3
Digital channel	1

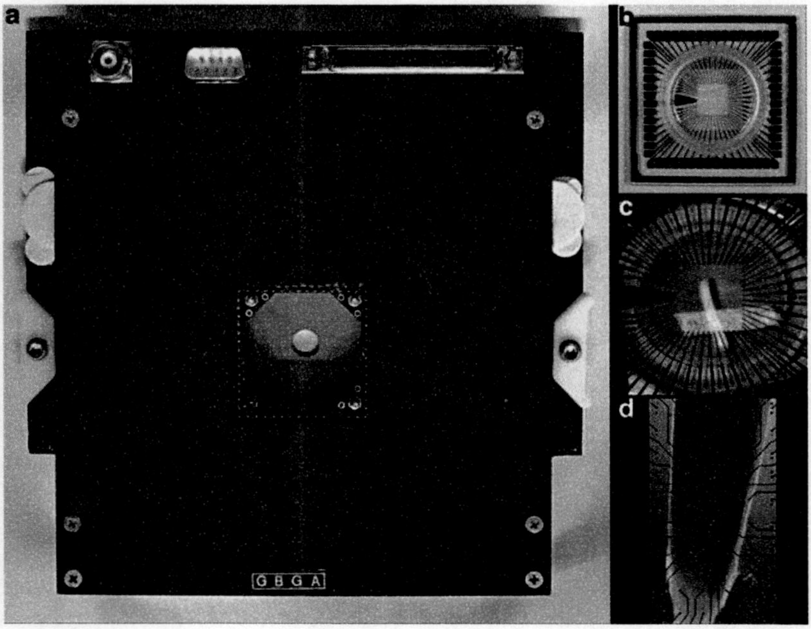

Fig. 3.2 MEA system (**a**); MEA chip with a 6 × 10 grid (500 μm IED) (**b**); a longitudinal slice of a root apex placed on a 6 × 10 MEA array (**c** and **d**) [Modified from Masi et al. 2009]

3.2.2 Recording Process

The sources of the monitored signals are blocks of single cells, which produce an extracellular field potential just after the local alteration of ion channel conductance. The time course of the response is approximately equal to the transmembrane current of the active cellular compartments, and the spatial distribution of the voltage in a thin layer of a conductive tissue sheet is recorded with respect to the reference electrode. The spontaneous or evoked electrical activity spreads from cells to cells into the whole tissue, and it is always accompanied by the flow of ionic current through the extracellular fluid. An extracellular voltage gradient, which varies in time and space according to the temporal activity, the spatial distribution and the orientation of the cells, is related to the current.

Slow field potentials and fast spikes arising from action potentials can be both recorded during any experiment. Spike activity with MEA electrodes can be detected and monitored at a certain distance from each electrode center: for example, Egert et al. (2002) detected spike activity with MEA electrodes at distances of up to 100 μm from a neuron. However, signal sources can be basically detected within a radius of 30 μm around the MEA electrode center. During the recording process, a Helmholtz double layer builds up at the metal–electrolyte interface, so producing an effective high-pass filter. The cut-off frequency of this high-pass filter is lowered by increasing the capacitance of the layer. This can be

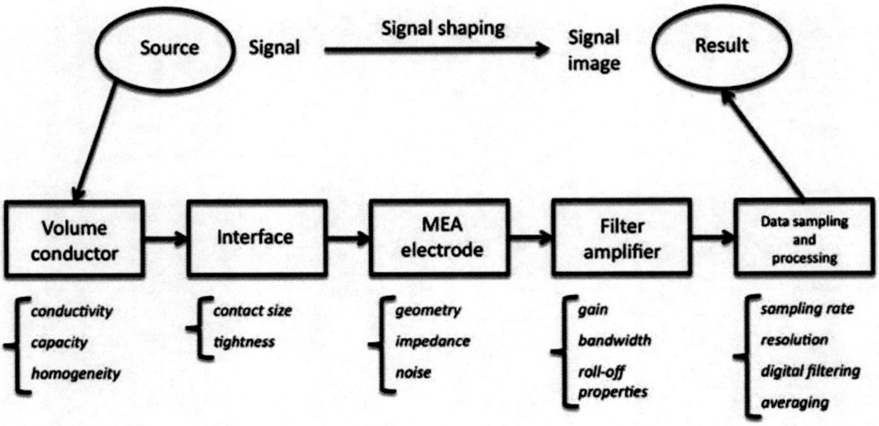

Fig. 3.3 The recorded, stored, and analyzed signal is an image of the original signal produced by the source (i.e., root apex), which is shaped by several parameters related to the components of the signal chain

achieved by the deposition of platinum or titanium nitride forming a stable columnar structure. The recording conditions for MEAs are defined by several biophysical and biological factors reported in Fig. 3.3.

3.3 Determination and Analysis of Spontaneous Electrical Activity in Roots

Masi et al. (2009) applied a 60-channels MEA to study and evaluate the spatiotemporal characteristics of the electrical network activity produced by thick root apex slices. Caryopses of *Zea mays* L. were soaked in distilled water and placed between damp paper towels in Petri dishes. Dishes were maintained in a vertical position, incubated at 26°C for 48 h; roots were used when length was approx. 3 cm (usually after 2 days). Longitudinal and transversal slices from the primary root tip were obtained by cutting the roots at 350 μm thickness with a tissue slicer (Mod. 51425, Stoelting, Illinois, USA) and then stored by submerging them for 2 h in $CaCl_2$ 5 mM (pH 6.5) at room temperature (22°C) before recording.

For recording, root samples were placed on a MEA consisting of 60 TiN microelectrodes (diameter 30 μm). We used two kinds of MEA chips: a 8 × 8 matrix for transversal slices (200 μm of IED) and a 6 × 10 matrix for longitudinal ones (500 μm IED) (Multi Channel Systems® MCS, Reutlingen, Germany; Fig. 3.2b). To achieve the best contact between the root tissues and the electrodes, the MEA was previously coated with nitrocellulose, and then root slices were fixed onto the MEA using an adhesive permeable and water resistant tape (3M Micropore Surgical Tape) that mediates close and flat adhesion of the slice to the MEA surface, allowing the concurrent superfusion of the tissue (Fig. 2c and d). Basically, recording was performed with tissues submerged in a bath solution containing $CaCl_2$ (5 mM, JT Baker Chemical Co, Deventer, The Netherlands) at a flow rate of 1 mL min^{-1} using a perfusion system at a temperature of 22 ± 1°C.

Data were acquired with a complete system for signal filtering, amplification and acquisition (MEA1060BC, MCS; Fig. 3.2a) at a sampling rate of 20 kHz (Table 3.2). A dedicated software such as MC Rack (MCS) was used to control the hardware and to record simultaneously field potentials from the 60 electrodes composing the MEA chip. Spike detection was performed using the virtual spike sorter instrument of MC Rack capable to individuate and extract spikes from the raw data using the method of the threshold. Waveforms that met the defined requirements (threshold at −30 μV, approx. 5 times the basal noise) were cut out from the raw data and used for further analysis (i.e., spike rate, cross correlations, etc.) with custom built MatLab applications. Signals were embedded in biological and thermal noise of ≈10–20 μV peak-to-peak, with S/N ratio of 8.7 ± 1.

Masi et al. (2009) mainly focused on the detection of individual spike activity produced by a single cell. When recording electrical activity from root apex slices, it is most likely to record the activity of single cells. A previous study demonstrated that an electrode with radius of 30 μm covers a detection area of no more than 100 μm in diameter (Lin et al. 2005). In the case of maize roots, the cells under investigation can be considered larger with respect to the electrode and homogeneous in morphology, with a diameter of 80–100 μm on average (Baluŝka et al. 1990). Therefore, by using an IED of 500 μm, it is reasonable to assume that the electrical signals generated by a single root cell can be detected by just one electrode.

Spikes, whose shape remind to those recorded in animal cells, were observed. It is important to check if the obtained results are compatible with what expected from other conventional and traditional electrophysiological techniques for the detection of electrical activity in the cells. In a typical plant AP recorded intracellularly with conventional microelectrodes, the current flow across the cell membrane that activates it, represents a point sink. Similarly to what already assessed about neurons and other excitable tissues in animals (Pine 2006), it is known that for a point current sink, a voltage is generated equal to

$$\frac{I\rho}{4\pi r} \tag{3.1}$$

where I is the current, ρ is the resistivity of the medium, and r is the distance from the center of the current sink (3.1). Extracellular recording with the MEA system would reasonably produce a signal about 50–100 times weaker than the voltage generated by (3.1).

Spikes amplitude varies from 30 to 400 μV, with duration of about 36–40 ms (Fig. 3.4). Amplitude greatly depends on the distance of the electrodes from the site of generation of the signal while the duration of the detected spikes has been estimated to be quite stable. It is also worth noting that, because of the noise, the

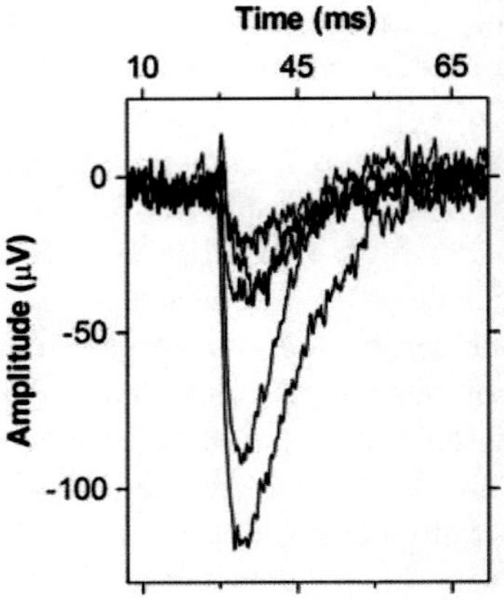

Fig. 3.4 Recordings of spontaneous spikes. Ten traces are superimposed to illustrate the typical shape and size. The signals are aligned at their maximum negative deflections. Data were digitized at 20 kHz

duration seems to be shorter for the spikes of small amplitude; in fact, in that case, the tail of the spike can be confused with the noise of the signal. In particular, the duration of the spikes turned out to be proportional to their amplitude: the smaller is the amplitude of the spike, the smaller is the duration due to the noise effect (Masi et al. 2009). This linear dependence suggests that the electrical events have a fixed temporal duration determined by the physiological features of the cell membrane.

3.3.1 Effects of Calcium and Glutamate

As expected, the presence of calcium in the bath solution is critical for the generation of spikes: no spikes were detected in bath solutions without calcium. The increase of external $CaCl_2$ concentration from 3 to 5 mM determined a significant ($P < 0.01$) increase in the rate of appearance of the spikes (from 3.0 ± 0.4 spikes sec^{-1} at 3 mM to 4.4 ± 0.7 spikes sec^{-1} at 5 mM). (Masi et al. 2009). The importance of calcium for the spike activity was indirectly confirmed by the use of gadolinium (Gd^{3+}), a commonly used blocker of plasma membrane calcium channels (Fig. 3.5a, b). Spikes were almost totally inhibited, at any external calcium concentration, by incubating the root for 1 h in presence of 1 mM of Gd^{3+} (Merck Pty Limited, Milan, Italy) Spike detection after overnight recovering of the root (the root was gently stirred in working solution) revealed that electrical network activity was restored (Fig. 3.5c).

In the complex and not yet fully understood mechanism that undergoes calcium signaling, glutamate receptors are possible candidates for an influx pathway (Dubos et al. 2003) especially in roots (Walch-Liu et al. 2006; Forde and Lea 2007). Externally applied L-glutamate (L-glu) induces membrane depolarization and cytosolic calcium spikes (Dennison and Spalding 2000; Qi et al. 2006). Accordingly, exogenous application of L-glu (0.5 mM, 1 mM, 2 mM; VWR, Milan, Italy) significantly ($P < 0.01$) increased spike rate (from 2 to 10 times greater; Fig. 3.5d); almost

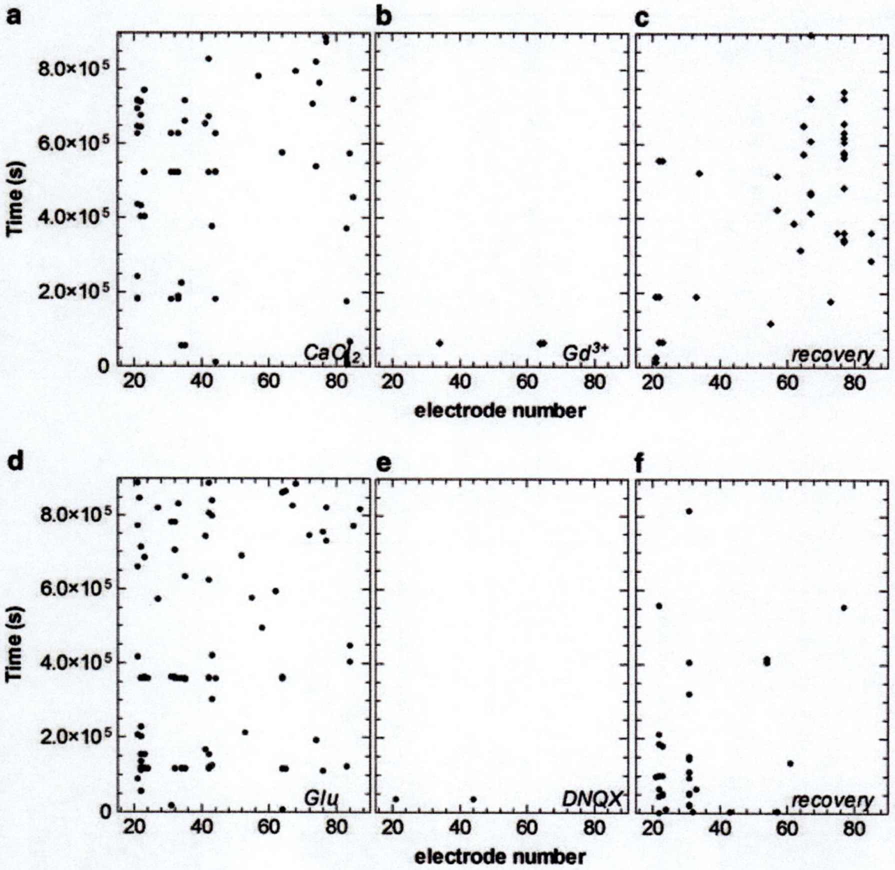

Fig. 3.5 Representative raster plots of spontaneous activity recorded from 60 electrodes (linear order) in maize root under the following conditions: (**a**) $CaCl_2$ 5 mM; (**b**) $CaCl_2$ plus Gd^{3+} 1 mM; (**c**) restored electrical activity after washing out of Gd^{3+} and overnight recovery; (**d**) L-glu 1 mM plus $CaCl_2$ 5 mM; (**e**) L-glu 1 mM plus $CaCl_2$ 5 mM plus DNQX 2.5 mM; (**f**) restored electrical activity after washing out of DNQX and overnight recovery [Modified from Masi et al. 2009]

likely, the effect of L-glu was not time-steady (data not shown). In all of the tested roots, the glutamate-evoked increase in activity was significantly ($P < 0.001$) attenuated (Fig. 3.5e) by the glutamate receptor antagonist 6,7-Dinitroquinoxaline-2,4-dione (DNQX, 2.5 mM, pH 6.7, 2 h incubation, Sigma, Milan, Italy) dissolved in 2.5% DMSO (Dimethyl sulfoxide, Sigma, Milan, Italy). Here, again it was possible to restore the electrical activity of the root after overnight recovery (Fig. 3.5f). In these experiments, each solution was adjusted to pH 6.5 and used, without root sample, as negative control to check the S/N ratio of the electrodes.

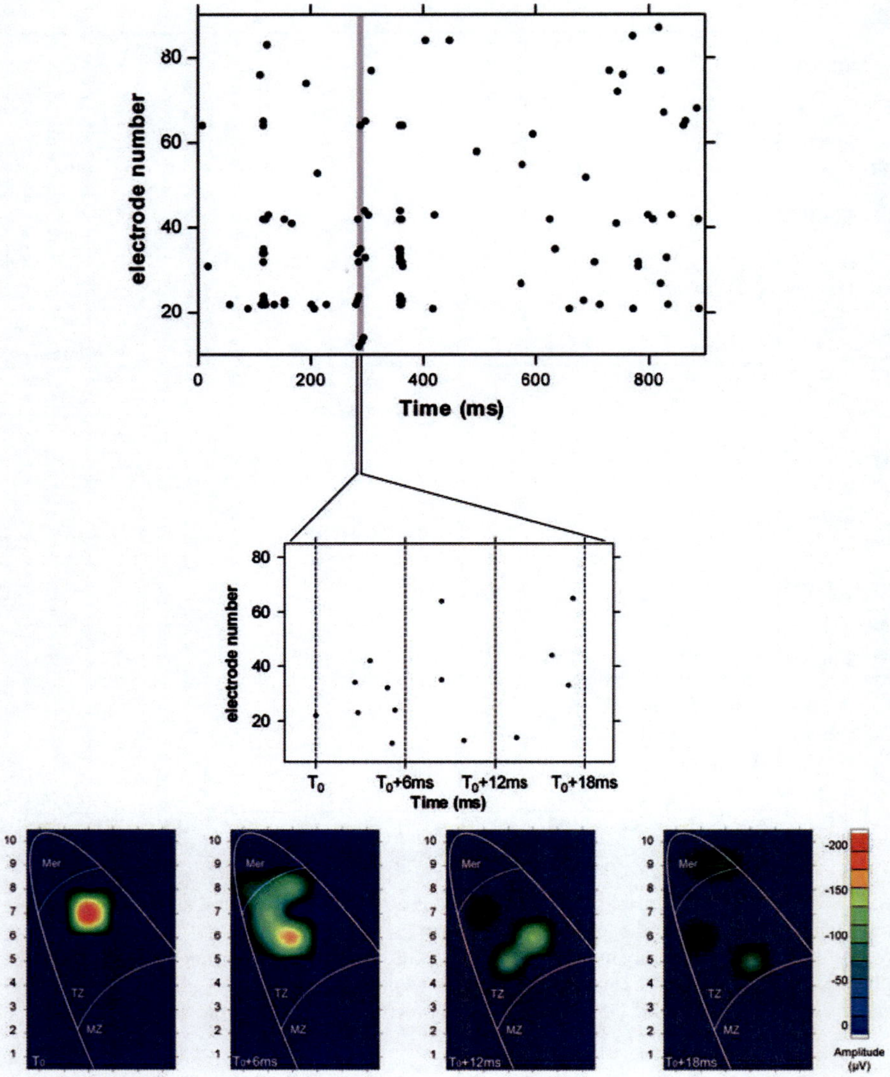

Fig. 3.6 Activity within synchronized periods. Raster plot of spontaneous activity (*Top*) shows correlated periods containing spatiotemporal patterns (*Middle*). A high-temporal-resolution map of the spatial propagation of a single impulse generated at the time 0 is shown (*Bottom*). A wide spread of the signal is evident [Modified from Masi et al. 2009]

3.3.2 Synchronization and Propagation of the Spikes

The electrical activity in the maize root apex followed a specific pattern of activity with spikes that developed spontaneously and concurrently on many electrodes, forming periods of synchronized activity separated by intervals of many seconds without any activity. This appears to be the first observation ever of synchronized electrical activity in plants (Fig. 3.6 Top). A detailed look to the activity shows that the recorded synchronized events are actually composed by a quite complex spatiotemporal pattern where the electrical signals did not appear from all the electrodes at the same exact time (Fig. 3.6 Middle). The propagation of the spikes (Fig. 3.6 Bottom) indicates the existence of excitable traveling waves in plants, similar to those observed in nonnerve electrogenic tissues of animals. The calculation of the speed of propagation of the signal from the center of excitation showed that both in the transversal and in the longitudinal direction the velocity of the occurring spikes decreased (Masi et al. 2009). The decrease was much more pronounced in the transversal direction. This supports the hypothesis that electrical signals propagate through plasmodesmata also in root cells; in fact, plasmodesmata are more frequent in transverse than in longitudinal wall of cells of the root apex, especially of the TZ. Consequently, the propagation of signals is easier in longitudinal direction, thus the speed decreased more slowly.

3.3.3 Localization of Prominent Electrical Activity in the Root Apex

The MEA approach for the study of electrical activity have provided data from a quite large area of the root apex, including meristem (Mer), transition zone, and mature zone (MZ). In each experiment performed, whatever the nature of the treatment, the most active root part was revealed to be the transition zone. In fact, this region showed a higher electrical activity than the neighboring root apex zones (Fig. 3.7) and a higher collective activity of groups of cells, that is, synchronization (Masi et al. 2009).

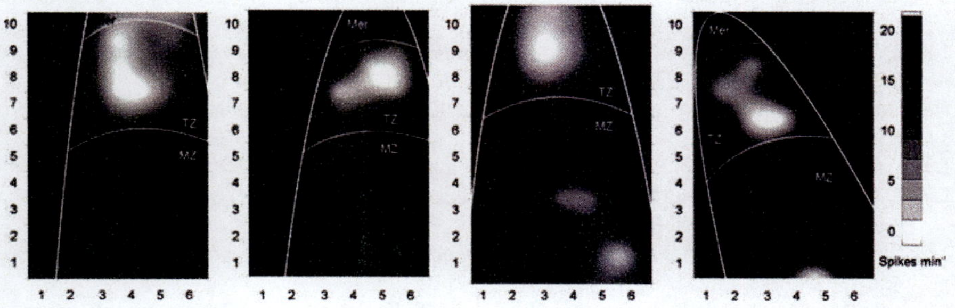

Fig. 3.7 Representative images of the spatial localization of the electrical activity in root apex of maize. Overall electrical activity measured as rate of appearance (spikes per minute) in four different roots/experiments. Root tips were always positioned in the upper part of the MEA array [Modified from Masi et al. 2009]

References

Baluŝka F, Kubica S, Hauskrecht M (1990) Postmitotic "isodiametric" cell growth in the maize root apex. Planta 181:269–274

Beilby M (2007) Action potential in charophytes. Int Rev Cytol 257:43–82

Bertholon ML (1783) De l'Electricité des Végétaux: Ouvrage dans lequel on traite de l'electricite de l'atmosphere sur les plantes, de ses effets sur l'economie des vegetaux, de leurs vertus medicaux (P.F. Didotjeune, Paris)

Blinks LR (1930) The direct current resistance of Nitella. J Gen Physiol 13:495–508

Burdon-Sanderson J (1873) Note on the electrical phenomena which accompany stimulation of a leaf of *Dionaea muscipula*. Proc R Soc London 21:495–496

Darwin C (1875) Insectivorous plants. Murray, London

Davies E (2004) New functions for electrical signals in plants. New Phytol 161:607–610

Dennison KL, Spalding EP (2000) Glutamate-gated calcium fluxes in Arabidopsis. Plant Physiol 124:1511–1514

Dubos C, Huggins D, Grant GH, Knight MR, Campbell MM (2003) A role for glycine in the gating of plant NMDA-like receptors. Plant J 35:800–810

Egert U, Heck D, Aertsen A (2002) Two-dimensional monitoring of spiking networks in acute brain slices. Exp Brain Res 142:268–274

Favre P, Krol E, Stolarz M, Szarek I, Greppin H, Trebacz K, Degli Agosti R (1999) Action potentials elicited in the liverwort Conocephalum conicum (Hepaticae) with different stimuli. Arch Sci 52:175–185

Felle H, Zimmermann MR (2007) Systemic signalling in barley through action potentials. Planta 226:203–214

Forde BG, Lea PJ (2007) Glutamate in plants: metabolism, regulation, and signalling. J Exp Bot 58:2339–2358

Friedman PA (2002) Novel mapping techniques for cardiac electrophysiology. Heart 87:575–582

Fromm J, Lautner S (2007) Electrical signals and their physiological significance in plants. Plant Cell Environ 30:249–257

Greenspan RJ (2007) An introduction to nervous systems. Cold Spring Harbor Laboratory, Cold Spring Harbor, USA

Hampson RE, Simeral JD, Deadwyler SA (1999) Distribution of spatial and nonspatial information in dorsal hippocampus. Nature 402:610–614

Heuschkel MO, Wirth C, Steidl EM, Buisson B (2006) Development of 3D multi electrode arrays for use with acute tissue slices. In: Taketani M, Baudry M (eds) Advances in network electrophysiology using multi-electrode arrays. Springer, Germany, pp 69–111

Heuschkel MO, Fejtl M, Raggenbass M, Bertrand D, Renaud P (2002) A three dimensional multi-electrode array for multi-site stimulation and recording in acute brain slices. J Neurosci Meth 114(2):135–148

Jimbo Y, Kasai N, Torimitsu K, Tateno T, Robinson HPC (2003) A system for MEA-based multisite stimulation. IEEE Trans Biomed Eng 50:241–248

Kerr JND, Denk W (2008) Imaging in vivo: watching the brain in action. Nat Rev Neurosci 9:195–203

Leys SP, Mackie GO (1997) Electric recording from a glass sponge. Nature 387:29–30

Lin Y, Chen C, Chen L, Zeng S, Luo Q (2005) The analysis of electrode-recording horizon. Conf Proc IEEE Eng Med Biol Soc 7:7345–7348

Madhavan R, Chao ZC, Potter SM (2007) Plasticity of recurring spatiotemporal activity patterns in cortical networks. Phys Biol 4:181–193

Maghribi M, Hamilton J, Polla D, Rose K, Wilson T, Krulevitch P (2002) Stretchable micro-electrode array [for retinal prosthesis]. In: Dittmar A, Beebe D (eds) Microtechnologies in Medicine and Biology, 2nd Annual International IEEE-EMB Special Topic Conference, Madison, p 80

Mancuso S (1999) Hydraulic and electrical transmission of wound-induced signals in *Vitis vinifera*. Aust J Plant Physiol 26:55–61

Masi E, Ciszak M, Stefano G, Renna L, Azzarello E, Pandolfi C, Mugnai S, Baluska F, Arecchi T, Mancuso S (2009) Spatiotemporal dynamics of the electrical network activity in the root apex. PNAS 106:4048–4053

Miljkovic-Licina M, Gauchat D, Galliot B (2004) Neuronal evolution: analysis of regulatory genes in a first-evolved nervous system, the hydra nervous system. BioSystems 76:75–87

Miller G (2009) On the origin of the nervous system. Science 325:24–26

Offenhausser A, Sprossler C, Matsuzawa M, Knoll W (1997) Field-effect transistor array for monitoring electrical activity from mammalian neurons in culture. Biosens Bioelectron 12(8):819–826

Pickard BG (1973) Action potential in higher plants. J Bot Rev 39:172–201

Pine J (2006) A history of MEA development. In: Baudry M, Taketani M (eds) Advances in network electrophysiology using multi-electrode arrays. Springer Press, New York, pp 3–23

Qi Z, Stephens NR, Spalding EP (2006) Calcium entry mediated by GLR3.3, an Arabidopsis glutamate receptor with a broad agonist profile. Plant Physiol 142:963–971

Rodger DC, Li W, Fong AJ, Ameri H, Meng E, Burdick JW, Roy RR, Edgerton VR, Weiland JD, Humayun MS, Tai Y (2006) Flexible microfabricated parylene multielectrode arrays for retinal stimulation and spinal cord field modulation. In: IEEE Engineering in Medicine and Biology Society Special Topic Conference on Microtechnologies in Medicine and Biology, Okinawa

Sahin M, Haxhiu MA, Durand DM, Dreshaj IA (1997) Spiral nerve cuff electrode for recordings of respiratory output. J Appl Physiol 83:317–322

Schuettler M, Stieglitz T (2000) 5th Annual International Conference of the International Functional Electrical Stimulation Society. Aalborg, Denmark, p 18

Schuettler M, Stiess S, King BV, Suaning GJ (2005) Fabrication of implantable microelectrode arrays by laser cutting of silicone rubber and platinum foil. J Neural Eng 2:S121–S128

Shew WL, Bellay T, Plenz D (2010) Simultaneous multi-electrode array recording and two-photon calcium imaging of neural activity. J Neurosci Meth 192:75–82

Stieglitz T (2001) Catalogue on available flexible, light-weighted microelectrodes. From www.ibmt.fraunhofer.de/gruppe_l/download/IBMT_Neuro%20Electrode%20Catalogue%20-%20July%202001.pdf

Stoppini L, Duport S, Correges P (1997) A new extracellular multirecording system for electrophysiological studies: Application to hippocampal organotypic cultures. J Neurosci Meth 72:23–33

Thomas CA, Springer PA, Loeb GE, Berwald-Netter Y, Okun LM (1972) A miniature microelectrode array to monitor the bioelectric activity of cultured cells. Exp Cell Res 74:61–66

Torborg CL, Hansen HA, Feller MB (2004) High frequency, synchronized bursting drives eye-specific segregation of retinogeniculate projections. Nat Neurosci 8:72–78

Tsay C, Lacour SP, Wagner S, Morrison B (2005) Architecture, fabrication, and properties of stretchable microelectrode arrays. Proc IEEE Sens, 1169–1172

Volkov AG (2006) Plant electrophysiology: theory and methods. Springer, Berlin

Volkov AG, Carrell H, Markin VS (2009) Biologically closed electrical circuits in Venus flytrap. Plant Physiol 149:1661–1667

Walch-Liu P, Liu LH, Remans T, Tester M, Forde BG (2006) Evidence that L glutamate can act as an exogenous signal to modulate root growth and branching in *Arabidopsis thaliana*. Plant Cell Physiol 47:1045–1057

Wildon DC, Thain JF, Minchin PEH, Gubb IR, Reilly AJ, Skipper YD, Doherty HM, O'Donnell PJ, Bowles DJ (1992) Electrical signalling and systemic proteinase inhibitor induction in the wounded plant. Nature 360:62–65

Chapter 4
The Vibrating Probe Technique in the Study of Root Physiology Under Stress

Camilla Pandolfi, Sergio Mugnai, Elisa Azzarello, Elisa Masi, Susanna Pollastri, and Stefano Mancuso

Abstract The present chapter introduces the use of a non-invasive vibrating probe technique to study membrane-transport processes in plant roots. Net ion fluxes and changes in the extracellular concentrations of ions (such as H^+, Ca^{2+}, K^+, Mg^{2+}, Cl^-, NH_4^+, Cu^{2+}) as well as gaseous molecules (such as O_2 and NO) and other compounds (IAA) are very important to understand the physiology of plants under stress conditions.

Here, we present an overview on the technique, focusing on background theory, main selective electrodes available, and the major outcomes obtained in the study of roots in order to show how in situ and real-time measurements of net ion and gas fluxes from root cells and tissues can provide insights into the functional genomics of plants and will significantly increase our understanding in plant adaptive responses to environmental stresses.

4.1 Introduction

Membrane transport have been widely studied in the last 25 years using a range of techniques, such as membrane potential, patch clamp, advanced image techniques, and ion selective microelectrode. The transport of gases and ions is central to the life of plants, and the measurement of specific ion or gas fluxes has contributed widely to the characterization of membrane transport system and gave an important

C. Pandolfi (✉) • S. Mugnai • E. Azzarello • E. Masi • S. Pollastri
Department of Plant, Soil and Environment, University of Florence, viale delle Idee 30, 50019 Sesto Fiorentino (FI), Italy
e-mail: camilla.pandolfi@unifi.it

S. Mancuso
Dpt. Plant, Soil & Environment, University of Firenze, Viale delle idee 30, Sesto Fiorentino (FI) – Italy

S. Mancuso (ed.), *Measuring Roots*,
DOI 10.1007/978-3-642-22067-8_4, © Springer-Verlag Berlin Heidelberg 2012

contribution to the knowledge at the functional level of all the transporters studied at a molecular and genetic level.

In recent years, ion flux measurements have become a popular tool in studying adaptive responses of plant tissues to a large number of biotic and abiotic stresses, and there are a growing number of laboratories around the world that employ this technique.

It has been applied successfully to study various aspects of membrane-transport processes in higher plants (Pharmawati et al. 1999; Newman 2001) and protoplasts derived from their tissues (Shabala et al. 1998). Nowadays, it can be coupled with a wide range of selective electrodes (commercial and custom-built) permitting the direct measurement of fluxes in a simple, quick, continuous, and non-invasive manner.

The first application is reported as vibrating self-referencing electrode technique (Jaffe and Nuccinelli 1974); later, it was adapted for the use with various ion-selective electrodes, allowing precise non-invasive measurements of ion fluxes in plant tissues (Kochian et al. 1992; Piñeros et al. 1998; Mancuso et al. 2000). A rigorous test of the theory of vibrating probe was performed by Newman et al. (1987). This technique has been successfully applied in many laboratories around the world, such as the National Vibrating Probe Facility at Woods Hale (USA) by Smith (1995); the University of Tasmania (Australia) by Newman (2001) who developed the MIFE system; the University of Florence (Italy) where an oxygen electrode was used for the first time with this technique and the first electrodes for Cu^{2+} and IAA were developed for the first time (Mancuso et al. 2000, 2005).

There are many features that provide a significant advantage of this technique: it allows in situ measurements of net ion fluxes in many physiological conditions; the electrode tip is very thin and makes possible the measure from single cells and protoplasts (Shabala et al. 1998); the small electrode tip also gives the possibility to map a tissue providing spatial distribution and functional expression of specific transporters; its high temporal resolution (0.1 Hz) is important to study rapid signalling events; thanks to its non-invasive characteristic its application is limited only by the lifetime of the electrodes, and it gives the possibility to follow the evolution of the phenomena, an important treat in the study of abiotic stresses; according to the experimental setup, it gives the possibility to measure simultaneously several ions or gas molecules to provide a combined explanation of the phenomena. Furthermore, the use of an extracellular and self-referencing probe, which averages signals over several seconds, minimizes external noise and interferences and allows very precise and reliable measurements (Newman 2001).

Cell membrane plays a fundamental role in cell biological processes: it is responsible for pH and ionic homeostasis, turgor, metabolite distribution, energy transduction and signalling. Changes in plasma membrane potential and modulation of ion flux are essential for plant adaptive responses and represent the earliest cellular events in response to light, temperature, osmotic stress, and salinity (Sanders et al. 1999; Zimmermann et al. 1999; Knight and Knight 2001). Taking salt stress as an example, change in plasma membrane and modulation of ion fluxes might be due to exclusion of toxic Na^+ from the cytosol via either the SOS1 plasma membrane Na^+/H^+ antiporter (Zhu 2003) or by compartmentalizing it into the vacuole by the NHX tonoplast Na^+/H^+ antiporter (Apse et al. 1999). Or, in the case of Al^{3+} toxicity, the adaptive response involves activation of anion channels for malate efflux and changes in the rhizosphere pH (Ryan et al. 2001). Oxygen and other compounds (such as auxin) have been measured along the root, and thanks to the vibrating probe technique it has been possible to determine the spatial distribution of fluxes. These peculiar characteristics have been used to assess changes in auxin transport in Arabidopsis mutants and changes in oxygen fluxes in stress situations. Such a central role of plant membranes and membrane transport processes in plant adaptive responses to environmental conditions makes them important targets, and the vibrating probe system is undoubtedly an easy and very informative electrophysiological approach.

4.2 The Technique

4.2.1 Theory

The main idea of the vibrating probe technique is that ions in solution tend to move from high to low electrochemical potential. If living tissues absorb an ion, its concentration close to the surface will be lower than that further away (Fig. 4.1). Vice versa, if the tissue releases the ion across the membrane, there will be a pronounced concentration gradient directed away from the tissue surface. The magnitude of this gradient will be proportional to the rate of ion movement across the tissue. In steady or at least slowly changing conditions, measurements of the net diffusive flux in solution close to the tissue give the net flux of the ion or gas molecule across the tissue surface.

The recorded voltage gradients ΔV between the two positions are converted into concentration differences (ΔC) using the calibrated Nernst slopes of the used electrodes (Nobel 1974). So, given the concentration of the ion in these two specific positions and the mobility and concentration of the specific ion in solution, it is possible to calculate the net flux of the tissue. This basic concept of the vibrating probe technique has been described in detail before (Kuhtreiber and Jaffe 1990;

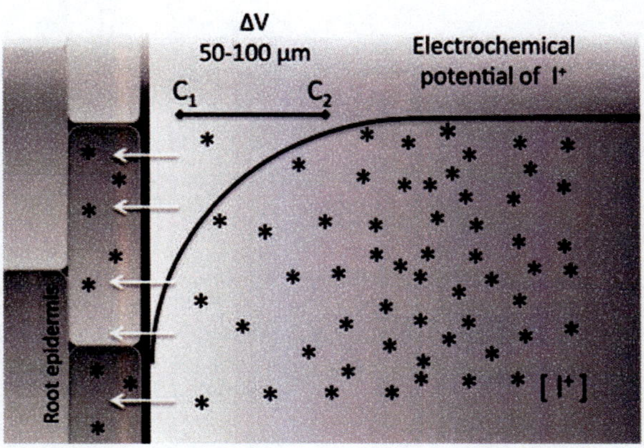

Fig. 4.1 Electrochemical gradient of a ion I^+ in the rhizosphere surrounding the root epidermis

Smith et al. 1994) and it is frequently called "self-referencing" because the electrode moves in a short time between the two positions, erasing all steady or drifting voltages in the measurement circuit (Smith 1995).

Ion and gaseous molecule fluxes can be measured as a net flux using Fick's law of diffusion (Smith 1995; Mancuso et al. 2000), or using the standard electrochemical theory (for an extensive review, see Newman 2001). Here, we give only a quick overview on the use of Fick's law of diffusion (4.1), which expresses the theory of the vibrating probe in a simplified way, and the results obtained match the same obtained by using the more detailed equation of electrochemical described by Dainty (1962) and Nobel (1974):

$$J = -D\frac{\Delta C}{\Delta x} \tag{4.1}$$

where J is the net influx or efflux from the tissue, ΔC is the differential value of the concentration of the ion between the two different positions C_2 and C_1, Δx is the vibrating distance, and D is the diffusion constant of the specific ion (Fig. 4.2).

By referencing the signals in this way, it is possible to minimize all the sources of interference caused by random drift and external noise and allowing the system to detect very small changes of concentration that otherwise would have been covered by these factors (Kuhtreiber and Jaffe 1990; Smith et al. 1994).

The time resolution of the measurement depends on how long the microelectrode is kept at each position (usually 2–10) and on the travel range of the electrode: in fact the calculation of ΔC requires the measurements in both positions. Considering an average time of 5 s in each position, with a travel range of 50 μm, the time resolution will be 0.1 Hz. Two practical considerations prevent this time being much shortened (a) the solution must not be disturbed with too fast movements, and (b) it is important to consider the internal response time of the electrode used, that

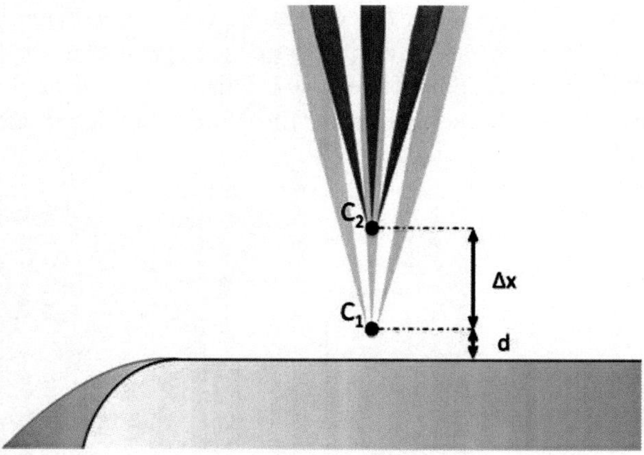

Fig. 4.2 Diagram of the vibrating probe theory, where three microelectrode tips are palced at a *d* distance from the root surface and moved along a known distance (Dx). Ion concentrations C_1 and C_2 are monitored, and flux calculated according to (4.1)

requires a specific time to settle to the new concentration value (Shabala et al. 1997).

4.2.2 *Selective Microelectrodes*

4.2.2.1 Custom-Built Ion Selective Microelectrode

Thanks to the wide range of selective ionophore produced by Fluka (Sigma-Aldich, USA), it is possible to build ion-selective microelectrodes in laboratory with many advantages: they are easy to make, can be produced everyday, and are cheaper than the ones available commercially.

Microelectrodes are usually made for use during the same day, and they are calibrated in solutions containing a series of concentrations covering the expected physiological range of the ion selected. Standard non-filamentous borosilicate glass capillaries used for patch clamp may be used. They are pulled with a puller apparatus, and micropipettes are coated with a hydrophobic material (the most often used is Tributylchlorosilane 90796, Fluka Chemicals) to hold the liquid ion exchanger (LIX). Electrode blanks are back-filled with appropriate solutions and afterwards by touching a drop of LIX it is possible to fill the tip. LIX are available commercially for a wide range of ions, and always new ones are being developed. The filling solution must contain chloride to provide a junction potential at the Ag/AgCl electrode inserted in the glass pipette, and a concentration of the ion of interest (typically 500 mM) to stabilize its electrochemical potential.

The shape of the capillary and the length of the LIX in the electrode tip are very important to get good response time and correlation of calibration curve: electrodes resistance should be less than 3 GOhm and correlation coefficients more than 0.999, otherwise electrodes must be discarded. The electrode resistance is critical for electrode signal-to-noise resolution, while low correlation coefficient led to absolute flux measurements not correct (Shabala et al. 2006). Reference electrode is fabricated in a similar way from a glass capillary and filled with 3 M KCl in 2% agar.

4.2.2.2 Solid State Microsensors

IAA Selective Microelectrodes

Recently, a carbon-nanotube-modified amperometric microelectrode has been developed by our laboratory that serves as a very reliable sensor of indole-3-acetic acid or IAA (Mancuso et al. 2005). For a detailed description of their fabrication refer to Mancuso et al. (2005). Briefly, a platinum wire, 10 mm long and 0.2 mm in diameter, is soldered to a silver connecting wire of the same diameter and mounted in a glass capillary (2 mm o.d., 1.12 mm i.d., World Precision Instruments (WPI),

USA). The wire is secured in position by melting the glass tip of the capillary. Etching of the protruding wire to form a sharp tip is accomplished galvanically by submerging the wire for 5 s in a mixture composed of KNO_3 and 5% $KHCO_3$ melted at 600°C. A current flow of 1.3 A is applied between the microwire and a platinum laminar anode of approximately 1 cm^2. With a little practice with this system, tips with diameters of approximately 7–10 μm can be easily obtained. Microwire insulation is achieved using an anodic electrodeposition paint (Glassophor ZQ 84-3225, BASF, Germany). The probe is completed with a platinum wire, 10 mm long and 0.5 mm in diameter, acting as an auxiliary electrode and with a reference half cell Ag/AgCl/KNO3 (0.1 M) and NaCl (0.08 M). The cell is arranged in a three-electrode polarographic configuration that maintains the working microelectrode at a polarizing voltage of 0.7 V, ensuring a condition of diffusion-limiting current for IAA oxidation. The diffusion coefficient for IAA at 25°C is 0.677×10^{-5} cm^2 s^{-1}, and this electrode is particularly suited to be used with the vibrating probe system.

Copper Electrodes

A copper Cu/CuSe was also developed by our laboratory, with the aim of using it to measure net fluxes of copper(II) ions in *Olea europaea* roots by the vibrating probe method (Papeschi et al. 2000). For the fabrication of the electrode, a copper wire 75 μm in diameter insulated with a thin layer of lacquer was utilized. It was soldered at one connector and mounted in a glass capillary used just as support. The tip of the wire was cut perpendicularly with a sharp surgical knife at a distance of about 3 mm from the capillary and the resulting copper disk was coated with a film of CuSe by cathodic deposition from a 0.1 M sodium selenite solution adjusted to about pH 6.0 with sulphuric acid.

The performance characteristics of the Cu/CuSe disk microelectrode in terms of its selectivity towards cupric ions, Nerstian response, and limit of detection were assessed and can be read elsewhere (Papeschi et al. 2000). The response time detected is shorter than 0.5 s, and this characteristic permits the use of this electrode with the vibrating probe system. The calibration can be performed by diluting a stock solution of $Cu(NO_3)_2$ 0.1 M in distilled water with a background of 50 μM $Ca(NO_3)_2$.

4.2.2.3 Commercial Microelectrodes

A wide range of commercial selective microelectrodes is now available. High response time (<3 s) and small electrode diameters (<50 μm) are the two main characteristics that these electrodes must have for flux measurements by using the vibrating probe technique. Here, we present only two commercial electrodes, which are the one that we commonly use in our laboratory: an oxygen sensor from Unisense (Denmark) and a NO sensor from WPI (USA).

Oxygen Microelectrode

Unisense produces high performance oxygen sensors for a large variety of tasks. They are Clark-type sensors, based on the diffusion of oxygen through a silicone membrane to an oxygen-reducing cathode, which is polarized against an internal Ag/AgCl anode. The flow of electrons from the anode to the oxygen-reducing cathode reflects linearly the oxygen partial pressure around the sensor tip in a pA range. The current is measured by a high quality picoammeter (Revsbech 1989). This kind of electrode is available in a wide range of tip dimension, from 6 to 60 μm and with resolution time less than 1 s. The electrode can be calibrated taking two points: measuring the microelectrode current in the solution used during the flux measurement with 100% N_2 (complete anoxia) and 100% O_2. The solubility of oxygen in water is strictly temperature-dependent so, given the temperature, it is possible to know the total amount of oxygen dissolved in water.

Nitric Oxide Microelectrode

WPI produces nitric oxide microelectrodes (ISO-NOP007) with tip diameter of 7 μm and a response time less than 3 s. It responds to successive additions of NO (100 nM) in a linear way and its lowest detection limit is 0.5 nM. Its main application is in animal science (Levine et al. 2001), but its characteristics make it suitable also for flux measurements with vibrating probe. It can be calibrated using different methods: *Method 1* is based on the decomposition of the S-nitrosothiol NO-donor (SNAP) using CuCl, while *Method 2* is similar to Method 1 using $CuCl_2$ as a catalyst.

4.3 An Example of a Vibrating Probe Set-up: The ViP System

The ViP system is the vibrating probe system developed by the LINV lab of the DiPSA department, University of Florence (Italy), in 2005 and since then it has been widely used for many physiological experiments.

The system is positioned on a vibration-isolated bench, inside a Faraday cage to prevent any external noise and vibrations. The microelectrodes are mounted on a manual micromanipulator providing three-dimensional positioning. During measurements, the distance between the root surface and the electrodes is changed with a three-way micromanipulator (Mercury DC-Motor controller, Physik Instrument, Germany) on which the measuring chamber is fixed. The manipulator is driven by the ViP software, developed by the group on Lab View 6 software (National Instruments, USA). The electrodes are connected by screened cables to various amplifiers, depending on the type of the electrode: oxygen electrodes are connected to a picoammeter (PA2000, Unisense, Denmark), nitric oxide electrodes are connected to an amplifier (Fee radical analyzer, TBR 4100, WPI, USA), whereas all the ion selective microelectrodes are connected to a dual channel differential electrometer (FD223a, WPI, USA). All the signals are successively converted to voltage, and connected via a BNC connector block (BNC 2090, NI National Instrument, USA) to a PC. This means that as far as there is a BNC connection, every kind of amplifier and electrode can be attached to the system.

The ViP software can handle up to three electrodes, and offer the unique possibility to combine oxygen, nitric oxide, and specific ions detections in one single measurement. Vibrating microelectrodes can be set perpendicularly to the root surface (Mancuso et al. 2000; Mancuso and Boselli 2002) or turned by an angle of 45° allowing undisturbed positive gravitropism of the root (Mancuso et al. 2005).

A differential signal is calculated from data obtained while oscillating the electrodes at 0.1 Hz between two points (5 s in each position). The electrode tip and the root are continuously monitored under a microscope throughout the experiment, and a videocamera will send the images to a monitor, so that the operator can continuously check the electrode position. Data are normally collected at a rate of 1,000 data points per second and filtered. For each electrode position, the first 2 s after the movement began are automatically discarded to eliminate stirring effects, movement artefacts, and to allow the reestablishment of the gradient; the remaining signal is then averaged (Fig. 4.3). The computer calculates the difference and records a measurement file that will be converted into flux using the calibration curves previously obtained.

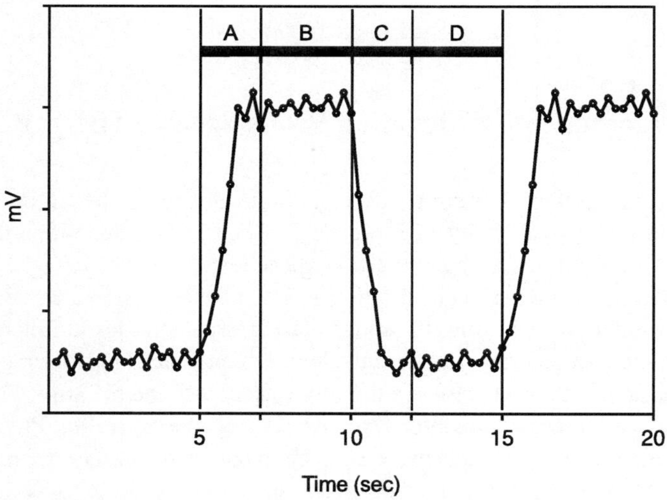

Fig. 4.3 Example of data averaging. Given the curve obtained from the reading of the electrode, the first 2 s in each new position (segment A and C) are discarded during the calculation. Segment B and segment D correspond to the steady fluxes in C_1 and C_2 (see Fig. 4.1) position: they will be averaged and used to calculate ΔC

4.4 Application of the Vibrating Probe Technique in Roots

4.4.1 Oscillation

Oscillations in membrane-transport activity are ubiquitous in the plant and have been widely studied using different vibrating probe techniques. Oscillations are observed at various levels of structural organization, from the molecular to the whole plant. In roots, there are numerous reports of different types of rhythmic processes: from circumnutations (Brown 1993; Shabala and Newman 1997) to oscillations in root intracellular and surface electric potentials (Scott 1957; Jenkinson 1962; Cortes 1997). At a membrane transport level, Shabala and Newman (1997) reported a strong association between H^+ and Ca^{2+} fluxes oscillations and root growth rate (Shabala et al. 1997; Shabala and Newman 1997). These oscillations were always observed in the root apex (Shabala 2003), and only occasionally in the mature zone (Shabala and Knowles 2002), pointing out the physiological importance of flux oscillation is such specific plant zone. In fact, the root apex has always been reported to be more sensitive to various stresses, including cadmium (Piñeros et al. 1998), aluminium stress (Ryan et al. 1993), and salinity (Chen et al. 2005), and it has been suggested that such oscillatory behaviour provides more efficient control and better stability needed for fine-tuning of physiological and metabolic processes in this dynamic region (Shabala et al. 2006).

Mancuso et al. (2000) reported another rhythmic pattern in root concerning O_2 influx, in agreement with H^+ and Ca^{2+} fluxes reported above, and proposed that the same mechanisms leading to these oscillations is responsible for the rhythmic uptake of oxygen at root level. Interestingly, oscillatory patterns in H^+, K^+, Ca^{2+}, and Cl^- uptake were also observed in different regions of the root using the vibrating probe technique (Shabala and Knowles 2002), confirming once more the importance of this technique in the study of root physiology.

4.4.2 Root Mapping

One of the most interesting features of the use of selective microelectrode with the vibrating probe system is the possibility to obtain "topographical" images of fluxes coming into and from the tissues. This opportunity is particularly important in the study of roots, where the different regions have always been found to have different sensitivity to various stresses.

For example, using an oxygen electrode, it was possible to examine O_2 fluxes at different positions along intact roots, and a spatial organization of O_2 influx was revealed (Mancuso et al. 2000). Two distinct peaks were recorded: one in the meristematic zone, and the second one in the distal elongation zone. This difference in the magnitude of the O_2 fluxes in different root regions suggests that distinct respiratory rates coupled to different energy requirements exist in the roots among the division zone (DZ), the elongation zone (EZ), and the mature zone (MZ).

In recent years, flux profiles for macroelements such as Ca^{2+} (Ryan et al. 1990) and Mg^{2+} (Grunes et al. 1993) or microelements such as Cd^{2+} (Piñeros et al. 1998) and Cu^{2+} (Papeschi et al. 2000) have been reported, revealing a marked spatial distribution along the root, with its maximum the elongation region.

Thanks to IAA-selective microelectrode, developed by Mancuso et al. (2005), the IAA flux along an intact root was examined, and in spite of some differences in the quantitative characteristics, all root apices showed the same spatial organization in net IAA influxes (Fig. 4.4). The peculiar spatial distribution of IAA fluxes along the root turned to be very useful in the study of auxin transport and its specific role in roots. In fact, this important plant hormone plays critical roles in a variety of processes related to plant growth and development as well as in numerous physiological processes, and the distribution of its fluxes, detectable with the vibrating probe technique, has been particularly useful in the study of its transport in roots. For example, it has been used to show some important traits of AtPGP4 transporter during early root development in addiction to other molecular techniques (Santelia et al. 2005). AtPGP4 loss-of-function plants revealed enhanced lateral root initiation and root hair lengths both known to be under the control of auxin; they showed altered sensitivities towards auxin and roots reveal elevated free auxin levels and reduced auxin transport capacities. The same procedure was applied to detect differences in IAA influx in root of *Arabidopsis* agravitropic mutants, revealing important insight in the role of polar transport of IAA (Bouchard et al. 2006). Furthermore, the effect of D'orenone in blocking polarized tip growth of root hairs

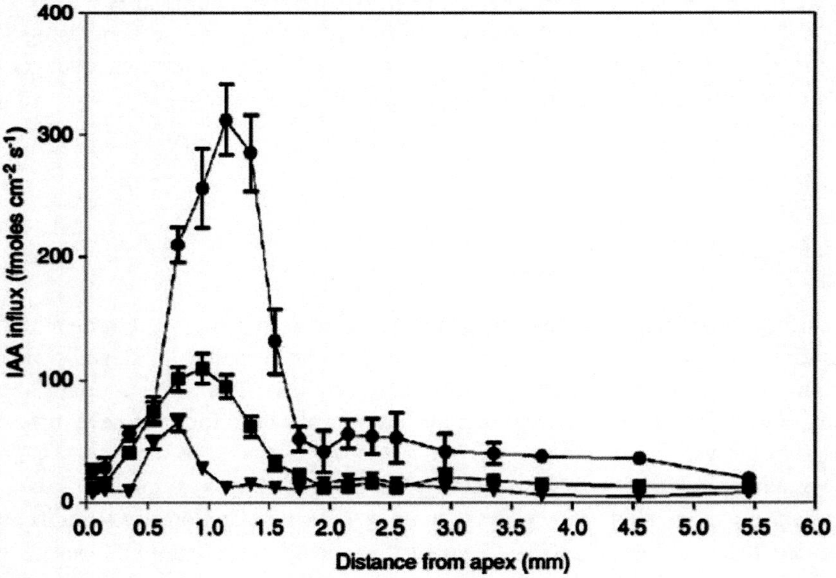

Fig. 4.4 Spatial distribution of IAA fluxes along the root surface (● = non treated, ■ = treated with TIBA, ▼ = treated with NPA). Modified from Mancuso et al. (2006)

by interfering with the PIN2-mediated auxin transport was once again validated thanks to the use of the vibrating probe technique coupled with molecular biology and fluorescence imaging (Schlicht et al. 2008).

4.4.3 Stresses

4.4.3.1 Oxygen Deprivation

When the soil become saturated with water under adverse environmental conditions (e.g., flooding or waterlogging), plants can experience anoxia because of the low solubility and diffusion rate of oxygen in water (Armstrong 1994). In these situations, oxygen is rapidly depleted with dramatic consequences in the root chemical environment and direct affects on root growth can be recorded due to a decline in acquisition of major nutrients such as N, P, K and Ca (Barrett-Lennard et al. 1999). It has been suggested that anoxia decreases membrane selectivity: for example the low concentration of oxygen decreases the selectivity of K^+/Na^+ uptake with a greater uptake of Na^+ (Armstrong and Drew 2002; Smethurst and Shabala 2003). More, loss of K^+ was reported after onset of anoxia presumably as a result of plasma membrane depolarization (Colmer et al. 2001; Mancuso and Marras 2006).

Recent studies showed that plants could also decrease their oxygen consumption in response to oxygen shortage, thus avoiding internal anoxia (Geigenberger 2003). The oxygen fluxes in the root of *Vitis* spp. with diverse tolerance to hypoxia/anoxia have been measured under low O_2 levels by using the vibrating probe technique (Mancuso and Boselli 2002).

4.4.3.2 Salinity

Salt tolerance is conferred by a large number of adaptive mechanisms, most of which are related to membrane-transport processes. Some important traits of salt tolerance were studied by the vibrating probe technique. Extensive research on the ionic mechanisms involved in the regulation of K^+ homeostasis and maintaining an optimal K^+/Na^+ ratio (critical to salt tolerance) in plants subjected to high levels of salinity were performed by Shabala and co-workers using the MIFE system (Microelectrode Ion Flux mEasurement). Two main adverse effects of salinity in non-tolerant plants are osmotic stress and ion (Na^+ or Cl^-) toxicity (Munns 2002; Zhu 2003; Tester and Davenport 2003). The above issue was addressed by measuring net ion flux responses to isotonic NaCl and mannitol solutions from root tissues (Chen et al. 2005), revealing that while NaCl promoted a net K^+ efflux, isotonic mannitol treatment induced a gradual increase in the net K^+ uptake, indicating that different ionic mechanisms are involved in perception of "ionic" and "osmotic" components of the salt stress.

The involvement of the plasma membrane ATP-dependent electrogenic H^+-pump in salt stress was also studied. A NaCl-induced increase in plasma membrane H^+-ATPase activity has been reported (Vera-Estrella et al. 1999), and there are evidences that the stimulation of H^+-ATPases by salt stress provide a driving force for a plasma membrane Na^+/H^+ exchanger to move Na^+ from the cytoplasm into the apoplast, providing a significant contribution to the salt adaptation of plant cells (Ayala et al. 1996). Using K^+ selective microelectrode, it was also possible to compare responses of some contrasting barley cultivars in their tolerance to the salt stress (Chen et al. 2005): a very strong correlation was reported between net K^+ fluxes measured from roots after NaCl treatment and plant physiological responses measured after a month. Interestingly, the K^+ efflux recorded from sensitive cultivars was much higher than that measured from the tolerant varieties (Chen et al. 2005).

4.4.3.3 Nutritional Disorders and Mineral Absorption

The application of ion selective microelectrodes to study various aspects of nutrition has been used to assess nutritional disorders, such as nitrogen and potassium (Colmer and Bloom 1998; Newman et al. 1987) and mineral absorption (Mugnai et al. 2008). The concentration dependence of NH_4^+ and NO_3^- fluxes around the roots was measured simultaneously revealing a complex heterogeneous nature of nutrient acquisition along the root axis, with some differences between functional root regions, and highlighting some interesting relationships between N and H^+ fluxes (Garnett et al. 2003). In another experiment (Mugnai et al. 2008), the capability of marine bioactive substances in enhancing mineral absorption by roots was assessed using ammonium and potassium selective microelectrodes, validating the positive effect of these substances on plant growth.

4.5 Conclusions

The use of the vibrating probe technique with ion-selective microelectrodes gives an unique opportunity to answer specific questions concerning how roots cope with environmental stresses and the processes for their adaptation to these changes. More, this technique can provide interesting insights into early events associated with stress perception and signalling. The key features shown above (non-invasiveness, high spatial and temporal resolution, and possibility to monitor several ions and molecules concurrently) make the vibrating prober technique a very useful tool to study membrane-transport processes in response to many environmental stresses, also giving the possibility to mimic a variety of natural situations and to evaluate the subsequent response of the roots.

Applications also include the possibility to perform a rapid screening of plants subjected to stress in order to evaluate their resistance in response to different treatments. The power of vibrating probe technique, however, strongly increases when combined with other advanced techniques such as imaging, biochemistry and molecular biology. These powerful combinations can allow the detection of the same phenomenon from different perspectives giving a wider and more complete picture of root response to environmental stresses.

References

Apse MP, Aharon GS, Snedden WA, Blumwald E (1999) Salt tolerance conferred by overexpression of a vacuolar Na^+/H^+ antiport in Arabidopsis. Science 285:1256–1258

Armstrong W (1994) Polarographic oxygen electrodes and their use in plant aeration studies. Proc Roy Soc Edinb 102:511–527

Armstrong W, Drew MC (2002) Root growth and metabolism under oxygen deficiency. In: Waisel Y, Eshel A, Kafkafi U (eds) Plant roots: the hidden half. Dekker, New York, pp 729–761

Ayala F, Oleary JW, Schumaker KS (1996) Increased vacuolar and plasma membrane H^+- ATPase activities in *Salicornia bigelovii* Torr in response to NaCl. J Exp Bot 47:25–32

Barrett-Lennard EG, van Ratingen P, Mathie MH (1999) The developing pattern of damage in wheat (*Triticum aestivum* L.) due to the combined stresses of salinity and hypoxia: experiments under controlled conditions suggest a methodology for plant selection. Aust J Agric Res 50:129–136

Bouchard R, Bailly A, Blakeslee J, Oehring SC, Vincenzetti V, Lee OK, Paponov I, Palme K, Mancuso S, Murphy AS, Schulz B, Geisler M (2006) Immunophilin-like TWISTED DWARF1 modulates auxin efflux activities of Arabidopsis P-glycoproteins. J Biol Chem 41:30603–30612

Brown AH (1993) Circumnutations: from Darwin to space flights. Plant Physiol 101:345–348

Chen Z, Newman I, Zhou M, Mendham N, Zhang G, Shabala S (2005) Screening plants for salt tolerance by measuring K^+ flux: a case study for barley. Plant Cell Environ 28:1230–1246

Colmer TD, Bloom AJ (1998) A comparison of NH_4^+ and NO_3^- net fluxes along roots of rice and maize. Plant Cell Environ 21:240–246

Colmer TD, Huang SB, Greenway H (2001) Evidence for down-regulation of ethanolic fermentation and K^+ effluxes in the coleoptile of rice seedlings during prolonged anoxia. J Exp Bot 52:1507–1517

Cortes PM (1997) Cortical intracellular electrical potential in roots of unstressed and stressed sunflower seedlings. I. Dependence on water status. Aust J Plant Physiol 24:643–649

Dainty J (1962) Ion transport and electrochemical potentials in plant cells. Ann Rev Plant Physiol 13:379–402

Garnett TP, Shabala SN, Smethurst PJ, Newman IA (2003) Kinetics of ammonium and nitrate uptake by eucalypt roots and associated proton fluxes measured using ion selective microelectrodes. Funct Plant Biol 30:1165–1176

Geigenberger P (2003) Response of plant metabolism to too little oxygen. Curr Opin Plant Biol 6:247–256

Grunes DL, Ohno T, Huang JW, Kochian LV (1993) Effects of aluminum on magnesium, calcium, and potassium in wheat forages. In: Golf S, Dralle D, Vecchiet L (eds) Magnesium. John Libbey & Co, London, pp 79–88

Jaffe LF, Nuccinelli R (1974) An ultrasensitive vibrating probe for measuring steady extracellular currents. J Cell Biol 63:614–628

Jenkinson IS (1962) Bioelectric oscillations of bean roots: further evidence for a feedback oscillator. Aust J Biol Sci 15:101–114

Knight H, Knight MR (2001) Abiotic stress signalling pathways: specificity and cross-talk. Trend Plant Sci 6:262–267

Kochian LV, Shaff JE, Kuhtreiber WM, Jaffe LF, Lucas WJ (1992) Use of an extracellular ion-selective vibrating microelectrode system for the quantification of K^+, H^+ and Ca^{2+} fluxes in maize roots and maize suspension cells. Planta 188:601–610

Kuhtreiber WM, Jaffe LF (1990) Detection of extracellular calcium gradients with a calcium-specific vibrating electrode. J Cell Biol 110:1565–1573

Levine DZ, Iacovitti M, Burns KD, Zhang X (2001) Real-time profiling of kindey tubular fluid nitric oxide concentrations in vivo. Am J Physiol Renal Physiol Renal 281:189–194

Mancuso S, Boselli M (2002) Characterisation of the oxygen fluxes in the division, elongation and mature zones of *Vitis* roots: influence of oxygen availability. Planta 214:767–774

Mancuso S, Marras AM (2006) Adaptative response of Vitis roots to anoxia. Plant Cell Physiol 47:401–409

Mancuso S, Marras AM, Magnus V, Baluska F (2005) Non invasive and continuous recordings of auxin fluxes in intact root apex with a carbon nanotube-modified and self-referencing microelectrode. Anal Biochem 341:344–351

Mancuso S, Papeschi G, Marras AM (2000) A polarographic, oxygen-selective, vibrating-microelectrode system for the spatial and temporal characterisation of transmembrane oxygen fluxes in plants. Planta 211:384–389

Mugnai S, Pandolfi C, Azzarello E, Salamagne S, Briand X, Mancuso S (2008) Ammonium and potassium root influxes enhancement by the application of marine bioactive substances positively affects *Vitis vinifera* plant growth. J Appl Phycol 20:177–182

Munns R (2002) Comparative physiology of salt and water stress. Plant Cell Environ 25:239–250

Newman IA (2001) Ion transport in roots: measurement of fluxes using ion-selective microelectrodes to characterize transporter function. Plant Cell Environ 24:1–14

Newman IA, Kochian LV, Grusak MA, Lucas WJ (1987) Fluxes of H^+ and K^+ in corn roots. Characterisation and stoichiometries using ion-selective microelectrodes. Plant Physiol 84:1177–1184

Nobel PS (1974) Introduction to biophysical plant physiology. WH Freeman, San Francisco

Papeschi G, Mancuso S, Marras AM (2000) Electrochemical behaviour of a Cu/CuSe microelectrode and its application in detecting temporal and spatial localisation of copper(II) fluxes along Olea europaea roots. J Solid State Electrochem 4:325–329

Pharmawati M, Shabala SN, Newman IA, Gehring CA (1999) Natriuretic peptides and cGMP modulate K^+, Na^+, and H^+ fluxes in Zea mays roots. Mol Cell Biol Res Commun 2:53–57

Piñeros MA, Shaff JE, Kochian LV (1998) Development, characterization, and application of a cadmium-selective microelectrode for the measurement of cadmium fluxes in roots of Thlaspi species and wheat. Plant Physiol 116:1393–1401

Revsbech NP (1989) An oxygen microelectrode with a guard cathode. Limnol Ocean 34:472–476

Ryan PR, Delhaize E, Jones DL (2001) Function and mechanism of organic anion exudation from plant roots. Annu Rev Plant Physiol Plant Mol Biol 52:527–560

Ryan PR, DiTomaso JM, Kochian LV (1993) Aluminium toxicity in roots: an investigation of spatial sensitivity and the role of the root cap. J Exp Bot 44:437–446

Ryan PR, Newman IA, Shields B (1990) Ion fluxes in corn roots measured by microelectrodes with ion-specific liquid membranes. J Membr Sci 53:59–69

Sanders D, Brownlee C, Harper JF (1999) Communicating with calcium. Plant Cell 11:691–706

Santelia D, Vincenzetti V, Azzarello E, Bovet L, Fukao Y, Düchtig P, Mancuso S, Martinoia E, Geisler M (2005) MDR-like ABC transporter AtPGP4 is involved in auxin-mediated lateral root and root hair development. FEBS Lett 579:5399–5406

Schlicht M, Šamajová O, Schachtschabel D, Mancuso S, Lichtscheidl I, Menzel D, Boland W, Baluška F (2008) D'orenone blocks polarized tip-growth of root hairs by interfering with the auxin transport network. Plant J 55:709–717

Scott BIH (1957) Electrical oscillations generated by plant roots and a possible feedback mechanism responsible for them. Aust J Biol Sci 10:164–179

Shabala S (2003) Regulation of potassium transport in leaves: from molecular to tissue level. Ann Bot 92:627–634

Shabala S, Knowles A (2002) Rhythmic patterns of nutrient acquisition by wheat roots. Funct Plant Biol 29:595–605

Shabala S, Newman I, Whittington J, Juswono U (1998) Protoplast ion fluxes: their measurement and variation with time, position and osmoticum. Planta 204:146–152

Shabala S, Newman IA, Morris J (1997) Oscillations in H^+ and Ca^{2+} ion fluxes around the elongation region of corn roots and effects of external pH. Plant Physiol 113:111–118

Shabala S, Newman IA (1997) Proton and calcium flux oscillations in the elongation region correlate with root nutation. Physiol Plant 100:917–926

Shabala S, Shabala L, Gradmann D, Chen Z, Newman I, Mancuso S (2006) Oscillations in plant membrane transport: model predictions, experimental validation, and physiological implications. J Exp Bot 57:171–184

Smethurst CF, Shabala S (2003) Screening methods for waterlogging tolerance in lucerne: comparative analysis of waterlogging effects on chlorophyll fluorescence, photosynthesis, biomass and chlorophyll content. Funct Plant Biol 30:335–343

Smith PJ, Sanger RH, Jaffe LF (1994) The vibrating Ca^{2+} electrode: a new technique for detecting plasma membrane regions of Ca^{2+} influx and efflux. Meth Cell Biol 40:115–134

Smith PJS (1995) Non-invasive ion probes – tools for measuring transmembrane ion flux. Nature 378:645–646

Tester M, Davenport R (2003) Na^+ tolerance and Na^+ transport in higher plants. Ann Bot 91:503–527

Vera-Estrella R, Barkla BJ, Bohnert HJ, Pantoja O (1999) Salt stress in *Mesembryanthemum crystallinum* L cell suspensions activates adaptive mechanisms similar to those observed in the whole plant. Planta 207:426–435

Zhu JK (2003) Regulation of ion homeostasis under salt stress. Curr Opin Plant Biol 6:441–445

Zimmermann S, Ehrhardt T, Plesch G, Muller-Rober B (1999) Ion channels in plant signalling. Cell Mol Life Sci 55:183–203

Chapter 5
The Use of Planar Optodes in Root Studies for Quantitative Imaging

Stephan Blossfeld and Dirk Gansert

Abstract The present state-of-the-art optical sensors – planar optodes – as a technology for noninvasive bioprocess analysis for rhizosphere research are presented. The measuring principle of planar optodes is described very briefly and their advantage over conventional techniques is discussed. The different ways of application of planar optodes for the quantitative imaging of rhizospheric parameters like pH, O_2, CO_2, or ammonium concentration are described. The core of this chapter is a review of the application of planar optodes in rhizosphere research since the first approach 5 years ago, highlighting the great potential of this technology.

5.1 Introduction

To improve the scientific understanding of root-associated biogeochemical processes in soils, the knowledge about the dynamics of the environmental parameters in the rhizosphere is one of the keys. Generally, the most important parameters are pH and oxygen content, because these two parameters dominate the so-called stability field of water, which forms the physic-chemical background for all life processes in soils (Blossfeld and Gansert 2007; Kirk 2004). Additionally, the knowledge about the carbon dioxide content is also a valuable parameter: it allows conclusions about the metabolic activities in the rhizosphere. But besides these three, many other parameters like nutrient or metal content are also of importance to understand the dynamic processes in the rhizosphere.

S. Blossfeld (✉)
Institute of Bio- and Geosciences, IBG-2: Plant sciences, Forschungszentrum Jülich GmbH, 52425 Jülich, Germany
e-mail: s.blossfeld@fz-juelich.de

D. Gansert
Department of Ecology & Ecosystem Research, University of Göttingen, Untere Karspüle 2, 37073 Göttingen, Germany

S. Mancuso (ed.), *Measuring Roots*,
DOI 10.1007/978-3-642-22067-8_5,

However, the conventional measuring techniques have some major drawbacks, especially when it comes to the question of suitability for quantitative noninvasive imaging solutions. For example, in the case of measurements of dissolved O_2 and CO_2, the standard measuring technique is the use of clark-type or severinghaus electrodes. But due to the electrochemical measuring principle of the electrodes (the current is proportional to the molecules consumed by the redox process in the electrode), the electrode has to be inserted into the biological system and an artificial sink for O_2 and CO_2 is created, hence the measurement is invasive. Furthermore, if the O_2 or CO_2 content is to be measured in the gaseous phase of the soil pores, methods and techniques for continuous measurements are missing. Beyond that, if a quantitative dynamic imaging of the parameter of interest is required, only a few techniques exist (e.g., videodensitometry for pH imaging; Jaillard et al. 1996), but none of them allow a *noninvasive* imaging of root–rhizosphere interactions in soil systems.

Hence, there is a strong need for sensors that show an independence of measurement from the state of aggregation as well as noninvasive characteristics for quantitative imaging studies in high spatial and temporal resolution. The state-of-the-art technology of quantification of physicochemical parameters for rhizosphere research are planar optodes, which overcome the limitations of conventional sensors like electrodes. For example, concerning the measurement of gases like O_2 or CO_2, planar optodes show an excellent measuring performance in the gaseous and the aqueous phase. Planar optodes use the change of the fluorescence properties of analyte-specific fluorophores as the carrier of information, and therefore, a decoupling between the analyte-specific sensor and the detector is possible, a major prerequisite for all noninvasive investigations of biogeochemical processes in soils. Moreover, the use of supporting materials like polymers allows the fixation of the analyte-specific indicator dyes as planar sensors, that is, sensor foils, from small sensor dots (millimeter range) to large sensor foils (decimeter range). Together with rather conventional optical setups, a dynamic noninvasive quantitative imaging of biogeochemical rhizosphere processes is possible.

This chapter provides basic information on the measuring principles of planar optodes as well as an overview of the available applications of planar optodes in rhizosphere research.

5.2 Measuring Principles of Planar Optodes

The measuring principle of planar optodes relies on the change of the fluorescence properties of analyte-specific fluorophores. During the measurements, these fluorophores are transferred into an excited state by the absorption of light of certain wavelengths. Depending on the analyte quantity, the fluorescence properties of the fluorophore are changed consequently and detected by an appropriate detector. The changed fluorescence properties include fluorescence decay time, luminescence intensity, polarization, quenching efficiency, radiative and nonradiative energy transfer as well as combinations of these properties (Gansert and Blossfeld 2008). Hence, the information about the analyte (CO_2, O_2, pH, etc.) is carried by the fluorescence light itself. Due to this, the sensor material and detector system can be decoupled and a noninvasive bioprocess analysis is possible.

Two fluorescence measuring approaches for planar optodes have been established.

(a) Fluorescence lifetime measurements (Glud et al. 1998; Holst and Grunwald 2001; Holst et al. 1998)
This approach is mainly used for oxygen optodes and utilizes short-term or sinusoidally modulated light pulses to excite the planar optodes. The fluorescence decay time is recorded either directly or indirectly via the phase shift between the modulated excitation light and the equally modulated fluorophore emission light.
(b) Ratiometric fluorescence measurements (Hakonen and Hulth 2008; Hakonen et al. 2010; Strömberg 2008)
For other planar optodes (e.g., pH, CO_2, or NH_4^+) the ratiometric measurements are generally applied: The ratio of the emission intensities of the planar optode at single excitation/dual emission, dual excitation/single emission, or dual excitation/dual emission wavelength measurements provides the information about the analyte quantity.

To perform the fluorescence measurements practically, the fluorophores are often immobilized in a polymer matrix, representing the active sensor membrane, that is, the planar optode. There exist several immobilization techniques that depend on the fluorophore used and the individual requirements of the experiments (Hakonen and Hulth 2010; Hulth et al. 2002; Oguri et al. 2006; Schröder and Klimant 2005; Schröder et al. 2007). For further details about the sensor chemistry and fluorophore immobilization, we refer to the specialized literature (e.g., Gansert et al. (2006), Orellana et al. (2005)).

The planar optode is positioned into the region of interest in the soil (e.g., rhizosphere), and the optical quantification of the analyte is done via an optical detector system from outside the soil system. Optodes can be fabricated almost transparent and therefore allow a further visual control of root growth (Holst and Grunwald 2001).

There is a key prerequisite to meet for planar optode measurements, that is, a transparent pathway for the excitation and emission light. In laboratory experiments, this is generally achieved by using so-called rhizoboxes, root containers with a transparent front side or at least a transparent window (Luster et al. 2009; Neumann et al. 2009). The planar optodes are attached to this front plate from inside of the rhizobox, and thus are brought in immediate contact with the rhizosphere (Blossfeld and Gansert 2007; Blossfeld et al. 2010; Strömberg 2008). In most cases, the rhizoboxes are tilted in order to force the roots to grow along the transparent window. A noninvasive optical detection system (CCD camera or stepper motor-based) is used to create a quantitative map of the analyte of interest. Due to the high spatial and temporal resolution of the applied system, visualization of the short- and long-term dynamics of the rhizosphere bioprocesses is possible. The most common rhizosphere parameters measured with planar optodes are the concentration of oxygen (Glud et al. 1998; Holst and Grunwald 2001; Holst et al. 1998), ammonium (Strömberg 2008; Strömberg and Hulth 2003), and the pH value (Hakonen and Hulth 2008; 2010; Hulth et al. 2002). Recently, CO_2 optodes are also available too (Schröder and Klimant 2005).

The CCD camera setup for mapping the chosen analyte consists of a light source with an optical filter that creates the excitation light and an adequate CCD camera with an optical filter for detecting the corresponding emission light (Frederiksen and Glud 2006; Glud et al. 1998; Holst et al. 1998). For illustration, Fig. 5.1 demonstrates the optical system and the sensor assembly for rhizospheric ammonium detection, originally described by Strömberg (2008). The spatial resolution depends on the camera properties, for example, Hakonen and coworkers achieved a spatial resolution of 277 × 277 μm pixel size over an effective optode area of 101 cm (Hakonen et al. 2010). The temporal resolution between two images lies in the second range and depends on the camera properties and on the data processing. Long-term measurements over days or weeks are possible, but result in large data sets.

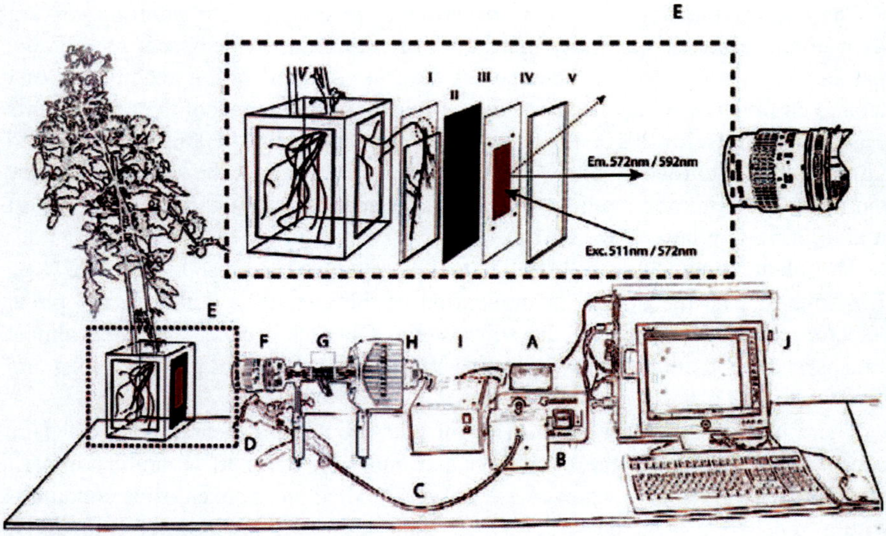

Fig. 5.1 Optical system and sensor assembly. The illustration describes the Xenon light source (A), the dual bandpass filter changer (B), the liquid light guide (C), the focusing lenses (D), the sensor assembly in the pot (E), the macrolens (F), the infinity region with bandpass filters for wavelength selection (G), the CCD camera (H), the dual filter control unit (I), and the image data transfer to computer (J). The close up of the sensor assembly (E) shows the transparency film (I) that isolated the studied root structure from the rest of the plant; the reservoir with 500 × 10–6 M ammonium solution (II), the optical isolation (III), the transparency film with ammonium sensor and alignment spots (IV), and the front removable glass (V) facing the illumination and the macrolens. Reprinted with permission from Strömberg 2008. Copyright 2008 American Chemical Society

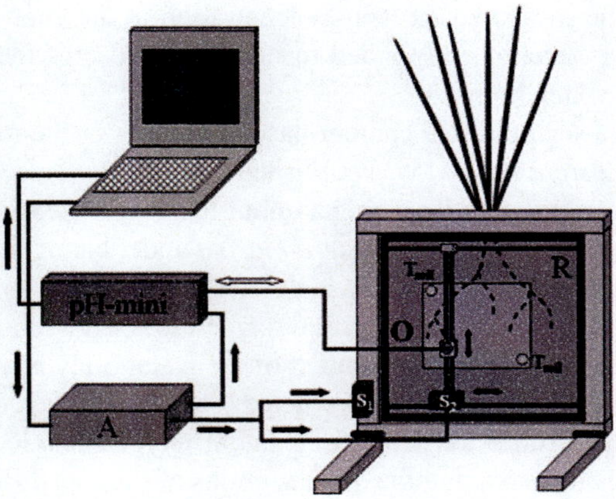

Fig. 5.2 A simplified setup of a stepper motor-based mapping of analyte concentrations in the rhizosphere. The planar optode is placed inside of a rhizobox (R) in direct contact with the roots and the soil. Two stepper motors (S) are controlled by a computer and a specific device (A) and move the optical fiber (O) across the sensor area from outside the rhizobox. The optical fiber is connected to the detector, which itself is connected and controlled by a computer. Reprinted with permission from (Blossfeld and Gansert 2007), copyright 2007 John Wiley and Sons

A different approach to the CCD camera setup for fluorescence measurements is the use of a stepper motor system that uses an optical fiber for light transmission, which is moved by a stepper motor in a defined raster across the region of interest in order to map the analyte concentration (Fig. 5.2; Blossfeld and Gansert 2007; Blossfeld et al. 2010). The spatial resolution depends on the step size and the diameter of the light spot. The reported resolutions range from 1.5 to 3.0 mm over an effective optode area of up to 80 cm. The temporal resolution between each step lies in the second range (2–3 s) and between each image in the minute range. Long-term measurements up to several weeks are possible and result in smaller data sets.

5.3 Examples of Application of Planar Optodes in Root Studies

Planar optodes represent a new tool for root studies, and thus only a few publications about the application of planar optodes exist so far. Nevertheless, the existing publications cover a wide field of root studies that demonstrate the great potential of this novel sensor technology.

The first rhizosphere application of planar optodes based on the CCD camera technique was done by Jensen and coworkers (2005). They created for the first time a quantitative map of the oxygen distribution in the rhizosphere of *Zostera marina*, which is a submerged growing marine grass. The quantitative data of this mapping revealed that up to 43% of the total oxygen loss from the roots of *Zostera marina* originated from the root tips, which is a key adaptation of submerged or waterlogged growing plants. The measured oxygen concentration at the root surfaces reached more than 175 μM which corresponded to more than 70% of the oxygen content in the overlaying water body.

One year later, a second CCD camera-based mapping of the radial oxygen loss from the roots of *Zostera marina* was published (Frederiksen and Glud 2006). This study presented detailed quantitative data about the two-dimensional *dynamics* of radial oxygen loss from the roots of *Zostera marina* under changing light conditions and oxygen concentrations in the overlying water body, as well as a function of the root age. The authors proved that the radial oxygen loss is governed by the diurnal change of light, but that the oxygen content of the water body is even more crucial for the radial oxygen loss. For example, when the oxygen content of the water body was increased above 100% air saturation, radial oxygen loss occurred not only along the root tips, but also along the other sections of the roots. Hence, the oxygen production via photosynthesis acts only as a supplementary pathway, whereas the plant internal gas transport is the driving force.

The first planar optode imaging study with crop plants focused on the ammonium turnover in the rhizosphere of a large fruit-bearing tomato plant (Strömberg 2008). For this study, a CCD camera-based system with a planar ammonium optode was used. Additionally, the use of so-called velocity/quiver plots allowed the creation of ammonium flow patterns in the rhizosphere of the tomato plant (Fig. 5.3). By this, Strömberg demonstrated that the entire root system of the investigated tomato plant is able to take up ammonium; however, young lateral roots showed a relatively stronger uptake rate as the main roots.

The first report about quantitative pH mapping in the rhizosphere of plants was based on the stepper motor approach (Blossfeld and Gansert 2007). The investigated plant species was *Juncus effusus*, growing under waterlogged alkaline soil conditions. The study proved that this species creates a diurnally fluctuating zone of rhizospheric acidification of up to 0.5 pH units, particularly along the elongation zone of young roots (Fig. 5.4). This acidification during daytime was attributed to the ammonia uptake and the equivalent proton release from the roots, because ammonia is the predominant nitrogen source in anoxic wetland soils. Furthermore, a long-term mapping of the rhizosphere pH revealed that the evolution of a complex root system causes a dynamic and heterogeneous pH field with spatio-temporal pH changes of up to 1.4 pH units. By this long-term mapping over a period of more than eight weeks, the dynamics interplay between root growth/root age and the rhizosphere pH field could be clearly visualized and quantified (Fig. 5.5).

Recently, the pH dynamics in the rhizosphere of trace-metal-contaminated soils was investigated in another stepper motor-based planar optode study (Blossfeld et al. 2010). The authors could demonstrate that two of the three investigated plant species (*Noccaea caerulescens* and *Lolium perenne*) strongly alkalinize their rhizosphere by up to 1.7 pH units above the bulk soil pH and that this raise of pH strongly reduces the bioavailability of trace metals. In the case of the third species (*Zea mays*), the authors detected an acidification of the rhizosphere. Most surprisingly, the authors could show that the strongest reduction of trace metal bioavailability occurred in the rhizosphere of the well-known trace metal hyperaccumulator

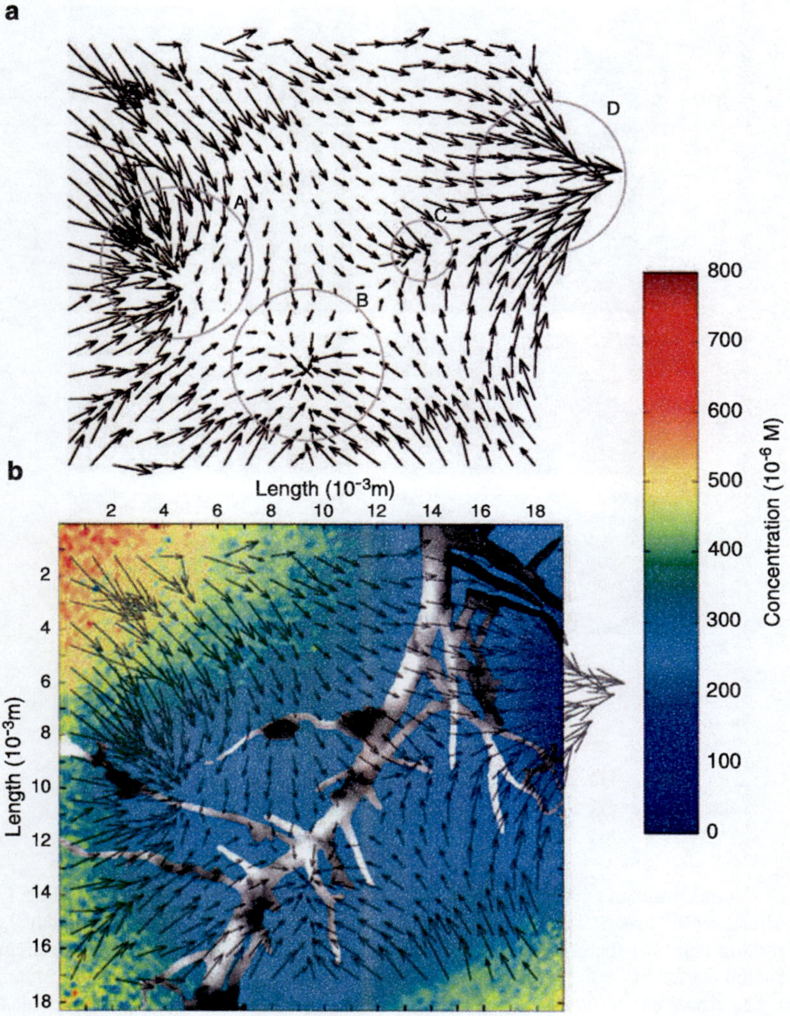

Fig. 5.3 Ammonium flow patterns (quiver plot) close to roots toward the end of the light period. (**a**) The quiver plot detected two local areas with consumption of ammonium (B and C) and a high supply region (A) and a high consumption region (D). (**b**) Region B was directly associated with a fresh root structure directed toward the sensor film, indicating that ammonium acquisition is highest at apical regions of roots. Region C is related to a small fresh outgrowth and is benefited by the supply to the high consuming region (D). The color bar indicate the ammonium concentration. Reprinted & adapted with permission from Strömberg 2008. Copyright 2008 American Chemical Society

Noccaea caerulescens. Hence, the originally expected process of rhizosphere acidification (which would increase the bioavailability) as the key process of *Noccaea caerulescens* roots for trace metal uptake is reversed. Therefore, it still remains unclear how the accumulation process in *Noccaea caerulescens* is operative.

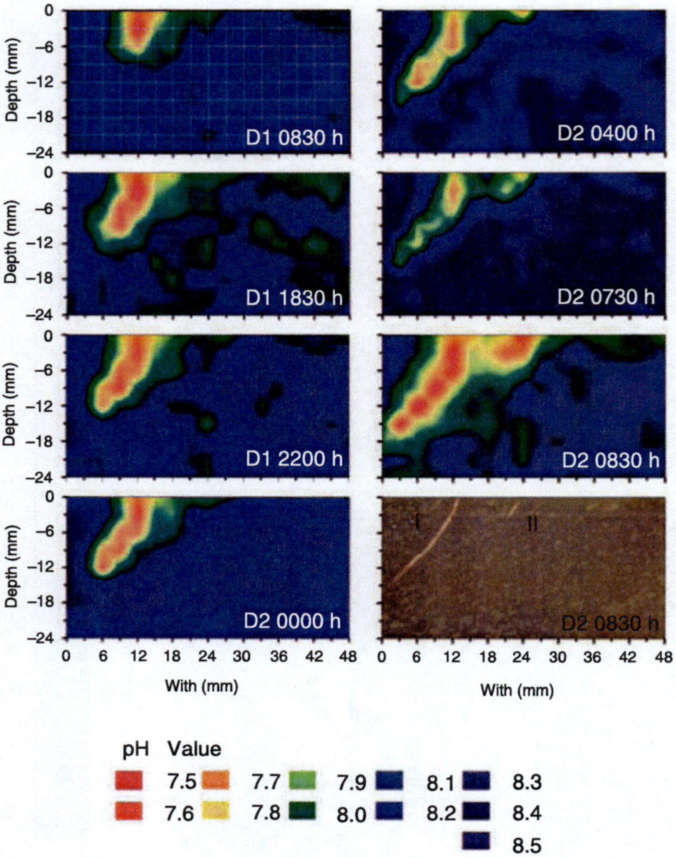

Fig. 5.4 Visualization of a 24 h series of pH profiles across growing roots of *Juncus effusus* L. The figure illustrates snapshots at different times during the time series from day 1 (D1) to day 2 (D2). The crossing points of the grid (*top left*) indicate the positions of fibre-optic pH measurements. Illumination started at 0800 h and ended at 2200 h each day. The colours indicate different pH values. The digital photo (*bottom right*) shows the investigated section of the planar optode at the end of the time series with two roots growing across the pH optode (roman numerals indicate different roots). Reprinted with permission from (Blossfeld and Gansert 2007), copyright 2007 John Wiley and Sons

5.4 Outlook

Planar optodes provide the unique opportunity for a broad range of rhizosphere studies. For example, the use of optodes in, for example, phenotyping screenings will certainly bring new fundamental insights with regard to the root–soil interaction under varying soil conditions. Especially the combination of planar optodes together with statistical methods like velocity plots is a very powerful step forward. Flux dynamics of, for example, ammonium, pH, and other key environmental

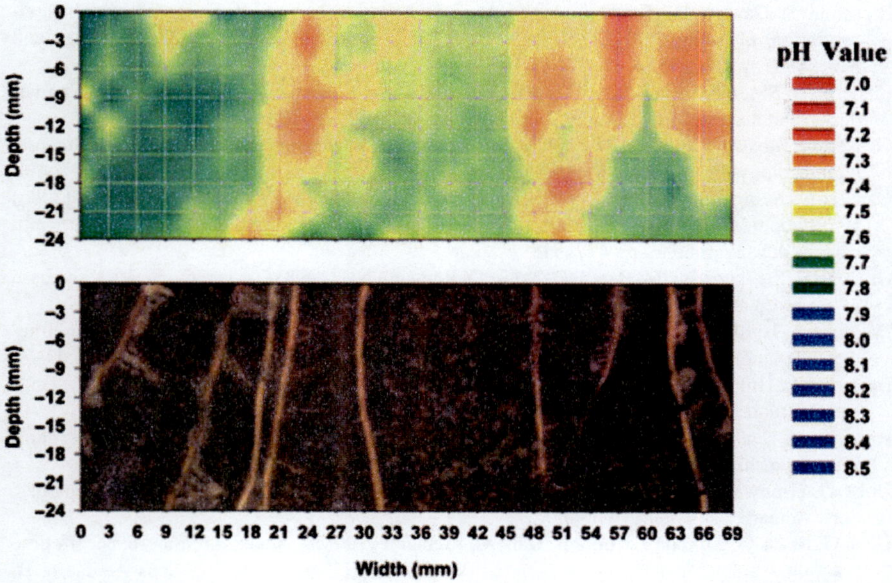

Fig. 5.5 Image of the acidification pattern of the roots of *Juncus effusus* L. 5 weeks after the onset of root growth across the sensor foil. Reprinted with permission from (Blossfeld and Gansert 2007), copyright 2007 John Wiley and Sons

parameters in the rhizosphere of plants could help to improve the application of fertilizers like urea depending on the specific soil conditions, which in turn could result in sustainable yields for agriculture. Alternatively, the prediction of the phytoremediation efficiency of the target plants growing on trace metal contaminated soil conditions will also be improved, if detailed proton flux models become available. Finally, multiple fluorescent sensors, that is, hybrid optodes are the latest state of optical chemical sensor technology (Stich et al. 2010; Blossfeld et al. 2011). These sensors are able to measure two to three parameters simultaneously at the same sensor spot (e.g., O_2 and temperature, pH and O_2, pH, O_2 and temperature or glucose, O_2 and temperature), and it seems that it is only a matter of time until new sensor types will become available. The use of these planar hybrid optodes will be a great opportunity for all kinds of rhizosphere research to unravel some of these highly complex processes that happen below our feet.

References

Blossfeld S, Gansert D (2007) A novel non-invasive optical method for quantitative visualization of pH dynamics in the rhizosphere of plants. Plant Cell Environ 30:176–186

Blossfeld S, Perriguey J, Sterckeman T, Morel J-L, Lösch R (2010) Rhizosphere pH dynamics in trace-metal-contaminated soils, monitored with planar pH optodes. Plant Soil 330:173–184

Blossfeld S, Gansert D, Thiele B, Kuhn AJ, Lösch R (2011) The dynamics of oxygen concentration, pH value, and organic acids in the rhizosphere of Juncus spp. Soil Biology and Biochemistry 43(6):1186–1197

Frederiksen SM, Glud RN (2006) Oxygen dynamics in the rhizosphere of *Zostera marina*: a two-dimensional planar optode study. Limnol Oceanogr 51:1072–1083

Gansert D, Blossfeld S (2008) The application of novel optical sensors (Optodes) in experimental plant ecology. Prog Botany 69:333–358

Gansert D, Stangelmayr A, Krause C, Borisov AY, Wolfbeis OS, Arnold A, Müller A (2006) Hybrid optodes (HYBOP). In: Popp J, Strehle M (eds) Biophotonics: visions for a better health care. Wiley, Weinheim, pp 477–518

Glud RNH, Santegoeds CM, Beer DD, Kohls O, Ramsing NB (1998) Oxygen dynamics at the base of a biofilm studied with planar optodes. Aquat Microb Ecol 14:223–233

Hakonen A, Hulth S (2008) A high-precision ratiometric fluorosensor for pH: implementing time-dependent non-linear calibration protocols for drift compensation. Anal Chim Acta 606:63–71

Hakonen A, Hulth S (2010) A high-performance fluorosensor for pH measurements between 6 and 9. Talanta 80:1964–1969

Hakonen A, Hulth S, Dufour S (2010) Analytical performance during ratiometric long-term imaging of pH in bioturbated sediments. Talanta 81:1393–1401

Holst G, Grunwald B (2001) Luminescence lifetime imaging with transparent oxygen optodes. Sens Actuators B: Chem 74:78–90

Holst G, Kohls O, Klimant I, König B, Kühl M, Richter T (1998) A modular luminescence lifetime imaging system for mapping oxygen distribution in biological samples. Sens Actuators B: Chem 51:163–170

Hulth SARC, Engström P, Selander E (2002) A pH Plate fluorosensor (Optode) for early diagenetic studies of marine sediments. Limnol Oceanogr 47:212–220

Jaillard B, Ruiz L, Arvieu J-C (1996) pH mapping in transparent gel using color indicator video densitometry. Plant and Soil 183:85–95

Jensen SI, Kühl M, Glud RN, Jorgensen LB, Priemé A (2005) Oxic microzones and radial oxygen loss from roots of *Zostera marina*. Mar Ecol Prog Ser 293:49–58

Kirk GJD (2004) The biogeochemistry of submerged soils. Wiley, Chichester, p 304

Luster J, Göttlein A, Sarret G, Nowack B (2009) Sampling, defining, characterising and modeling the rhizosphere – the soil science toolbox. Plant Soil 321:457–482

Neumann G, George T, Plassard C (2009) Strategies and methods for studying the rhizosphere – the plant science toolbox. Plant Soil 321:431–456

Oguri K, Kitazato H, Glud RN (2006) Platinum octaetylporphyrin based planar optodes combined with an UV-LED excitation light source: an ideal tool for high-resolution O_2 imaging in O_2 depleted environments. Mar Chem 100:95–107

Orellana G, Moreno-Bondi M, Guillermo O, Garcia-Fresnadillo D, Marazuela M (2005) The interplay of indicator, support and analyte in optical sensor layers. Front Chem Sens 3:189–225

Schröder CR, Klimant I (2005) The influence of the lipophilic base in solid state optical pCO2 sensors. Sens Actuators B: Chem 107:572–579

Schröder CR, Polerecky L, Klimant I (2007) Time-resolved pH/pO_2 mapping with luminescent hybrid sensors. Anal chem 79:60–70

Stich MIJ, Fischer LH, Wolfbeis OS (2010) Multiple fluorescent chemical sensing and imaging. Chem Soc Rev 39:3102–3114

Strömberg N (2008) Determination of ammonium turnover and flow patterns close to roots using imaging optodes. Environ Sci Technol 42:1630–1637

Strömberg N, Hulth S (2003) A fluorescence ratiometric detection scheme for ammonium ions based on the solvent sensitive dye MC 540. Sens Actuators B: Chem 90:308–318

Chapter 6
Applications of Confocal Microscopy in the Study of Root Apparatus

Susanna Pollastri, Elisa Azzarello, Elisa Masi, Camilla Pandolfi, Sergio Mugnai, and Stefano Mancuso

Abstract Confocal microscopy can be considered as one of the most important progress in optical microscopy within the last decades and has become a powerful investigation tool for molecular, cellular, and development biologist. In this chapter, the authors analyze the main uses of confocal microscopy and fluorescent dyes for studying many aspects of root physiology.

The developments of confocal laser scanning microscopy and fluorescent probes that can be applied, in vivo, to plant root cells, have improved new possibilities for imaging cellular components and activities. Moreover, the opportunity to create transgenic plants and cells permits the visualization of fluorescently labeled components in cells, with minimum invasive manipulation. The combination of all this techniques provides more information for the comprehension of physiological and developmental processes in plant roots. Finally, the use of confocal laser microscopy in these types of studies has several advantages: (1) it is simple, rapid, and accurate; (2) it is nonhazardous; (3) a proper use of all the equipment is helpful to not obtain artifacted images but reliable ones.

6.1 Introduction

Fluorescence microscopy is one of the main tools of investigation available for life science researches. One of its greatest benefits is represented by its minimal invasiveness, which allows to obtain images with good spatial resolution of organisms,

S. Pollastri (✉) • E. Azzarello (✉) • E. Masi • C. Pandolfi • S. Mugnai
Department of Plant, Soil and Environment, University of Florence, viale delle Idee 30, 50019 Sesto Fiorentino (FI), Italy

S. Mancuso
Dpt. Plant, Soil & Environment, University of Firenze, Viale delle idee 30, Sesto Fiorentino (FI) – Italy

S. Mancuso (ed.), *Measuring Roots*,
DOI 10.1007/978-3-642-22067-8_6,

cells, and biomolecules, without alteration of their architecture, in an environment as close as possible to their biological reality.

Because of the high resolution limit of conventional optical microscopy techniques (around 200 nm), researchers have an excellent opportunity to study the subcellular level of organisms. With fluorescence microscopy it is also possible to make a functional analysis, for example, providing the opportunity to characterize the kinetic parameters of biochemical reactions. In contrast, electron microscopy has a resolution that can reach the molecular scale, although the conditions of observation with this type of microscopy are incompatible with life.

Fluorescence microscopy has undergone a revolution with the advent of Confocal Laser Scanning Microscopy (CLSM), introducing the possibility to obtain a microscopic world vision closer to reality, because of its capacity to analyze the samples in a tridimensional way. Nowadays, CLSM is an essential tool in the studies of life science. It is developed from the necessity to visualize, in a more clear and defined way, the processes that occur at cellular level. The main characteristics of this kind of microscope are the presence of a laser as a light source, sensitive photomultiplier tube detectors (PMTs) which allows to amplify the signal and a pinhole which permits to eliminate the out-of-focus light. In particular, laser scans the sample, collecting emitted light with a PMT and builds up an image, pixel by pixel. It is possible to obtain a high optical resolution and a serial optical sections that can be reassembled to form 3D images of the samples being studied and also collect images during time, obtaining time-lapse imaging. This temporal dimension enriched the applications in biological cell sciences of a fundamentally important parameter with the possibility to observe dynamically specific biological events (Fig. 6.1).

The CLSMs can be used in transmission and reflection modes, although their extensive use is overall linked to fluorescent probes for imaging either fixed or living tissues labeled. The use of fluorescent dyes and fluorescent proteins has become an important tool for cellular imaging of CLSM. There are many fluorescent probes specific and selective to different organelles and structures, but also for physiological parameters, such as pH, Ca^{2+}, metal ions, protons, etc., with different excitation and emission spectra. To obtain a multifluorescence image it is also possible to combine different dyes together for staining cells, and use the laser at different wavelengths.

Many cell wall stains are existing, such as propidium iodide, Hoechst 33258 (aka bis-benzimide), lectin conjugates, Calcofluor, and all of them localize to the polysaccharide components. For staining cell membrane it is possible to use the FM dyes which are nonfluorescent in aqueous media and not toxic to cells; these probes become fluorescent only when they reached the membranes and bind with. This class of compounds is almost used to study endocytosis and vesicle trafficking (Fig. 6.1 a,b). Also Lucifer Yellow, Texas Red dye, carboxyfluorescein, and sulforhodamine G can be used for endocytosis and transport studies in plant cells, in fact these dyes are taken up by endocytosis and can be moved to different intracellular compartments.

ER-Tracker Green and ER-Tracker Red stains, conjugate of glibenclamide with BODIPY fluorophores, are the most commonly used and efficacious selectively dyes for endoplasmic reticulum (ER) labeling in plants. There are different fluorescent stains for mitochondria in vivo imaging as MitoTracker Green FM and MitoTracker

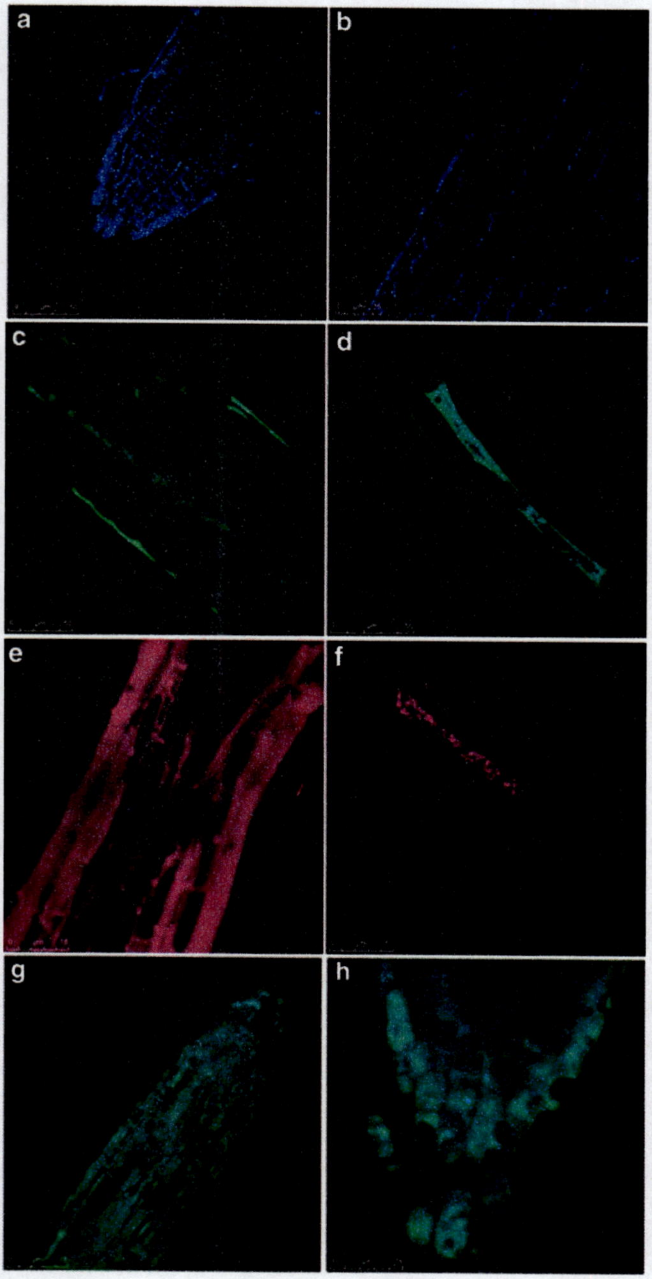

Fig. 6.1 (continued)

Red FM. A plethora of dyes are also available to stain nucleic acid in plants: 7-AAD, DAPI, Hoechst 33258, YOYO-1 iodide, YO-PRO-1 iodide, acridine orange, ethidium bromide, and propidium iodide. To study the cytoskeleton structure it is possible to

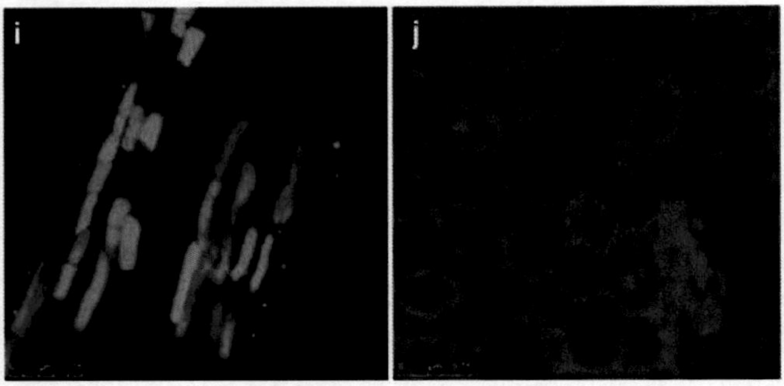

Fig. 6.1 Confocal Laser Scanning Microscopy Images of Roots using different dyes. (**a, b**) Roots of *Arabidopsis* stained with the endocytotic tracer FM4-64. In image **b** it is possible to see endocytotic vesicles. (**c, d**) Confocal images of Fluo-3 fluorescence showing the presence of free cytosolic calcium in *Arabidopsis thaliana* root cells. (**e, f**) Representative images illustrating the CLSM detection of zinc, with the FluoZin dye in *Nicotiana tabacum* (**e**) and *Arabidopsis thaliana* roots (**f**), grown in hydroponic conditions with the addition of 100 μM and 500 μM of zinc, respectively. (**g, h**) Detection of endogenous NO in *Arabidopsis* root with the use of DAF-FM as fluorescent probe. (**i**) Images illustrating the CLSM in vivo detection of sodium in *Olea europaea* root. Sodium Green AM was the dye used for the visualization of sodium in the cytosol of root cells. (**j**) Section of *Gossypium* spp. infected by *Fusarium oxysporum* (*red arrow*) and visualized with trypan blue dye. (*Scale bars* are illustrated in each picture)

stain the actin filaments with a selective probe as phalloidin. Those are only a small fraction of fluorophores used to label and to study the dynamics of different organelle systems in plant cells. Moreover, to monitor in vivo cell dynamics and to visualize specific cellular compartments, genetic transformation of plants with a series of Green Fluorescent Protein (GFP)-tagged organelle proteins is also possible (Lipka and Panstruga 2005). Many different fluorescent proteins are available, allowing the visualization of different intracellular compartments to which the tag is associated.

The opportunities for using CLSM in plant science are many and varied. For example, confocal microscopy is a powerful tool to study root physiology, mainly because of the possibility to use living tissues and to monitor, in real time, changes in root physiology, responses to stresses, and in vivo interaction with plant microbes. These physiological studies usually try to understand the spatial distribution, the localization and the quantification of the ion or molecule under inspection, and also if these ions or molecules change in level or distribution during a developmental or physiological event (Hepler and Gunning, 1998). The use of CLSM in conjunction with fluorescent probes, based on small organic molecules, helped the researchers to have one more possibility to understand the behavior of such important organ of the plant, also without the need of genetic engineering of the sample (Nagano, 2010).

The topics included in this chapter are selected to illustrate the utility of CLSM in plant root studies. In this section, we report newly developed live root cells imaging methods and provide some example of how CLSM, in addition with fluorescent probes, has advanced our understanding of plant roots. Surely, the improvement of these techniques will permit to make even more comprehensive how plant roots response in different situations.

6.2 $[Ca^{2+}]$

Chemical fluorescent dyes are widely used for in vivo calcium (Ca^{2+}) imaging in plants. Calcium is reported to be essential for plant life; in fact it is involved in various physiological processes and acts as a second messenger in responses to most external stimuli in plants (Bush 1995; Webb et al. 1996). To measure changes in cytoplasmic free calcium activity it is essential to understand the role of calcium in cellular processes, as transduction of external stimuli (Bush 1995).

Developments of calcium-sensitive fluorescent probes and CLSM offer the possibility to monitor the spatial differences in intracellular Ca^{2+} concentrations within the root (Read et al. 1992; Webb et al. 1996).

In 2009, Sobol and Kordyum used Fluo-4, a calcium fluorescent probe, to study the distribution of calcium ions in cells of *Beta vulgaris* roots. Fluo-4 is a visible light-excitable Ca^{2+} indicator. The advantages of using this fluorophore are linked to a reduced sample auto-fluorescence interference, a lower cellular photodamage, and a strong absorption by the dyes. For these reasons, it is possible to use a lower dye concentration which is also less phototoxic for living cells. Fluo-4 is available as cell-impermeant potassium salt or as its cell-permeant acetoxymethyl (AM) ester. The AM esters enter in the cell but they may translocate to intracellular compartments, where they are still fluorescent but they are not indicator of cytosolic ion levels. For this reason, the duration of the measurements of calcium cytosolic levels is limited because of sequestration of the indicator into organelles. The authors observed Ca^{2+} distribution and relative content in gravisensing plant cells, using clinorotation, reproducing partly the biological effects of microgravity. Because of the use of confocal microscopy and of Fluo-4 the authors demonstrated that the amount of calcium differs in root of seedlings growth in different conditions.

In Leshem et al. (2006), the vacuolar Ca^{2+} dynamics in *Arabidopsis* root cells during salt stress were assayed with Fluo-4. In the nonstressed seedlings, the dye was seen predominantly in the vacuoles, which occupy most of the mature cell volume, while after 20 h from the beginning of the salt stress, Ca^{2+} was detected mostly into the cytosol. Other available options to monitor calcium are Fluo-3, Fura Red, Fura-2, PBFI AM, and Calcein AM (Fig. 6.1 c,d).

6.3 $[Na^{+}]$

To visualize sodium (Na^{+}) in cellular and subcellular compartments, different Na^{+} indicators are commercially available. Sodium detection is essential in plants subjected to salt (NaCl) stress, in particular to understand which molecular mechanisms are used by the same plants to resist to NaCl toxicity. In fact, under salt stress, plants must regulate sodium and potassium (K^{+}) cellular concentrations, adjusting activities of membrane transporters, channels, and co-transporters (Zhu, 2003). In particular, root cells are the first to suffer salt stress because the influx of sodium is related to the transpiration of water (Läuchli et al. 2008). The major mechanism, which regulates the cytosolic ion level, is to compartmentalize toxic ions into vacuole (Munns and Tester 2008). In a recent research, Choi et al. (2011) identified and characterized NKS1, an ER protein, as an element involved in the control of Na^+ and K^+ homeostasis pathway, also because of the visualization of intracellular Na^+ by the sodium-specific fluorescent dye CoroNa-Green, in *Arabidopsis* roots. CoroNa-Green was also used by Oh et al. in 2010 to visualize sodium in root cells of *Arabidopsis* SOS1 knockout mutant (*sos1-1*), where SOS1 is a very important plasma-membrane Na^+/K^+ antiporter, determinant in the exclusion of sodium ions from cells (Shi et al. 2000, 2002). They were able to see differences in CoroNa-Green intensity between *sos1-1* and control. Particularly, sodium was detected in the vacuole of control root cells, while in *sos1-1* there was a leakage of sodium out of the vacuoles.

Sodium Green is another probe for the fluorometric determination of Na^+ concentration that works with sufficient selectivity in the presence of other monovalent cations. There are the cell-impermeant as well as the cell-permeant forms for monitoring intracellular levels of sodium (Fig. 6.1 i).

6.4 pH

pH is one of the most important factors that regulate protein activities and all the physiological processes and biochemical reactions in which they are involved. In plants, it is also suggested that pH may be involved in many processes, such as plant growth and development, and acts as a signal development (Fasano et al. 2001; Felle and Hanstein 2002; Dodd et al. 2003; Jia and Davies 2007). Nowadays, it is not easy to measure change of pH in some subcellular compartments of plant cell, and for this reason the progress of pH signaling in plants is very limited.

The confocal microscope with pH fluorescent indicators has become an important tool to determinate cellular pH. The fluorescent intensity of some dyes may change in a pH-dependent manner, and because of this property, the intensity of fluorescence gives a measure of pH. This technique is well established in animal rather than in plant because there are some difficulties that inhibit the right use of this kind of probes; in fact the fluorescent intensity is related to the pH but also to the concentration of the dye and the probe must be loaded to the right compartment. There are several types of fluorescent pH indicators, with different features, that may be suitable for some types of cells instead of others.

The fluorescent pH dyes are generally divided into two groups, depending on the range of pH measurements, one is to use in environment with near-neutral pH and the other in relatively acidic environment. In the cytosol, pH is generally between 6.8 and 7.4, and in cell's acidic organelles it is between 4.5 and 6.0. Usually for pH measurement in roots, common pH indicator conjugates are used.

Yu et al. (2001) measured the apoplastic pH in situ of tap roots of *Lupinus angustifolius* L. using confocal microscopy. In higher plants, many transport processes across the plasma membrane are influenced by the ionic composition and concentration of the apoplast, and for this reason the pH of root apoplast is related to nutrient uptake (Canny 1995, Sattelmacher et al. 1998). This study was focused on *Lupin* that grows poorly on alkaline soils; in fact high pH decreases cell elongation and reduces cell volume in the epidermis and the outer cortex (Tang et al. 1993). For a better understanding of such dynamic in determined region and tissue, the employment of CLSM in combination with specific fluorescent dyes is essential. They used pH indicator conjugated to a dextran, Cl-NERF-dextran in combination with Texas Red. The properties of pH indicators are not affected upon conjugation to these relatively inert polysaccharides, but it gives higher water solubility and remains inside the cells for longer period. The authors were able to measure apoplastic pH of the epidermis in the elongation zone of lupin tap roots and the difference of pH in relation to the pH of the solution where the plants were grown.

Fasano et al. (2001) measured the dynamics of cytoplasmic pH in root cells of *Arabidopsis* after gravistimulation, in order to better understand the events at cellular level that mediates gravity signaling. They measured pH of apoplast using CBD-OG CBD peptide of *Clostridium cellulovarans* in combination with succinimidyl ester cellulose, loaded into the apoplast by pressure injection. For measurement of cytoplasmic pH, root cells were pressure microinjected with BCECF linked to 10,000 molecular weight dextran. Moreover, they used transgenic *Arabidopsis* plants expressing cytoplasmic pH sensor (GFP H148D) (Elsliger et al. 1999). These procedures allow to monitor the pH changes in the wall and cytoplasm of the columella cells of the root cap in response to gravistimulation.

6.5 Metal Ions

Metal ions, such as iron (Fe), manganese, zinc (Zn), copper (Cu), etc., are essential in many biochemical functions because of their incorporation into or association with proteins. An excess of those metals, however, could be toxic to cells and cell organelles functions. For example, the heavy metal cadmium can inhibit enzyme activity (Nieboer and Richardson 1980) and mitochondrial function (Kesseler and Brand 1994; Smiri et al. 2009) or high levels of Fe, Zn, and Cu can interfere on plant respiratory functions (Kampfenkel et al. 1995; Padua et al. 1996, 1999). Many studies were conducted to evaluate the effects of these metals on plant function and to understand how many plants can maintain cell homeostasis of essential metals. Because of techniques such as atomic absorption spectroscopy, ICP, radioactive isotopes, energy dispersive X-ray fluorescence (EDXRF) (Zaichick et al. 1999), laser ablation ICP-MS (Becker et al. 2005), and micro-proton-induced X-ray emission (micro-PIXE) (Vogel-Mikus et al. 2007) metals can be measured and localized in plant tissues. Also fluorescence sensors were recently used to examine metals in vivo in cells and subcellular compartments. The ability to visualize metal ions in plant cells would be a real value to the analysis of homeostasis or detoxification strategies.

Numerous researches have been carried out, for example, on zinc absorption and accumulation in plants. Zinc is present in biological systems at relatively high concentrations as compared with other micronutrients (Wilson 1988). Natural levels of total Zn in soils are 30–150 ppm (Mulligan et al. 2001) with available fraction as low as < 10 ppm (Ebbs and Kochian 1997). At these concentrations, Zn is an essential element for plant metabolism and growth, but human activities such as mining operations have led, in numerous sites contaminated with Zn, to concentrations potentially harmful for the environment. Zinc fluorophores have been used for imaging Zn in vivo in animal cells, particularly in neural cell biology (Woodroofe et al. 2004; Kay et al. 2006). The most common fluorescent probes to analyze accumulation of zinc in plant cells are (6-methoxy-(2-*p*-toluenesulfonamido) quinoline (TSQ), Zinquin, Zinpyr and the ZnAF family (Thompson 2005), FluoZin (Fig. 6.1 e,f). The Zinpyr-1, a membrane permeable fluorophore highly selective for Zn (Burdette et al. 2001), was used by Sinclair et al. (2007) and collaborators to localize zinc in root tissue of two *Arabidopsis* mutants. They concluded that Zinpyr-1 is an efficient fluorophore to study and localize zinc in root tissue. The same conclusion can be reached by using the fluorophore Zinquin. Zinquin has never been used to show the accumulation of Zn in plant roots, even if in Sarret et al. (2006) it was strategic to discover deposition of zinc at the base of trichomes in *Nicotiana tabacum* leaf.

Aluminum (Al), unlike zinc, is toxic to many plants even at low concentrations (Ma et al. 2001). Acid soils are particularly rich in aluminum, which is the most limiting factor for plant growth in these environments (Jones and Kochian 1997; Matsumoto 2000). Just only after 1 h of exposure to this metal an inhibition of root elongation can be observed (Llugany et al. 1995). However, some plant species can live in these soils, because of strategies that they have evolved over the time.

Several methods can be used to assess the uptake and accumulation of Al in root tissues, but the use of staining techniques with specific dyes is relatively simple and rapid.

There are different dyes to detect the presence of aluminum in plant samples, as hematoxylin (Cançado et al. 1999), eriochrome cyanine R (Aniol 1983), aluminon, and solochrome azurin (Denton et al. 1984), but all of them are not enough sensitive to study Al cellular distribution.

Many authors used Morin, a pentaprotic acid, that form a fluorescent compound with Al (Browne et al. 1990), but the results using this acid to detect Al in root cells are contradictory (Eticha et al. 2005; Ahn et al. 2002; Vitorello and Haug 1996; Tice et al. 1992) especially with regard to the main sites of Al accumulation. The use of lumogallion, a fluorescent reagent developed by Kataoka et al. in 1997, seems to have the greatest chance of success to detect Al in plant roots using confocal microscopy (Henry et al. 2005; Kataoka and Nakanishi 2001). In Kataoka and Nakanishi (2001), the researchers were able to detect aluminum in the middle of the root tissue of soybean (*Glycine max* (L.) Merr. cv. Tsurunoko) exposed to short-term aluminum treatment. Due to the low sensitivity of the methods utilized before, it was not possible to understand how aluminum accumulates in root tips after a short time treatment. Silva et al. in 2000 made some experiments to determine whether aluminum accumulated in cell nuclei after an exposition to relatively low level of aluminum. They also used the fluorescent stain lumogallion, confirming that aluminum can interfere with nuclear activities causing decrease of root growth.

In addition to zinc and aluminum indicators, other fluorescent probes exist and are useful to detect various metal ions such as copper, iron, manganese, etc., in root cells.

6.6 Nitric Oxide

The role of nitric oxide (NO) has become relevant as a key signaling molecule in plants, and in the regulation of many growth and developmental processes. Two different enzymatic pathways are evolved in the production of NO, the Arg-dependent nitric oxide synthase (NOS) pathway and the nitrite-dependent nitrate reductase (NR) pathway, although, also nonenzymatic processes contribute to the synthesis of NO in plants (Neill et al. 2003, 2008; Wilson et al. 2008).

To analyze NO production in plants, many different methodological approaches have been developed, but most of them detect NO outside the cell.

Confocal microscopy, in association with the appropriate dye, can be a useful tool to analyze NO production inside the cell. The molecule of NO is unstable and it is readily oxidized to NO^+. These reactive nitrogen oxides can be trapped by aromatic amines; for this reason, DAF-FM (4-amino-5-methylamino-2′,7′-difluorofluorescein) and DAF-FM Diacetate are the most used indicators to visualize intracellular NO (Fig. 6.1 g,h).

Corpas et al. (2009) demonstrated the importance of *Arabidopsis* peroxisomes for NO accumulation in the cytosol under salt stress conditions in root. For the visualization of endogenous NO in primary roots, they used 4-aminomethyl-2′,7′-difluorofluorescein diacetate (DAF-FM DA) as fluorescent probe. The use of transgenic *Arabidopsis* plants expressing GFP with peroxisomal targeting signal 1 (PST1) (Mano et al. 2002) allowed the authors to visualize NO inside these organelles in association with the specific dye for NO in vivo. To test the response in terms of NO production after gravistimulation, Hu et al. (2005) analyzed the distribution of NO on the lower side of the apical region of soybean (*G. max*) seedling roots. They found out that gravistimulation of soybean roots induced asymmetric NO accumulation with the use of confocal laser microscopy and DAF-2DA as specific dye.

6.7 ROS

Reactive Oxygen Species (ROS), including superoxide anion, hydroxyl radical, and hydrogen peroxide, are highly reactive molecules containing oxygen. They are generally produced during normal cellular metabolism, but their formation can be enhanced by external environmental conditions. These molecules, if present at low concentrations, are useful to various cellular metabolism functions; on the contrary, an excessive production of ROS determines a condition of oxidative stress that could cause damages to cells (Asada and Takahashi 1987). Although over the past years it has been shown that ROS play a fundamental role in sensing and signal transmission pathways in response to environmental and biological stresses (Foyer and Noctor 2003; Apel and Hirt 2004), their physiological roles are not yet well defined. A precise determination and localization, because of fluorescent probes and to the use of CLSM, could be helpful to define these roles.

There are several probes that can be used to detect O_2^- and H_2O_2 in plant roots. For example, nitroblue tetrazolium (NBT) is a widely used indicator that is reduced by O_2^- to form an insoluble blue formazan precipitate (Maly et al. 1989; Flohe and Otting 1984; Beyer and Fridovich 1987). The blue color denotes the presence of O_2^-, which was found, for example, especially in the elongation zone of *Arabidopsis* roots subjected to different chemical treatments, by Dunand et al. (2007), after staining them in a solution of NBT.

Hydrogen peroxide is involved in the growth of root hairs (Foreman et al. 2003), in the peroxidase-mediated formation of lignin (Ros Barcelò 1997), and in other many developmental and physiological processes (Gapper and Dolan 2006; Kwack et al. 2006). Di-amino benzidine (DAB) (Yan et al. 2010), Amplex Red, and Hydroxyphenyl fluorescein (HPF) are dyes which form compounds with hydrogen peroxide. After the reaction with DAB and H_2O_2, brown polymerization products are formed, while the reaction product of Amplex Red and H_2O_2 is resorufin, a colored and fluorescent compound. Hydroxyphenyl is not fluorescent until it reacts with ROS, as it was shown by Dunand et al. (2007) to detect H_2O_2 in *Arabidopsis* roots. They found out that 2 min of incubation was sufficient to observe the samples using confocal microscopy and obtain fluorescent images useful to detect the presence of H_2O_2 especially in expanding root hairs. In Rodríguez-Serrano et al. (2006), the authors evaluated cadmium effect on oxidative metabolism of pea roots. H_2O_2 was localized, because of the use of DCF-DA (2′,7′-dichlorofluorescin diacetate), especially in root tips of cadmium-treated roots, because oxidative damages caused by the presence of heavy metal. DCF-DA becomes fluorescent after cellular oxidation and after removal of acetate groups by cellular esterases. CM-H2DCFDA (5-(and-6)-chloromethyl-2′,7′-dichlorodihydrofluorescein diacetate, acetyl ester) is a derivative of DCF-DA and was used by Shin et al. (2005) to visualize ROS species in *Arabidopsis* roots after potassium, nitrogen, or phosphorus deprivation. Because of confocal microscopy, they found ROS accumulation in the epidermis in roots grown in absence of potassium and nitrogen, while ROS accumulates more in the cortical tissue when roots were deprived of phosphorus. This visualization helped them to formulate the hypothesis that ROS produced, in the absence of phosphorus, in the cortex, may be involved in signaling, and also in the stimulation of lateral root growth from adjacent pericycle cells. H2DCFDA has the ability to bind intracellular components, but it has some problem of autoxidation in the presence of light and it is not as specific as other dyes.

More studies can be carried out on visualization, localization, and quantification of ROS and their physiological roles in plant roots using these specific dyes. A possible and important application of this technique could be the visualization of ROS production in root tips under various stresses.

6.8 Plant Microbes

The starting point of plant–microbe interactions is at cellular level. It is well known that these cellular interactions are important susceptibility and resistance to microbial growth, and for these reasons. The ability to visualize and monitor at cellular level the interactions between plant and microbes with high resolution images technique and imployment of fluorescent probes, is important in order to understand the susceptibility and resistance to microbial growth. Biochemical studies of tissue extracts do not provide considerable information.

New opportunities to understand the cellular and subcellular responses to microbes in living plant samples have been opened by CLSM together with fluorescent tags; also three-dimensional (3D) reconstructions of subcellular objects allow us to understand the spatial organization of the components at contact sites.

For these reasons, confocal microscopy is a fundamental tool that can permit to understand in which way plant and microbe recognize each other and begin their relationship.

Although interactions between fungi and plants have been studied at microscopic level, it is very interesting to clarify and to monitor the subcellular events that occur (Fig. 6.1 j).

Genre and coworkers in 2005 investigated, at microscopic level, the Arbuscular mycorrhizae (AM), which are specialized endosymbiotic associations formed between a restricted group of biotrophic soil fungi (the Glomeromycota) and the large majority of vascular land plants.

Arbuscules, the main sites of metabolic exchange between the two organisms, are created when these fungi colonize root (Harrison 2005). The development of extraradical hyphae permits the exchange of important nutrients between fungi and plants. The development of extraradical hyphae permits the exchange of important nutrients between fungi and plants, but the cellular and molecular mechanisms which characterize this process are still not well understood. To monitor in vivo intracellular responses and remodeling during all the stages of the AM–plant association, the use of confocal microscopy was successfull. Transformed *M. truncatula* root, expressing specific GFP-labeled markers to observe plant cytoskeleton and the ER, was used. Confocal microscope in combination with fluorescent molecules allows them to discover the prepenetration apparatus, a novel but transient intracellular structure.

Not always the interactions within plants and fungi produce benefit, and for this reason it is also important to understand molecular mechanisms underlying infection by pathogen. Marcel et al. (2010) conducted a beautiful work on a rice blast disease caused by *Magnaporthe oryzae*. Although these fungi generally infect leaves, the authors test its ability to infect roots, even though the distribution in the soil is not well characterized. Using confocal microscope and specific fluorescent dyes they provide a detailed characterization of root infection. To follow the infection they used propidium iodide to reveal plant cell walls and FM4-64 dye to stain membranes.

Propidium iodide is commonly used as a nuclear or chromosome counterstain and as a stain for dead cells, it is low membrane permeant and it is generally excluded by viable cells. In plant it is used to detect plant cell walls. The FM dyes (FM4-64 and FM1-43) are probes selective for membrane which stain the plasma membrane and most organelle membranes in cells and can be used to detect both in plant and fungal cells. The FM dyes fluoresce strongly only when located in a hydrophobic environment. For example, when used in fungal cells FM dyes enter fungal cells primarily by endocytosis, and then are distributed to different organelle membranes and different organelles [reviewed in Hickey et al. (2004)].

In another study, fluorescent-labeled nonpathogenic and nonsymbiotic strains of a bacterium and a fungus where used to demonstrate how microbes enter root cells and how plant used them as a source of nutrients (Paungfoo-Lonhienne et al. 2010). From this study emerges the evidence of the fact that *Arabidopsis* (*Arabidopsis thaliana*) and tomato (*Lycopersicum esculentum*) are able to take up nonpathogenic *E. coli* and *S. cerevisiae* into root cells modifying the walls. Confocal images of root cross-sections show that microbes are in different parts of the root as epidermis cells, cortex cells, and the apoplastic space. This kind of approach can be used to study plant–soil–microbes ecology.

Kumar et al. (2008) used trypan blue for staining arbuscular mycorrhizal fungal spores, fungal hyphae, arbuscules, and vesicles, as a fluorescent dye in association with the use of confocal microscope. The study was conducted on roots of three different types of mangroves. The use of CLSM in association with trypan blue let them to visualize, without sectioning and with high resolution, arbuscular mycorrhizal structure during roots infection.

Another interesting way to study the interaction between plant and pathogen is to analyze the production of NO by the pathogen during the infection. Prats et al. (2008) reported that NO is also generated by *Blumeria graminis* f.sp. hordei as a pathogenesis determinant on barley. Infection by *B. graminis* f.sp. hordei is dependent on appressorium formation in order to penetrate the host. Using fluorescent dye diaminofluorescein-2 diacetate (DAF-2DA) and CLSM, transient NO generation was detected within the *B. graminis* f.sp. hordei appressorium during its maturation, indicating that NO plays a key role in formation of *B. graminis* f.sp. hordei appressoria.

References

Ahn SJ, Sivaguru M, Chung GC, Rengel Z, Matsumoto H (2002) Aluminium-induced growth inhibition is associated with impaired efflux and influx of H^+ across the plasma membrane in root apices of squash (*Cucurbita pepo*). J Environ Qual 53:1959–1966

A (1983) Aluminium uptake by two winter wheat varieties of different tolerance to aluminium. Biochem Physiol Pflanz 178:11–20

Apel K, Hirt H (2004) Reactive oxygen species: metabolism, oxidative stress, and signal transduction. Annu Rev Plant Biol 55:373–399

Asada K, Takahashi M (1987) Production and scavenging of active oxygen in photosynthesis. In: Kyle DJ et al. (eds) Photoinhibition (Topics in Photosynthesis) Vol. 9. pp 227–287

Becker JS, Zoriy MV, Pickhardt C, Palomero-Gallagher M, Zilles K (2005) Imaging of copper, zinc and other elements in thin section of human brain samples (hippocampus) by laser ablation inductive coupled plasma mass spectrometry. Anal Chem 77:3208–3216

Beyer W, Fridovich I (1987) Assaying for superoxide dismutase activity: some large consequences of minor changes in conditions. Anal Biochem 161:559–566

Browne BA, McColl JC, Driscall CT (1990) Aluminium speciation using morin. I. Morin and its complexes with aluminium. Journ Env Qual 19:65–72

Burdette SC, Walkup GK, Spingler B, Tsien RY, Lippard SJ (2001) Fluorescent sensors for Zn^{2+} based on a fluorescein platform: synthesis, properties and intracellular distribution. JACS 123:7831–7841

Bush DS (1995) Calcium regulation in plant cells and its role in signaling. Annu Rev Plant Physiol Plant Mol Biol 46:95–122

Cançado GMA, Loguercio LL Martins, PR, Parentoni SN, Paiva E, Borém A, Lopes MA (1999) Hematoxylin staining as a phenotypic index for aluminum tolerance selection in tropical maize (Zea mays L.). Theor Appl Genet 99:747–754

Canny MJ (1995) Apoplastic water and solute movement: new roles for an old space. Annu Rev Plant Physiol Plant Mol Biol 46:215–236

Choi W, Baek D, Oh DH, Park J, Hong H, Kim WY, Bonhert HJ, Bressan RA, Park HC, Yun DJ (2011) *NKS1*, Na^+-and K^+-sensitive 1, regulates ion homeostasis in a SOS-independent pathway in Arabidopsis. Phytochemistry 72:330–336

Corpas FJ, Hayashi M, Mano S, Nishimura M, Barroso JB (2009) Peroxisomes are required for in vivo nitric oxide (NO) accumulation in the cytosol following salinity stress of Arabidopsis plants. Plant Physiol 151(4):2083–2094

Denton J, Freemont AJ, Ball J (1984) Detection and distribution of aluminium in bone. Journ Chem Pathol 37:136–142

Dodd IC, Tan LP, He J (2003) Do increases in xylem sap pH and/or ABA concentration mediate stomatal closure following nitrate deprivation? J Exp Bot 54:1281–1288

Dunand C, Crèvecoeur M, Penel C (2007) Distribution of superoxide and hydrogen peroxide in *Arabidopsis* root and their influence on root development: possible interaction with peroxidases. New Phytol 174:332–341

Ebbs SD, Kochian LV (1997) Toxicity of zinc to Brassica species: implication for phytoremediation. J Environ Qual 26(3):776–781

Elsliger MA, Wachter RM, Hanson GT, Kallio K, Remington SJ (1999) Structural and spectral response of green fluorescent protein variants to changes in pH. Biochem 27:5296–5301

Eticha D, Staß A, Horst WJ (2005) Localization of aluminium in the maize root apex: can morin detect cell wall-bound aluminium? J Exp Bot 56:1351–1357

Fasano JM, Swanson SJ, Blancaflor EB, Dowd PE, Kao T, Gilroy S (2001) Changes in root cap pH are required for the gravity response of the *Arabidopsis* root. Plant Cell 13:907–922

Felle HH, Hanstein S (2002) The apoplastic pH of the substomatal cavity of *Vicia faba* leaves and its regulation responding to different stress factors. J Exp Bot 53:73–82

Flohe L, Otting F (1984) Superoxide dismutase assays. Meth Enzymol 105:93–104

Foreman J, Demidchik V, Bothwell C, Mylona P, Miledema H, Torres MA, Linstead P, Costa S, Brownlee C, Jones JDG, Davies JM, Dolan L (2003) Reactive oxygen species produced by NADPH oxidase regulate plant cell growth. Nature 422:442–446

Foyer CH, Noctor G (2003) Redox sensing and signaling associated with reactive oxygen species in chloroplasts, peroxisomes and mitochondria. Physiol Plant 119:355–364

Gapper C, Dolan L (2006) Control of plant development by reactive oxygen species. Plant Physiol 141:341–345

Genre A, Chabaud M, Timmers T, Bonfante P, Barker DG (2005) Arbuscular Mycorrhizal Fungi Elicit a Novel Intracellular Apparatus in Medicago truncatula Root Epidermal Cells before Infection. Plant Cell 17:3489–3499

Harrison MJ (2005) Signaling in the arbuscular mycorrhizal symbiosis. Annu Rev Microbiol 59:19–42

Hepler PK, Gunning BES (1998) Confocal fluorescence microscopy of plant cells. Protoplasma 201:121–157

Hickey PC, Swift SR, Roca MG, Read ND (2004) Live-cell Imaging of filamentous fungi using vital fluorescent dyes and confocal microscopy. Meth Microbiol 34:63–87

Hu X, Neill SJ, Tang Z, Cai W (2005) Nitric oxide mediates gravitropic bending in soybean roots. Plant Physiol 137:663–670

Irving HR, Gerhing CA, Parish RW (1992) Changes in cytosolic pH and calcium of guard cells precede stomatal movements. PNAS 89:1790–1794

Jia WS, Davies WJ (2007) Modification of leaf apoplastic pH in relation to stomatal sensitivity to root-sourced abscisic acid signals. Plant Physiol 143:68–77

Jones DL, Kochian LV (1997) Aluminium interaction with plasma membrane lipids and enzyme metal binding sites and its potential role in citotoxicity. Febs Lett 400:51–57

Ma JF, Ryan PR, Delhaize E (2001) Aluminium tolerance in plants and the complexing role of organic acids. Trends Plant Sci 6(6):273–278

Kampfenkel K, Van Montagu M, Inze D (1995) Effect of iron excess on *Nicotiana plumbaginifolia* plants. Implications to oxidative stress. Plant Physiol 107:725–735

Kataoka T, Nakanishi T (2001) Aluminum distribution in soybean root tip for a short time Al treatment. J Plant Physiol 158:731–736

Kataoka T, Iikura H, Nakanishi TM (1997) Al distribution and viability of plant root and cultured cells. Soil Sci Plant Nutr 43:1003–1007

Kay AR, Neyton J, Paoletti P (2006) A startling role for synaptic zinc. Neuron 52:572–574

Kesseler A, Brand MD (1994) Localisation of the sites of action of cadmium on oxidative phosphorylation in potato tuber mitochondria using top-down elasticity analysis. Eur J Biochem 225:897–906

Kwack JM, Nguyen V, Schroeder JI (2006) The role of reactive oxygen species in hormonal responses. Plant Physiol 141:323–329

Kumar T, Majumdar A, Das P, Sarafis V, Ghose M (2008) Trypan blue as a fluorochrome for confocal laser scanning microscopy of arbuscular mycorrhizae in three mangroves. Biotech Histochem 83(3–4):153–159

Läuchli A, James RA, Huang CX, McCully M, Munns R (2008) Cell-specific localization of Na^+ in roots of durum wheat and possible control points for salt exclusion. Plant Cell Environ 31:1565–1574

Leshem Y, Melamed-Book N, Cagnac O, Ronen G, Nishri Y, Solomon M, Cohen G, Levine A (2006) Suppression of *Arabidopsis* vesicle-SNARE expression inhibited fusion of H_2O_2-containing vesicles with tonoplast and increased salt tolerance. PNAS 103(47):18008–18013

Lipka V, Panstruga R (2005) Dynamic cellular responses in plant–microbe interactions. Curr Opin Plant Biol 8:625–631

Llugany M, Poschenriedr C, Barceló J (1995) Monitoring of aluminium induced inhibition of root elongation in four maize cultivars differing in tolerance to Al and proton toxicity. Physiol Plant 93:265–271

Maly FE, Nakamura M, Gauchat JF, Urwyler A, Walker C, Dahinden CA, Cross AR, Jones OTG, De Weck AL (1989) Superoxide-dependent nitroblue tetrazolium reduction and expression of cytochrome b-245 components by human tonsillar lymphocytes and B cell lines. J Immunol 142:1260–1267

Mano S, Nakamori C, Hayashi M, Kato A, Kondo M, Nishimura M (2002) Distribution and characterization of peroxisomes in Arabidopsis by visualization with GFP: dynamic morphology and actin-dependent movement. Plant Cell Physiol 43:331–341

Marcel S, Sawers R, Oakeley E, Angliker H, Paszkowski U (2010) Tissue-adapted invasion strategies of the rice blast fungus *Magnaporthe oryzae*. The Plant Cell 22:3177–3187

Matsumoto H (2000) Cell biology of aluminium toxicity and tolerance in higher plants. Int Rev Cytology 200:1–46

Moyer-Henry K, Silva I, Macgall J, Johannes E, Allen N, Goldforb B, Rufty T (2005) Accumulation and localization of aluminium in root tips of loblolly pine seedlings and the associated ectomycorrhiza *Pisolithis tinctorius*. Plant Cell Env 28:111–120

Mulligan CN, Yong RN, Gibbs BF (2001) Surfactant-enhanced remediation of contaminated soil: a review. Engin Geol 60:371–380

Munns R, Tester M (2008) Mechanisms of salinity tolerance. Annu Rev Plant Biol 59:651–681

Neill S, Barros R, Bright J, Desikan R, Hancock J, Harrison J, Morris P, Ribeiro D, Wilson I (2008) Nitric oxide, stomatal closure, and abiotic stress. J Exp Bot 59:165–176

Nagano T (2010) Development of fluorescent probes for bioimaging applications. Proc Jpn Acad Ser B 86:837–847

Neill SJ, Desikan R, Hancock JT (2003) Nitric oxide signalling in plants. New Phytol 159:11–35

Nieboer E, Richardson DHS (1980) The replacement of the nondescript term "heavy metals" by a biologically and chemically significant classification of metal ions. Environ Pollut Ser B 1:3–26

Oh DH, Lee SY, Bressan RA, Yun DJ, Bonhert HJ (2010) Intracellular consequences of SOS1 deficiency during salt stress. J Exp Bot 61:1205–1213

Padua M, Aubert S, Casimiro A, Bligny R (1996) Arrest of mitochondrial biogenesis in copper-treated sycamore cells. FEBS Lett 398:248–252

Padua M, Aubert S, Casimiro A, Bligny R, Millar AH, Day DA (1999) Induction of alternative oxidase by excess copper in sycamore cell suspensions. Plant Physiol Biochem 37:131–137

Paungfoo-Lonhienne C, Rentsch D, Robatzek S, Webb RI, Sagulenko E, Näsholm T, Schmidt S, Lonhienne TGA (2010) Turning the table: plants consume microbes as a source of nutrients. PLos ONE 5(7):11915

Prats E, Carver TLW, Mur LAJ (2008) Pathogen-derived nitric oxide influences formation of the appressorium infection structure in the phytopathogenic fungus Blumeria graminis. Res Microbiol 159:476–480

Read ND, Allan WTG, Knight H, Knight MR, Malho R, Russell A, Shacklock PS, Trewavas AJ (1992) Imaging and measurements of cytosolic free calcium in plant and fungal cells. J Microsc 166:57–86

Rodríguez-Serrano M, Romero-Puertas MC, Zabalza A, Corpas FJ, Gómez M, del Río LA, Sandalio LM (2006) Cadmium effect on oxidative metabolism of pea (*Pisum sativum* L.) roots. Imaging of reactive oxygen species and nitric oxide accumulation in vivo. Plant Cell Environ 29:1532–1544

Ros Barcelò A (1997) Lignification in plant cell walls. Intern Rev Cytol 176:87–132

Sarret G, Harada E, Choi YE, Isaure MP, Geoffroy N, Fakra S, Marcus MA, Birschwilks M, Clemens S, Manceau A (2006) Trichomes of tobacco excrete zinc as Zn-substituted calcium carbonate and other Zn-containing compounds. Plant Physiol 141:1021–1034

Sattelmacher B, Mühling KH, Pennerviß K (1998) The apoplast – its significance for the nutrition of higher plants. Zeitschrift für Pfanzenernährung und Bodenkunde 161:485–498

Shi H, Ishitani M, Kim C, Zhu JK (2000) The *Arabidopsis thaliana* salt tolerance gene SOS1 encodes a putative Na+/K+ antiporter. PNAS 97:6896–6901

Shi H, Quintero FJ, Pardo JM, Zhu JK (2002) The putative plasma membrane Na+/K+ antiporter SOS1 controls long-distance Na+ tran sport in plants. Plant Cell 14:465–477

Shin R, Berg RH, Schachtman DP (2005) Reactive oxygen species and root hairs in Arabidopsis root response to nitrogen phosphorus and potassium deficiency. Plant Cell Physiol 46:1350–1357

Silva IR, Smyth TJ, Moxley DF, Carter TE, Allen NS, Rufty TW (2000) Aluminium accumulation at nuclei of cells in the root tip. Fluorescence detection using lumogallion and confocal laser scanning microscopy. Plant Physiol 123:987–996

Sinclair SA, Sherson SM, Jarvis R, Camakaris J, Cobbet CS (2007) The use of zinc-fluorophore Zinpyr-1, in the study of zinc homeostasis in *Arabidopsis* roots. New Phytol 174:39–45

Smiri M, Chaoui A, El-Ferjani E (2009) Respiratory metabolism in the embryonic axis of germinating pea seed exposed to cadmium. J Plant Physiol 166:259–269

Sobol M, Kordyum E (2009) Distribution of calcium ions in cells of the root distal elongation zone under clinorotation. Microgravity Sci Technol 21:179–185

Tang C, Kuo J, Longnecker NE, Thomson CJ, Robson AD (1993) High pH causes disintegration of the root surface in *Lupinus angustifolius* L. Ann Bot 71:201–207

Thompson R (2005) Studying zinc biology with fluorescence: ain't we got fun? Curr Opin Chem Biol 9:526–532

Tice K, Parker DR, DeMason DA (1992) Operationally defined apoplastic and symplastic aluminium fractions in root tips of aluminium-intoxicated wheat. Plant Physiol 100:309–318

Vitorello VA, Haug A (1996) Short-term aluminium uptake Bt tobacco cells: growth dependence and evidence for internalization in a discrete peripheral region. Physiol Plant 97:536–544

Vogel-Mikus K, Pongrac P, Kump P, Necemer M, Simcic J, Pelicon P, Budnar M, Povh B, Regvar M (2007) Localization and quantification of elements within seeds of Cd/Zn hyperaccumulator Thlapsi praecox by micro-PIXE. Environ Pollut 147(1):50–59

Webb AAR, McAinsh MR, Taylor JE, Hetherington AM (1996) Calcium ions as intracellular second messengers in higher plants. Adv Bot Res 22:45–96

Wilson ID, Neill SJ, Hancock JT (2008) Nitric oxide synthesis and signalling in plants. Plant Cell Environ 31:622–631

Wilson J (1988) The cost of heavy metal tolerance: an example. Evolution 42:408–413

Woodroofe CC, Masahla R, Barnes KR, Frederickson CJ, Lippard SJ (2004) Membrane-permeable and -impermeable sensors of the Zinpyr family and their application to imaging of hippocampal zinc in vivo. Chem Biol 11:1659–1666

Yan W, Ye S, Jin Q, Zeng L, Peng Y, Yan D, Yang W, Yang D, He Z, Dong Y, Zhang X (2010) Characterization and mapping of a novel mutant sms1 (senescence and male sterility 1) in rice. J Genet Genomics 37:47–55

Yu Q, Kuo J, Tang C (2001) Using Confocal Laser Scanning Microscopy to measure apoplastic pH change in roots of *Lupinus angustifolius* L. in response to high pH. Ann Bot 87:47–52

Zaichick V, Ovchjarenko N, Zaichick S (1999) In vivo energy dispersive X-ray fluorescence for measuring the content of essential and toxic trace elements in teeth. Appl Radiat Isot 50:283–293

Zhu JK (2003) Regulation of ion homeostasis under salt stress. Curr Opin Plant Biol 6:441–445

Chapter 7
High-Throughput Quantification of Root Growth

Andrew French, Darren Wells, Nicola Everitt, and Tony Pridmore

Abstract In this chapter, we describe and discuss our experience of developing high-throughput imaging systems within the Centre for Plant Integrative Biology (CPIB), at the University of Nottingham. After an introductory literature review, a description of the development of robot imaging hardware for high-throughput digital imaging of *Arabidopsis thaliana* roots inside a growth room is presented. This is followed by a description of the imaging software and techniques that have been developed to analyse such images. The intention of the chapter is to provide an unfamiliar reader with a practical overview of the subject, from the perspective of a team of engineers and computer scientists who have worked together to produce a high-throughput imaging system in active use by plant biologists.

A. French (✉) • D. Wells
The Centre for Plant Integrative Biology, University of Nottingham, Sutton Bonington Campus, Loughborough LE12 5RD, UK
e-mail: andrew.p.french@nottingham.ac.uk

N. Everitt
The Centre for Plant Integrative Biology, University of Nottingham, Sutton Bonington Campus, Loughborough LE12 5RD, UK

Materials, Mechanics and Structures Division, Faculty of Engineering, The University of Nottingham, University Park, Nottingham NG7 2RD, UK
e-mail: nicola.everitt@nottingham.ac.uk

T. Pridmore
The Centre for Plant Integrative Biology, University of Nottingham, Sutton Bonington Campus, Loughborough LE12 5RD, UK

School of Computer Science, University of Nottingham, Jubilee Campus, Wollaton Road, Nottingham NG8 1BB, UK

S. Mancuso (ed.), *Measuring Roots*,
DOI 10.1007/978-3-642-22067-8_7,

7.1 Background and Literature Review

Quantitative data describing the structure and development of root systems is central to the understanding of plant growth and function. Traditional root bioassays, however, are based on at best a small number of measurements, and often only end-point analyses (Parry et al. 2001). These are informative but have the limitation of only examining long-term effects on root growth. Transient events and subtle temporal changes can be missed.

Image acquisition and analysis provide a potential solution. Image sequences provide a rich source of data on plant growth. Each image contains a detailed description of a plant's state of development at the time of acquisition, and images can be captured at high speeds. Once captured, they can be stored and re-examined to extract further information, perhaps for a different scientific purpose, at a later date.

Time-lapse photography was used as early as the 1930s (Van der Laan 1934; Michener 1938) to measure the heights of seedlings at various times after application of the phytohormone ethylene, providing important information about the timing of the effects of this hormone on growth regulation. Today, a wide variety of image acquisition devices are available which can be deployed to analyse root growth. Confocal laser microscopy provides high quality digital images at the molecular and cellular scale. Standard light microscopes can be used to detail the development of individual roots in high resolution (again digital) images. Digital cameras are now of sufficient quality that even domestic devices can be used to gather data on sets of plant roots growing together on growth room plates (French et al. 2009).

Imaging modalities are also being developed and deployed which allow plant roots to be assessed in their natural growth medium – soil. Traditional methods of measuring the development of plants grown in soil are destructive, requiring root systems to be removed from their environment and washed. These processes disrupt the topology of the root, limiting the architectural features that can be detected and measurements that can be made. Recently, X-ray computed tomography (Gregory 2003; Perret et al. 2007; Tracy et al. 2010), NMR (Jahnke et al. 2009) and magnetic resonance imaging (van As 2007) have been used to visualise the 3D structure of plant roots in soil.

This rapid expansion in the range of image data available begs the question: how is it to be analysed? The human analysis of images employed in early time-lapse photography work is time consuming, subjective and prone to error. Analysis "by eye" can produce measurements that are difficult to replicate and may result in subtle phenotypes, such as a delayed response, being missed. As biological experiments often require large numbers of samples to be examined, a key requirement of any tools providing data on plant growth is that they be high-throughput. High-throughput systems can process large numbers of samples in short time periods with minimal user involvement. To achieve high-throughput recovery of data on root growth, automatic image analysis methods are required.

A number of software packages exist that aim to automate aspects of the kinematic analysis of growing roots. Many of these software tools attempt to measure growth parameters from high-resolution microscope images. RootFlow (van der Weele et al. 2003) and relative elemental growth rate (REGR) analysis (Walter et al. 2002) have made use of optic flow-based techniques (Barron et al. 1994; Barron and Liptay 1994) to recover the motion of texture features through a sequence of root images. Intensity features are identified and matched between frames, and corresponding velocity flow fields are calculated. From these vector fields, estimates of growth can be made across any part of the image, provided that reliable image features are available in the areas of interest.

Following a review of the value of related methods in computer vision, Roberts et al. (2007, 2010) proposed an optic flow approach for the analysis of confocal microscope images which combines a robust multi-frame likelihood model and a technique for estimating uncertainty. Quelhas et al. (2010) used optic flow data extracted from confocal image sequences to identify cell division events; the key cue being the difference in velocity of pixels arising from growing and splitting cells. Sethuraman et al. (2009), in contrast, analysed confocal image sequences at the tissue level, focusing on groups of cells which are tracked using a modified network snake approach.

The analysis of rhizotron images attracts considerable attention from those interested in assessing root system development. Rhizotrons occur in different forms (Taylor et al. 1990), all of which involve placing a camera inside a transparent casing and burying it in soil. As plant roots grow against the casing wall, images of their structure and development may be captured. Though widely used, the presence of the rhizotron may disrupt root growth and may affect the results obtained. A good review of rhizotron-oriented image analysis methods is provided by Le Bot et al. (2010).

Attempts are underway in a number of laboratories to automatically extract quantitative descriptions of the 3D structure and growth of root systems from X-ray tomography, NMR and PET images. The imaging devices involved are, however, costly. X-ray tomography is limited in the size of containers that can be scanned and, though high resolution data can be obtained, this comes at the cost of increased imaging time. NMR is also expensive and complex to maintain, and very sensitive to the media in which the plants are grown. Though they hold much promise for the future, none of these devices can support high-throughput analysis at the time of writing.

In what follows, we describe the high-throughput system developed to measure the growth and curvature of populations of plated *Arabidopsis thaliana* roots within the Centre for Plant Integrative Biology (CPIB) at the University of Nottingham. Sections 7.2 and 7.3 discuss the hardware and software components of the system, respectively, before lessons learnt are summarised and conclusions drawn in Sect. 7.4.

7.2 Hardware Approaches

Within CPIB, many of the day-to-day growth studies of Arabidopsis roots involve examining the growth of multiple roots (perhaps 20 or 40) on square petri dishes in growth rooms. Often, it is desirable to study growth on multiple plates at once, perhaps to study contrasting treatments or simply because a large number of plants are required. Previously, single plate time-lapse imaging has been possible using digital cameras and timed capture software. The motivation for a robotic imaging system is to allow, with only one or two cameras, the imaging of up to 20 plates at a time, without the need for a researcher to spend time moving plates or cameras. This section will detail the development of the hardware side of such a device, built from the ground up with both commercially available and easily constructed custom parts. This approach allowed the development of a system which was both able to meet the requirement of imaging using the current growth room setup, and could be tailored to the budget available.

7.2.1 Basic Configuration Options

A number of different hardware setups are possible to achieve high-throughput imaging in the laboratory, each with advantages and drawbacks. The simplest approach is to provide a separate camera and lighting arrangement for each sample. This has the advantage that neither the samples nor the camera is moved, but hardware costs can be prohibitive when imaging large numbers of plants. Thus, hardware solutions for high-throughput imaging are usually motorised and furthermore can be categorised into two approaches, based on which component is to be moved: the sample, or the imaging equipment itself.

Static camera systems require the translocation of samples to an imaging area, typically achieved by using motorised carousels, turntables or conveyor belts. This approach is advantageous in that a single camera and lighting setup is required but care must be taken to ensure that movement to the imaging area does not disturb the samples, especially when imaging delicate roots growing on the surface of agar.

Static sample systems, which move the imaging equipment to each sample in turn, allow many such samples to be imaged in situ using a single acquisition system. The simplest design is to mount a camera on a turntable; more flexible designs employ linear actuators either singly or in gantry arrangements. The positional accuracy required in linear systems can be achieved by using either belt driven or ball screw actuators whilst turntables can be positioned by stepper motors.

Whichever approach is used, the temporal resolution of such systems is dictated by the speed of the motorised handling equipment and the number of samples being imaged. All systems also require accurate and repeatable positioning, especially if time-lapse images are to be acquired over an extended period. Automated software registration of collected images can correct for minor displacements but such post-collection processing can be avoided if suitable hardware is used; it is always desirable to try and avoid such errors at the point of capture if at all possible.

7.2.1.1 The CPIB Imaging Robot

The system currently in use at CPIB has been developed to image root growth and root system architecture in seedlings of the model plant *A. thaliana* grown vertically on agar plates. The system was designed to be used in standard growth rooms, with no alterations to the conditions under which the plants are usually grown. A linear actuator moving a pair of cameras along existing illuminated growth shelves best met the experimental demands and has proved a robust and reliable solution for laboratory-scale high-throughput imaging (Fig. 7.1).

The system comprises a 1.8 m belt-driven linear actuator which is floor mounted in front of a standard growth room shelving unit. The actuator translocates a tower supporting two cameras whose vertical position is adjustable to the height of individual shelves. The tower is positioned by a stepper motor driven via a stepper driver board controlled via a dedicated microcontroller. Commands are sent to the microcontroller via a serial connection to a personal computer running the control software. Control software was written using the LabVIEW Graphical Development Platform (National Instruments Corp., Austin, USA) to provide a graphical user interface for setting motion, time lapse and image acquisition parameters. With two cameras mounted, the system can image twenty × 125 mm square growth plates with a minimum interval between consecutive plate images of two min.

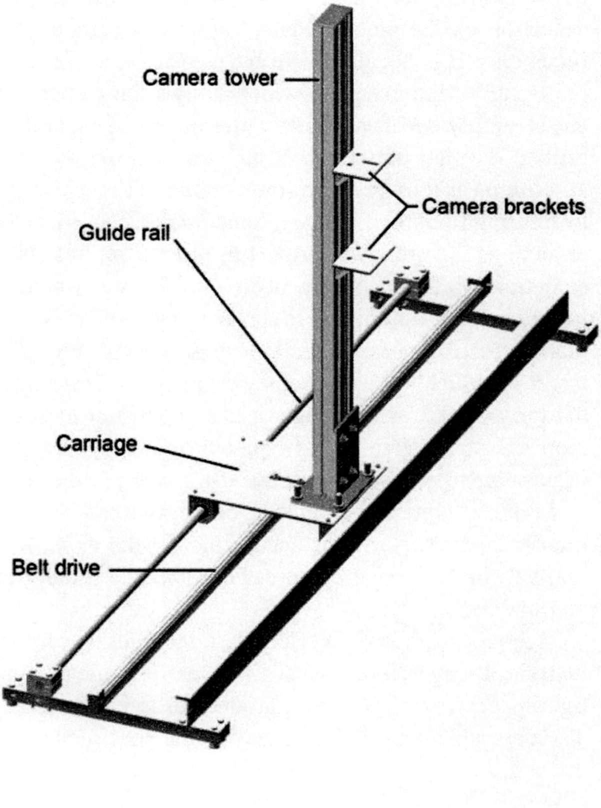

Fig. 7.1 The CPIB Image Robot. 3D model of the imaging system. Full schematics, software and video are available for download from the CPIB website (CPIB 2010)

Images are collected and stored in individual directories for each plate for later analysis. Actual control of the cameras is handled externally to, but controlled by, LabView using the PSRemote (Breeze Systems Ltd, Surrey, UK) software to control two × Canon G9 high-end consumer compact cameras. We will discuss more about the analysis of the captured images in Sect. 7.3.

7.2.2 *Capturing Scientifically Useful Images*

When cameras are used to take images which are later going to be processed using an automated image analysis approach, it is important to spend time making sure the images are of the best possible quality to start with. The term "best quality" here is often largely dependent on the scientific task at hand, but there are some unwanted effects when capturing images which are common problems across different applications, be they microscope images of roots, CCD camera images of individual roots, images of plated plants, etc. Most are easily resolved with some prior knowledge and planning. Below is a list of some common points to check when setting up an image capture process:

Focus. It may sound obvious, but is the plant in focus? If a partially reflective target is used, such as a Petri dish, is the camera focusing on the plants or the reflection on the plates? Many auto-focus cameras have the ability to lock their focus once it is acquired. We have found it useful to use a dedicated focus target built onto a dummy plate which allows the camera to easily acquire focus. This target simply consists of many alternating black and white lines of various scales, printed on matt paper, which aids the camera's auto-focusing algorithms. These algorithms usually perform some search of the focus space (e.g., Wright et al. 2009) to find high contrast images, hence having an ideal, high contrast target without reflections can greatly improve this procedure; it is also for this reason that focusing sometimes fails on poorly lit or uniformly textured surfaces. The focus is then locked and the dummy plate replaced with the experimental plate. Of course, with manual focus lenses, this is less of a concern, but it is still useful to have a clear target on which to focus. Achieving good focus is important, as recovering detail is hard in an unfocussed image; it is easy to later increase the "blur" on the image if required (to remove high frequency features, for example, like small lines), but impossible to recover fine detail lost at the point of capture.

Lighting. Optimising lighting conditions can be a challenge, as clearly there is a trade-off between what is acceptable for the experimental conditions and what is optimal for the image capture. For cameras, usually brighter conditions are more favourable.

If the intensity and wavelength of the lighting cannot be altered for experimental reasons, it may still be possible to alter the angle of illumination. Backlighting, side lighting and front lighting all produce different types of image. The image produced also depends on the transparency of the specimen. Side-lighting opaque roots will produce a different image to translucent roots. Clearly, the growth medium used also has an impact here; backlighting obviously is not possible with an opaque medium.

Various light sources (fluorescent, tungsten, daylight, etc.) all cast a different colour onto the image, which can cause the image to appear tinted blue or yellow. It is possible to correct for this by selecting the correct white balance setting in the camera, to match the lighting conditions. Post-capture correction is possible if a known white target is placed in the field of view. Software can then correct the colour balance of the image to make the appearance of this target once again colour-neutral.

It may be that whatever requires imaging needs to be kept dark. In this situation, imaging may still be possible using invisible wavelengths, such as infrared (IR). IR light extends beyond the visible light spectrum, further past the red light wavelengths we can see. Most common infrared imaging uses near infrared. The purchase of a dedicated IR camera is a possibility, or it may be possible to convert existing cameras by removing the visible light filter over the sensors or on the lens; often the light sensor in the camera is responsive to a wider range of light than we can see, and is narrowed to visible bands using a physical filter. Consult the camera manual for information, as some of these filters are either user-swappable or, as is sometimes the case, the filter is clear glass and does not block any wavelengths to start with.

Camera type. When setting up a high-throughput system, choice of camera is clearly important as it determines the ultimate quality of your images. We have worked with £250 consumer digital cameras and £2,500 industrial grade cameras. Both can ultimately produce high enough quality images for most automated processing applications, but the ease of image capture and the kind of images that can be captured does vary. Using an off-the-shelf compact camera forces use of whichever lens the camera is factory-fitted with (precluding the switching of custom lenses, which is possible with most industrial cameras, digital SLRs, etc.). Choosing a custom lens is useful if you want to take a very highly magnified image of a root, for example, or if the capturing of a wide field of view is essential. Another factor when deciding on a consumer camera is that they are often limited to working in visible light wavelengths; some more expensive cameras have removable filters over the sensor which allow them to function as infrared cameras.

Cameras are commonly either of USB or FireWire connection type. FireWire allows for faster data transfer between the camera and the computer, and can supply power directly to the camera along the same cable. However, a dedicated FireWire board may be required in the host PC (although these are inexpensive). USB cameras usually require an external power source. The maximum length of connecting cable that can be used without repeaters also varies dependant on the connection type.

The cost and availability of cameras obviously feeds into the choice of robotic imaging system outlined earlier, i.e. whether it is economical to replicate the imaging system multiple times.

File type. Saving images as JPEGs compresses them in a lossy way; in other words, the size is much smaller than an uncompressed file, but at the cost of losing information (particularly colour information) which cannot be recovered. Care must be taken with JPEGs as they are designed to lose information that we cannot perceive very well in order to reduce the file size. This means that they may look high quality to the human eye, when in fact much data has been destroyed; this is good news for photograph storage, but bad news for scientific analysis. Dependant on the processing required though, a high quality JPEG image may be sufficient, especially considering the trade-off with the large file size required for an equivalent uncompressed image file.

7.3 Software Approaches

The phrase "high-throughput" suggests the generation of large amounts of data. This data generation can take the form of an automated process like time-lapse image capture, and thus many thousands of digital images can be produced with little user interaction required. The ability to generate such raw resources with ease is thanks to recent leaps forward in technology and the corresponding decrease in the associated costs, both of capturing the data and storing it. This dream rate of data production does have two obvious drawbacks, however. The bottleneck in analysing the data is shifted from the capturing process to the next step in the sequence: processing the stored data. The second drawback lies with the storage itself: producing huge volumes of data with little difficulty means the data needs to be stored in a safe way (i.e. backed up), and catalogued using a system which means people can simply *find* their, and other people's, data again, and can link it to the subsequent processing steps and analytical results.

In the following sections, we will describe approaches to solving both of these problems.

7.3.1 Automating Measurement Processes

Traditional, methods of root growth measurement are really only suited to low-throughput data capture rates. Such traditional methods can take the form of marking plates at fixed intervals to indicate growth position and subsequent manual measuring, or measuring root length in image editing software such as ImageJ (Abramoff et al. 2004), etc.

Our solution to extracting root growth data from these large, time-series image datasets was to develop some software in-house which adopts a top-down approach to the image analysis. The software, RootTrace (French et al. 2009), combines user knowledge of where each root lies, with various levels of modelling and tracking techniques which allow the path of the root to be automatically extracted from the image.

Taking a top-down approach means some prior model is used to interpret the image. This is in contrast to a bottom-up approach, which starts with basic operations on pixels and builds up, by combining pixels into increasingly larger groups, to higher level segmentations. The main advantage of top-down approaches is that they tend to be more robust to noise and distracting clutter in the images. The high-level knowledge we have about the scene can be used to make sense of confusing areas within the image, where a typical bottom-up process might fail. Take, for example, a bright reflection on an image of plated roots. Using prior knowledge of where seeds were placed and having a model of what we expect a root to look like and how we think it might grow, a top-down approach might never need to go near the area of image with the reflection, and, if it does obscure a root, the models of appearance and growth of the root gives the system a fair chance of still correctly tracing the root. Conversely, a bottom-up approach might be based on thresholding pixel intensity values; this means selecting all pixels in the image above this threshold value. The pixels of which the reflection image is comprised would be selected as they are bright, and the existence of a root might wrongly be inferred from the bright patch. Further processing, such as looking at the shape of the patch or the number of pixels in it would be needed to discard the unwanted reflection, and even this might not be possible if the reflection "looks" like a root.

EZRhizo (Armengaud et al. 2009) is a good example of an effective bottom-up approach to root measurement. The software begins by thresholding the image, selecting the brightest pixels. Several processing steps are then used to fill in gaps in the root, remove non-root regions and skeletonise the root outline to a structure one pixel wide. The user must guide the process at each stage. The result can be a good representation of a root and the architecture of the lateral roots, in a format well suited for further analysis; indeed from the final structure, summary statistics can be automatically measured and saved in a database. The interaction required, though, means this approach is not well suited for use in high-throughput systems.

To automatically analyse thousands of roots across multiple images, both of different plates and different time points, several features are needed in the software. The first, and most obvious, is image management. The tool must accept images for analysis as a batch and must understand how the images relate to one another, for example as a time series. After supplying prior information to the software, such as where the system must start measuring each root, the tool should be able to be left to process the images without further interaction.

One of the dangers of a fully automated approach is if the software does manage to perform a measurement on each image, but that measurement is wrong. How can such mistakes be detected? In some cases, it may be possible to put limits in the software to detect unusual measures. However, this still requires a judgement call on what is permissible and what is not, with the added danger that if the user believes the software is checking the validity of its own results, the user might not bother to check for themselves. A sensible alternative is to present each measured result to the user in a way which allows them to be easily verified, such as on a graph or a series of images. Ideally multiple visualisations of the data should be available. Dubious measurements can be interrogated in detail in the raw data. Furthermore, adopting an analysis procedure which expects there to be some inaccurate measurements in a sequence and robustly handles them is a sensible approach. All of this relies on the software providing the user with raw data to process rather than analysing it itself; presenting summary statistics within the tool may be useful, but may remove some of the necessary quality assurance work from the user, which can be undesirable.

7.3.1.1 RootTrace

Met with the problem of needing to analyse increasing numbers of plated roots to measure growth over time, development on the RootTrace software (French et al. 2009) was started. This is an open source software tool, continually in development driven by the needs and suggestions of biologist users. Here, detail will be presented about its motivation and use. Focusing on this high-throughput tool here is not intended to suggest it is the only tool capable or the best tool in all situations (see the introduction for examples of more). Indeed, as is the case with image analysis tools, they are often constrained to quite specific application areas, and well-defined images. But it is a tool which is in frequent use and active development, and, even if not directly applicable to everyone's datasets, has practical lessons to be learnt from the general way high-throughput software can be implemented and used.

There are two main reasons why the data being collected justified the development of a new software tool: the quantity of images (and number of roots on each image) and the quality of the images themselves. The aim of RootTrace was that the processing of time series datasets should progress automatically, without input from the user, and that the quality and type of images should be that produced using a consumer digital camera under growth room conditions, as in Fig. 7.2a.

Use of RootTrace is as follows. First, an image dataset must be captured. Originally, this involved setting up a camera on a tripod in the growth room, and using time-lapse image capture software. Today, multiple plates can be imaged at once using a robotic imaging setup (Sect. 7.2). RootTrace is designed to be run once per plate (i.e. once per time series sequence). The user is required to click on the uppermost part of each root and a small example background region. This setup is carried out once per plate; this seems acceptable as the overhead to the user is only perhaps two minutes per time series; the rest of the processing is carried out by the image analysis software itself. One area of current research is concerned with automated initialisation, but as the workload to the user is so low anyway, certainly compared to fully manual methods, this has not been found to be a necessity. Once all this necessary prior knowledge has been added by the user, the software is instructed to begin processing. Processing continues until all roots in the time series have been traced. This can take anything from a few minutes to several hours, dependant on the dataset size.

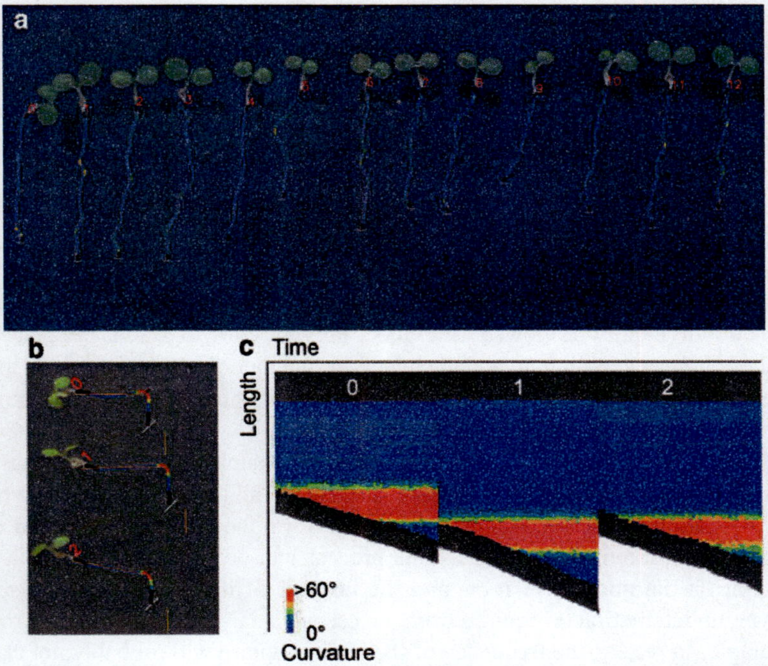

Fig. 7.2 (**a**) Plated roots as traced by RootTrace. (**b**) Roots responding gravitropically, and the resulting curvature traces (**c**) of these roots. The overlaid *colours* in (**a**) and (**b**) indicate curvature, and the detected angle of the root tip can be seen represented by the *offset lines* in (**b**)

7.3.1.2 Output Data

Presenting the results of processing large, multi-faceted datasets to the user in an understandable yet powerful way is a challenge. On the one hand, the software should export raw data, ready for further processing in statistics packages, for example. On the other hand, it can be productive and helpful to show the user summary charts displaying various aspects of the data as soon as processing is finished, or even as it is happening.

The most complete pictures of these datasets contain both time and space dimensions, displaying, for example, the local curvature along the root length as the root grows. RootTrace currently does this on a per root basis; see Fig. 7.2b, c.

A true summary would collate information across roots to present a representative picture of how the roots as a group grew. Some high-throughput tools do present this information to the user; with RootTrace currently this particular analysis must be performed by the user from the raw data results.

Presenting growth information is easily done, as for each time point the length of each root on the plate is recorded. Thus, growth charts can be easily constructed to show lengths over time, the gradient of which is the rate of growth.

Some higher level processing is performed by RootTrace, to detect, for example gravitropic bend points when large angles are formed in the root. When found, the distance along the root can be determined, and from the growth information a time when this bend was formed can be estimated. Other interesting analysis that could potentially be performed with curvature information includes Fourier analysis of frequencies of local curves and the total amount of curvature produced. Later versions of RootTrace also specifically measure tip angle as well as local curvature along the root.

RootTrace aims to help the user detect errors by showing the processing as it happens, and saving a time series of annotated output images which can be examined individually or viewed as a video. Preview growth graphs are generated as analysis happens, giving an added chance for the detection of outliers. Also, the raw data for each root in every frame is saved in CSV (comma separated value) text files, ready for further analysis in statistics packages.

The actual processing of the data takes place in a batch mode. The user has told the software which set of images requires processing, and where to start each trace. RootTrace will then load each image in turn, perform the image analysis and write the various measurements to a file. This process can be time consuming, largely dependent on the number of roots and the number of images in the time series. However, no interaction is required from the user so it can be left running overnight, for example. In reality, the frequency of the image capture will probably not exceed one every ten minutes, as a minimum. However, a frequency of one image every two minutes has been used before. The frequency makes no difference to RootTrace, and it will simply take longer to process the data. The advantage of oversampling the data is that if some errors occur during the tracing, discarding problematic frames then becomes a possibility, whilst still achieving the desired time resolution between images.

Of course, with robotic imaging taking multiple plates, it is desirable to process all these images in one batch. This will still need the user to initialise each plate sequence; with the potential of automatic initialisation, this "parallel" processing of all images in a robot-derived time capture becomes plausible. The time to then process this dataset increases correspondingly. With fully automated multiple image capture and automated image processing, the bottleneck in processing becomes not the image capture, nor the user performing the measuring. Instead, simply accessing and processing the raw data becomes the limiting factor. Dividing the image analysis into processing batches of plates breaks down nicely into a parallel processing arrangement, whereby the processing of batches of images could be farmed out to separate computers, and the results collated at the end of the processing. Although not yet implemented, the independent processing of images lends itself to this kind of arrangement, if it is desirable to decrease the processing time.

With this much data, storage space needs consideration. A typical 10 megapixel compressed (JPEG) image might use 5 MB of disk space. Multiply this by perhaps six per hour for 5 days, and you have 720 images per plate (or about 3.5 GB). If you are imaging 20 plates, then the data produced jumps to about 70 Gb for 1 week's worth of imaging data. Where should this data be stored? Typically, as the images are taken, the software controlling the cameras will store them on a local computer close to the camera. It may be possible to store them directly on a network drive which is backed up, although the transfer speed of the network, especially low strength wireless networks, may mean the camera control software is unable to do this. If not stored directly on backed up media, the images should be backed up as soon as possible, as losing the data at this stage means at least 1 week, plus the preparation time, of research is lost. This is especially true if the images are stored on a local computer in a growth room, which may be in environmental conditions which make the computer storage more likely to fail (due to temperature or humidity, for example). Although 70 Gb for one dataset may sound a lot, modern hard drives are extremely competitively priced. Even so, be aware that a backup system may end up storing around 3× this amount of data as part of the redundancy and versioning strategies.

7.3.2 *Pre-processing Images for Analysis*

While this section may not be relevant to all readers, it may be useful or of interest to those wishing to develop their own systems. Here, details will follow of the foundational image processing techniques which can improve the quality of images before further analysis in automated tools such as RootTrace.

Before any image analysis actually occurs, the image may need to be tidied up to remove artefacts such as noise and lens distortion. Lens distortion is caused by the optical components of the camera warping the image, often appearing as a "barrel distortion" effect, leading to the edges being more compressed than the centre of the image. This may not be an issue and is more of a problem in wide angle lenses. But if it is an issue, there are standard ways to remove it; for example, it is possible to estimate the parameters needed to correct the image if a known straight line in the real world can be traced in the image (it will appear as an arc).

If the camera has moved between frames of a time series, the images may need to be aligned to each other, a process known as registration. This ensures that features that should not move between frames do indeed actually remain static throughout the time series.

All imaging devices produce noise, which takes the form of erroneous measurements in addition to the true measure of light intensity at each pixel. The most common form of noise in digital camera images is called Gaussian image noise. This name simply means that noise drawn from a Gaussian distribution (which, of course, is constructed from a mean value and a standard deviation describing the likely spread of numbers surrounding it) is added to the true measurement at each pixel.

Correcting for Gaussian noise simply involves blurring the image using a Gaussian blur. For each pixel, a weighted average of the surrounding pixels is taken, as defined by a Gaussian function so that pixels close to the pixel under consideration have higher weight than those further away. The "width" of this blur can be adjusted; higher noise levels can be attenuated with more blur, but consequently sharp features in the image are lost, so a trade-off is required.

Although not directly used in RootTrace, additional processing may sometimes be required to correct for uneven lighting across an image. This is commonly required in software which uses some form of thresholding (selecting all pixels above a certain intensity value, either set globally or calculated locally). For example, plates lit from above may be significantly brighter at the top of the image than the bottom. A basic procedure for background illumination correction is simple. Construct an image with the foreground objects (roots in our case) removed, by blurring the image to remove these high-frequency components. A large Gaussian blur will typical remove all details and just leave an approximation of the general brightness and colour of the image, like looking at a photo through heavily misted glass. Subtracting this image from the original image effectively cancels the low-frequency background out, removing large-scale variations in lighting, but leaving the high-frequency components, like the roots, untouched.

7.3.3 Data Storage and Organising Data

We touched earlier on the large quantities of data that will need to be stored in these high-throughput imaging systems. Organising this data and storing it safely is key in any high-throughput system. We cannot present more than the briefest introduction here, but be aware that the storage and organisation of such scientific data electronically is a burgeoning and fast moving research field (e.g. OMERO 2010).

It is good practice to store metadata with the images, i.e. information about the images themselves. This might be as simple as tagging or associating words from a controlled vocabulary to the images. Normally image files are automatically tagged with information about the camera type and settings of the camera for a particular photo. This information is usually stored within the image file, in a section reserved for text rather than image data. Further tags can be added, either to the text data in the image file itself or associated with the file in a database. The kinds of tags that are useful relate the image to the experiment and content of the image. For example, storing who was running the experiment, who took the images, particular information about the plant in the image, etc. means a good chance of finding the image set again using searches at a later date. You might, for instance, search for all images of wild-type roots, or all images taken by a particular researcher between certain dates.

Although some of this data can be harvested automatically, much of this metadata must currently be entered manually, and there lies a problem: people's work procedures will need to be changed to make adding metadata easy, error free and most importantly as unobtrusive as possible in their workflow. Really, there should not be much work involved in adding information to all the images in one time series, but tools need to be designed to make it easy to add this data. Success here demands that researchers are willing to put in the effort and time required to "mark up" their data with metadata, and it is the software developer's job to make this as user friendly as possible.

It may be possible to manage the image store itself using a database, which can then be easily searched and images retrieved using a database front end, for example via a website. If this is not possible, then at least the directory structure of the folders where images are stored should be considered, ideally before the capture of any data, and adhered to by all experimenters for all experiments. In practice, this may be hard to achieve, but again enforcing consistency from the start means it should be very easy to find images at a later date, perhaps long after the original researchers have left: this is the goal of any data organisation system.

7.4 Lessons Learnt from High-Throughput Imaging at CPIB

In this chapter, we have presented our experience of developing a high-throughput image acquisition and analysis system within our lab. It has been very much driven by the needs of the biologists, and was developed as a collaboration between those biologists, engineers and computer scientists. Obviously, the system we have presented, as with all tailor-built setups, is specific for our needs. However, there are some general lessons which have been learnt and are worth passing on to others considering work in this field. As a form of summary, we shall present them here to conclude this chapter.

- *Consider carefully the biology you wish to research.* Any system such as this is likely to be built with a particular use in mind. This may sound obvious, but it is important to make sure right at the start that it is capable of the job. Do you need to image in the dark as well as the light? If so, you may need to investigate infrared lighting and cameras for imaging in darkness. Will you be studying architecture as well as basal root growth? There are clear differences in both the quality of the images required and the image analysis process used. What species will you be looking at? This can determine the plate size required and consequently the distance between the camera and the plate in order to fill the image frame.
- *Do you want to move the plants or the cameras?* Choosing to move the plants allows potentially more plants to be imaged as they can be stored in an efficient arrangement when not being photographed. However, moving plates may vibrate the plants, which might have an unwanted effect on growth or the positioning of the plants on the plate. There is clearly a cost trade-off in how many cameras are used versus the engineering required to position each plate in front of the cameras available.
- *Building fully automated analysis software is a challenge.* Many processing packages for biological images rely, quite rightly, on a sizeable amount of interaction from the user. There are some challenges in image analysis which we as people perform effortlessly, yet which are very hard problems indeed to solve algorithmically. RootTrace currently relies on user interaction to

essentially show the software where the plants are on the plate. For the user, this is such a trivial task, but to identify reliably the required position automatically, across a wide range of plating possibilities is a challenge. With high-throughput software, this interaction phase is best handled at the start of the process. If the user's attention is required every 5 min during processing, then it is doubtful the software could be considered truly high-throughput.

- *Image analysis methods invariably use models.* Many image analysis methods and systems have been described in the literature. These address problems commonly found in plant phenotyping, such as segmentation and object tracking. To produce a solution to these problems, each method must make assumptions about the input images and the output required. In RootTrace, these assumptions are made explicit in the various models employed within the tracker. Other methods may not state their assumptions (models) so clearly, but must use them – it is simply not possible to segment an image without some model of what constitutes a segment. If, however, the models used do not match the task at hand the results obtained may be incorrect. It is important to clearly understand the models embedded in any image analysis software and/or methods used.
- *Think about how to store the large amounts of data.* Time-series image datasets can quickly fill a lot of disk space. Thinking about local and long-term backed-up storage is something that must be considered on day one. Simply moving the data around can be problematic, although with the adoption of gigabit networking this has become a lot faster. USB hard disks are an option, but can feel slow when copying tens of gigabytes of files (not uncommon amounts with the image sets generated here).
- *Improving the images at the point of capture makes later analysis much easier.* Spending time optimising the image capture process pays dividends when it comes to analysing them later. Problems like condensation on the plates, which can disrupt automated analysis, can often be eliminated, or at least greatly reduced, in the growth room. Not investing the time and effort to do this at the start of the process means every image captured may be affected. Developing a high quality capture process on day one allows you to reap the rewards of high quality data throughout your experiments.

In conclusion, we hope this chapter has shown how it is possible to build custom high-throughput imaging hardware and software from the ground up. This arrangement has proved popular with the plant biologists at the University of Nottingham, and has found use in day-to-day image capture and root growth analysis. Additionally, we hope we have provided some useful thoughts and tips for those wishing to investigate high-throughput root imaging in their own laboratory.

Acknowledgements The authors would like to thank colleagues and collaborators at CPIB and the University of Nottingham, including Chris Payne, Oya Aksoy, Charles Long and Michael Pearce (robot development); Tara Holman, Susana Ubeda-Tomás (sample imaging and RootTrace development); Michael Wilson and Michael Stout (data management). CPIB is funded by the BBSRC and EPSRC Systems Biology Initiative.

References

Abramoff MD, Magelhaes PJ, Ram SJ (2004) Image processing with ImageJ. Biophotonics International 11(7):36–42

Armengaud P, Zambaux K, Hills A, Sulpice R, Pattison RJ, Blatt MR, Amtmann A (2009) EZ-Rhizo: integrated software for the fast and accurate measurement of root system architecture. Plant J 57:945–956

Barron JL, Liptay A (1994) Optical flow to measure minute increments in plant growth. Bioimaging 2:57–6

Barron JL, Fleet DJ, Beauchemin SS (1994) Performance of optical flow techniques. Int J Comput Vis 12:43–77

CPIB (2010) CPIB imaging robot resources, available from http://www.cpib.ac.uk/tools-resources/imaging-robots/. Cited 22 Aug 2011

French A, Ubeda-Tomás S, Holman TJ, Bennett MJ, Pridmore T (2009) High-throughput quantification of root growth using a novel image-analysis tool. Plant Physiol 150(4):1784–95

Gregory PJ, Hutchinson DJ, Read DB, Jenneson PM, Gilboy WB, Morton EJ (2003) Non-invasive imaging of roots with high resolution X-Ray micro-tomography. Plant and Soil 255:351–359

Jahnke S, Menzel MI, Dusschoten D, Roeb GW, Buhler J, Minwuyelet S, Blumer P, Temperton VM, Hombach T, Streun M, Beer S, Khodaverdi M, Ziemons K, Coenen HH, Schurr U (2009) Combined MRI-PET dissects dynamic changes in plant structures and functions. The Plant Journal 59:634–644

Le Bot J, Serra V, Fabre J, Draye X, Adamowicz S, Pagè L (2010) DART: a software to analyse root system architecture and development from captured image. Plant Soil 326:261–27

Michener H (1938) The action of ethylene on plant growth. Am J Bot 25:711–720

OMERO (2010). The OMERO Platform http://www.openmicroscopy.org/site/products/omero. Cited 22 Aug 2011

Parry G, Delbarre A, Marchant A, Swarup R, Napier R, Perrot-Rechenmann C, Bennett MJ (2001) Novel auxin transport inhibitors phenocopy the auxin influx carrier mutation aux1. Plant J 25:399–40

Perret JS, Al-Belushi ME, Deadman N (2007) Non-Destructive visualization and quantification of roots using computed tomography. Soil Biology & Biochemistry 39:391–399

Quelhas P, Mendonça A, Campilho A (2010) Optical flow based Arabidopsis thaliana root meristem cell division detection. Lect Notes Comput Sci 6112:217–226

Roberts T, McKenna SJ, Wuyts N, Valentine T, Bengough G (2007) Performance of low-level motion estimation methods for confocal microscopy of plant cells in vivo. IEEE Workshop on Motion and Video Computing (WMVC'07): 13

Roberts T, McKenna SJ, Du C-J, Wuyts N, Valentine T, Bengough AG (2010) Estimating the motion of plant root cells from in vivo confocal laser scanning microscopy images. J Machine Vision Appl 21:921–939

Sethuraman V, Taylor S, French A, Wells D, Pridmore T (2009) Segmentation and tracking of confocal images of Arabidopsis thaliana root cells using automatically-initialised network snakes. In: Proceedings of 3rd International Conference on Bioinformatics and Biomedical Engineering, 2009

Taylor HM, Upchurch DR, McMicheal BL (1990) Applications and limitations of rhizotrons and minirhizotrons for root studies. Plant Soil 129:29–35

Tracy SR, Roberts J, Black CR, McNeill A, Davidson R, Mooney SJ (2010) The X-factor: visualizing undisturbed root architecture in soils using X-ray computed tomography. Journal of Experimental Botany 61:311–313

van As H (2007) Intact plant MRI for the study of cell water relations, membrane permeability, cell-to-cell and long distance water transport. Journal of Experimental Botany 58(4):743–756

Van der Laan P (1934) Der Einfluss von Aethylen auf die Wuchsstoffbildung be aena und Vicia. Rec Trav Bot Neerl 31:691–74 van der Weele CM, Jiang HS, Palaniappan KK, Ivanov VB, Palaniappan K, Baskin TI (2003) A new algorithm for computational image analysis of deformable motion at high spatial and temporal resolution applied to root growth: roughly uniform elongation in the meristem and also, after an abrupt acceleration, in the elongation zone. Plant Physiol 132:1138–114

Walter A, Spies H, Terjung S, Küsters R, Kirchgeßner N, Schurr U (2002) Spatio-temporal dynamics of expansion growth in roots: automatic quantification of diurnal course and temperature response by digital image sequence processing. J Exp Biol 53:689–69

Wright EF, Wells DM, French AP, Howells C, Everitt NM (2009) A low-cost automated focusing system for time-lapse microscopy. Meas Sci Technol 20:027003

Chapter 8
Flat Optical Scanner Method and Root Dynamics

Masako Dannoura, Yuji Kominami, Naoki Makita, and Hiroyuki Oguma

Abstract This chapter outlines those methods for assessing root dynamics in situ based on flat office optical scanners inserted into the ground. The main advantages of these scanners are (a) they take wide images of root developing on a flat surface, (b) they can be easily coupled to PC via USB or wireless LAN connection for automatic image acquisition at high frequency, and (c) their low price allows researchers to burry several scanners for several months or years. This new method was successfully used to describe the dynamics of mycorrhizal tips formation on *Pinus pinaster* root and the development of individual root tips of *Quercus serrata*.

8.1 Introduction

Instead of disruptive methods such as digging, soil core sampling and monoliths, development of non-disruptive methods is needed to better understand fine root dynamics in the soil.

Non-disruptive methods such as rhizotrons and minirhizotrons are mainly used for continuous observation of root extension or to investigate the distribution of live roots in the field. Rhizotron or root window allows root to grow and develop along a transparent board inserted along a trench (Green et al. 2005; Sword 1998;

M. Dannoura (✉) • N. Makita
Graduate School of Agricultural, Kyoto University, Kitashirakawa oiwake-cho, sakyo-ku, Kyoto 606-8502 Japan
e-mail: dannoura@kais.kyoto-u.ac.jp

Y. Kominami
Kansai Research Center, Forestry and Forest Product Research Institute, 68, Nagaikyutaro, Momoyama, Fushimi, Kyoto 612-0855 Japan

H. Oguma
Center for Environmental Measurement and Analysis, National Institute for Environmental Studies, 16-2 Onogawa, Tsukuba, Ibaraki 305-8506 Japan

S. Mancuso (ed.), *Measuring Roots*,
DOI 10.1007/978-3-642-22067-8_8,

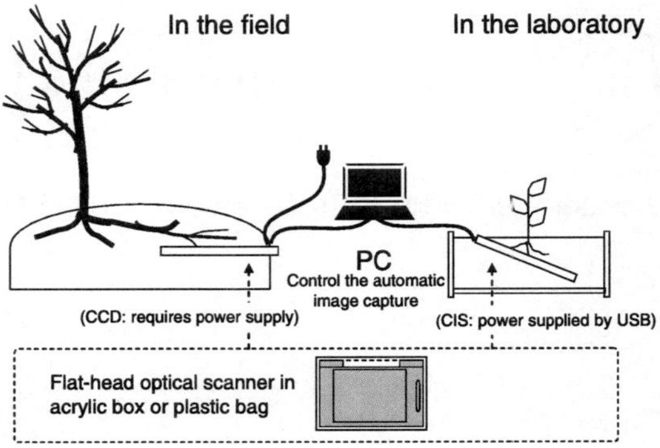

Fig. 8.1 Outline of the scanner system set up in the field (*left*) or in the laboratory (*right*) (Dannoura et al. 2008)

Thongo M'bou et al. 2008). Images are collected manually by drawing or using camera, or scanner (Dauer et al. 2009). Additionally, the root development of plants growing in root box, i.e. container having at least one transparent face, has also been developed for greenhouse or growth cabinet experiments (Epron et al. 1999; Pagès 1992; Riedacker 1974). Minirhizotron method is a well-established technique in long-term studies of root dynamics using a transparent tube installed into the soil (Johnson et al. 2001, Satomura et al. 2006). We have alternatively developed a method using flat optical scanners from which we obtained automatically wide image of root developing on a flat surface (Dannoura et al. 2008; Fig. 8.1).

8.2 The Characteristic of Scanner Methods

The first advantage of flat office scanner method is to get wide flat images of developing roots in contrast to minirhizotrons that are made of transparent tubes (around 65 mm in diameter), thus capturing small round shaped images. Using minirhizotron, a succession of single images is captured by moving either a camera (Minirhizotron BTC-100X; Bartz Technology Corporation, USA, image size 16 × 18 mm) or a scanner (CI-600, CID Bioscience, USA, image size 216 × 196 mm, up to 1,200 dpi) along the tube. In contrast, the method presented here uses office scanner that has flat observational surface with a typical size of 216 × 297 mm depending on the size of commercially available scanners. The observation surface will face one direction. The quality of image is from 96 to 4,800 dpi depending on the specification of product; actually 300–600 dpi is realistic for avoiding handling excessive memory capacities. Additionally, flat scanners fit into observation of rhizo-boxes. The large observed area is well suited for observing the distribution and branching of root systems, as are root window methods.

The second advantage is that the optical scanner can be set underground during all the entire experimental period. In the case of minirhizotron, observers bring the camera or scanner to the place where tubes have been buried to take images. Thus, the observation intervals limited typically at monthly to weekly. However, If scanners and PC are connected via USB or through web connection, images can be collected automatically by the PC, allowing frequent image captures (ex, hourly intervals) suitable for studying the dynamics of fine root growth or decomposition.

The third advantage is the low cost. One scanner costs less than 10,000 JPY (more or less 100 euros). It allows observers to set a respectable number of scanners permanently in the field, yet it is also possible to use only one scanner manually by inserting it in several transparent boxes set up in the soil.

The main disadvantage of this scanner method is the disturbance at the time of setting up the scanner box in the soil when roots are cut and soil structure altered. In addition, the scanner box creates an artificial environment, the environmental conditions on the glass or on the acrylic board are not the same to soil. This problem is not specific to the flat scanner method but also concerns the minirhizotron tubes and root window method as well.

8.3 How to Set Up the Scanner in the Field

8.3.1 Selection of Scanner

There are two types of flat office scanners: Contact Image Sensor (CIS) and Charge Coupled Device (CCD). The CIS-type scanner is relatively thin, its power can be supplied through a USB cable from a PC and warm-up time is eliminated. The CCD-type scanner is thicker than the CIS and it reads the image through the lens after reflected light is consolidated in a mirror. It differs from CIS in its clear interpretation of rugged objects. The selection of the scanner type will depend on the thickness of the water-proof protection. The CIS scanner is suitable when thin protection material is used (mainly when experiments are setup in the lab). In the field, the thickness of protection material will often require the use of CCD scanners because of their capacity to capture depth in the objects.

8.3.2 Installation in the Field

It is not recommended to burry scanner directly, because water can enter the scanner system. The scanner can be protected using plastic bags (suitable for laboratory experiments) or transparent boxes made of glass or acrylic materials.

Holes in the protection box or bag are needed for electric and USB connection but waterproofing should be most carefully done to prevent scanner failure, especially if it is setup for a long period in the field. Additional holes may be useful to connect a pump circulating dry air to prevent or remove condensation. Desiccating agent such as silica gel could alternatively be put in the box.

The scanner can be set up from vertical to horizontal position depending on the purpose. However, a small departure from horizontal position is recommended to avoid accumulation of water after rainfall. Note also that the soil mass over horizontally displayed scanner boxes might induce some deformation of the box.

8.3.3 Collecting the Images

The images could be obtained automatically using software or a written program. For example, UWSC free software records actions with mouse and keyboard into a macro allowing the PC to playback the sequence iteratively and automatically. The images are stored in PC, and the frequency of image collection depends on the aim of the experiment. It is possible to connect several scanners to one PC. The distance between scanner and PC is determined by the length of USB cable (up to 10 m) but it can be increased using wireless LAN connection. In the site with no electricity in the field, scanners and PC can be supplied by car battery connected to solar panels.

8.4 The Image Analysis

8.4.1 Image Observation

Figure 8.2 shows an example of the quality of images of *Pinus Pinaster*'s root in a pine plantation which has been established in 1998 (12 year old) at the INRA domain of Pierroton (44°45′ N, 0°42′ W, France) on a sandy spodosol. The scanners were inserted vertically in the field in March 2009. These images were taken from 19th of May to 14th July with 300 dpi resolution. The frequency of captured images was every 1.5 h. The development of mycorrhiza was observed in this measurements period. Monthly observation would have not been enough to record the detail of the kinetics of both root and mycorrhiza growth, while weekly observation gave a better description of root growth kinetics. Daily observations gave a nice view of the formation of mycorrhizal tips that occurred when the root growth was finished. We were also able to track the change of the colour of tips, which is important to know the function and the healthiness of the fine roots.

8.4.2 Elongation Analysis

Figure 8.3 shows an image of a *Quercus serrata* Thunb.'s root at Yamashiro experimental forest (34°47′ N, 135°51′ E, Japan). The scanner was inserted almost horizontally in the field in February 2007. Images were taken every 30 min with

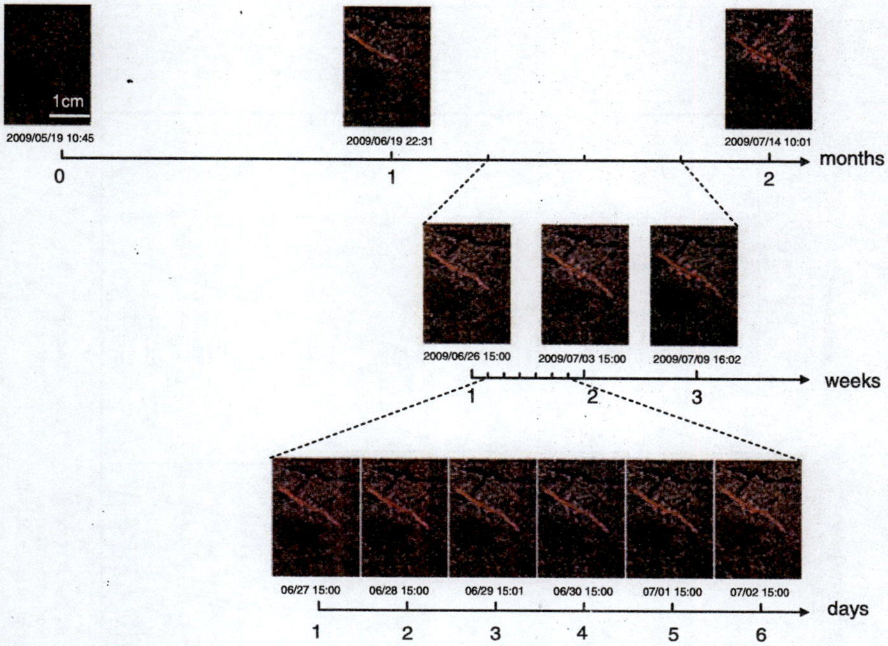

Fig. 8.2 Images of a single fine root of *Pinus pinaster* at monthly interval (*top*), weekly interval (*middle*) and daily interval (*bottom*). The development of each mycorrhizal root tip is shown with time at daily interval

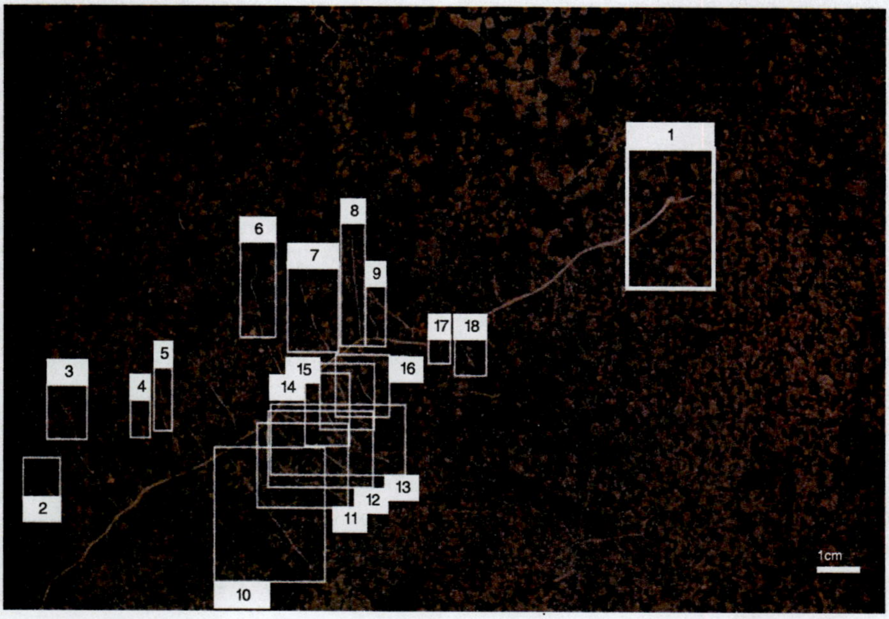

Fig. 8.3 Image of a well-developed branched fine root system of *Quercus serrata*. The number refers to the different root tips that were identified and for which root growth rate has been calculated (see Fig. 8.4)

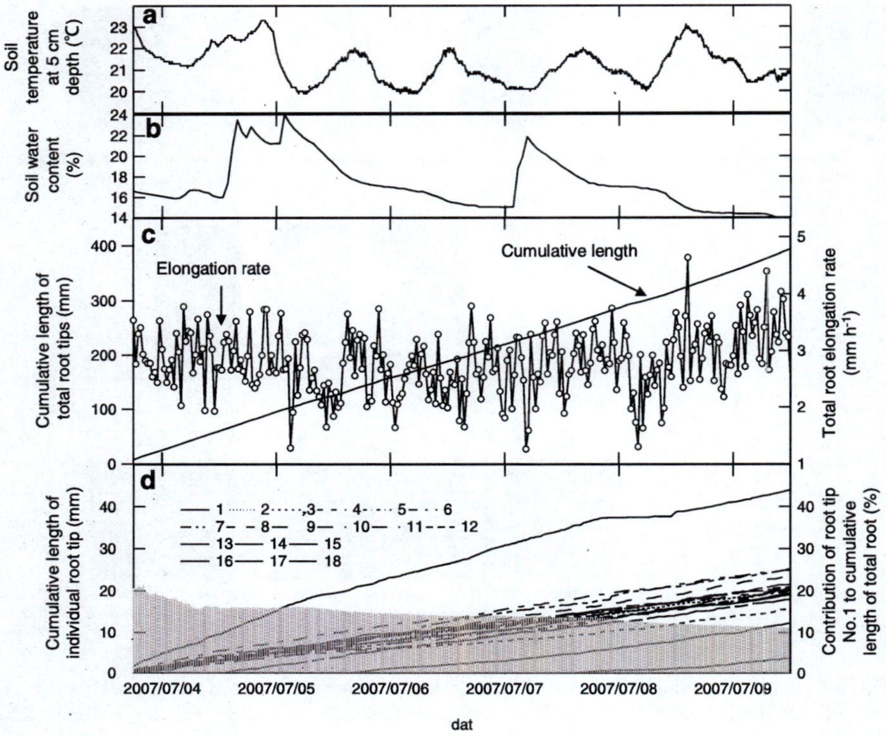

Fig. 8.4 Temporal change of soil temperature at 5 cm depth (°C), soil water content (%), total root elongation rate (mm h^{-1}) and accumulate length of each root (mm) calculated from the well-developed branched fine root system of *Quercus serrata* exhibited in Fig. 8.3

300 dpi resolution. Using A-root program (Nakano et al., in preparation), the elongation rate of each root tip was calculated every 30 min from July 3, 2007 to July 9, 2007. Soil temperature and soil water contents (ECH2O probes, Decagon, USA, Fig. 8.4a, b) were taken during the same period. The top root No.1 (Fig. 8.3) shows highest growth rate of all the branched roots (Fig. 8.4d). The total growth rate had high frequency fluctuation (Fig. 8.4c). Growth of root tips was blocked for short period by grain of sands before they began again to grow. The fluctuation in growth rate seems to be due to this movement.

Total root elongation rate was 2.8 mm h^{-1} (Fig. 8.4c) and root tip No.1 was 0.31 mm h^{-1} on average during the measurement period. After 6 days of measurements, cumulative length of total root tips reached more than 400 mm. Moreover, the ratio of root tip No.1 decreased from 20% to 11% of that of the total root. These high frequency image capture allows to show such phenomena as the decrease of elongation velocity of top root (No.1) when the 'new' branched roots (No.16, 17 and 18) start to grow.

8.5 Conclusion

In conclusion a flat office scanner coupled to PC for automatic image acquisition is a promising tool to study the dynamics of fine root and mycorrhiza. Although commercially available software can be used to analyse the images, development of new program is needed to better describe the growth dynamics of individual root types.

References

Dannoura M, Kominami Y, Oguma H, Kanazawa Y (2008) The development of an optical scanner method for observation of plant root dynamics. Plant Root 2:14–18

Dauer JM, Withington JM, Oleksyn J, Chorover J, Chadwick OA, Reich PB, Eissenstat DM (2009) A scanner-based approach to soil profile-wall mapping of root distribution. Dendrobiology 62:35–40

Epron D, Toussaint ML, Badot PM (1999) Effects of sodium chloride salinity on root growth and respiration in oak seedlings. Ann For Sci 56:41–47

Green J, Dawson L, Proctor J, Duff E, Elston D (2005) Fine root dynamics in a tropical rain forest is influenced by rainfall. Plant Soil 276:23–32

Johnson MG, Tingey DT, Phillips DL, Storm MJ (2001) Advancing fine root research with minirhizotrons. Environ Exp Bot 45:263–289

Pagès L (1992) Mini-rhizotrons transparents pour l'étude du système racinaire de jeunes plantes. Application a la caractérisation du développement racinaire de jeunes chênes (*Quercus robur*). Can J Bot 70:1840–1847

Riedacker A (1974) Un nouvel outil pour l'étude des racines et de la rhizosphère: le minirhizotron. Ann For Sci 31:129–134

Satomura T, Hashimoto Y, Koizumi H, Nakane K, Horikoshi T (2006) Seasonal patterns of fine root demography in a cool-temperate deciduous forest in central Japan. Ecol Res 21:741–753

Sword MA (1998) Seasonal development of loblolly pine lateral roots in response to stand density and fertilization. Plant Soil 200:21–25

Thongo M'bou A, Jourdan C, Deleporte P, Nouvellon Y, Saint-André L, Bouillet J-P, Mialoundama F, Mabiala A, Epron D (2008) Root elongation in tropical Eucalyptus plantations: effect of soil water content. Ann For Sci 65:609–615

Chapter 9
3D Quantification of Plant Root Architecture In Situ

Suqin Fang, Randy Clark, and Hong Liao

Abstract Root systems play important roles in plant nutrient and water uptake. The spatial distribution and structure of root systems can affect many physiological functions, carbon distribution, and plant anchorage. The accurate measurement of root systems is necessary for better understanding of plant growth and responses to biotic and abiotic stress. Due to their underground growth habitat, root systems are usually excavated from the soil before root measurements are taken. This process can destroy the root system architecture and result in the lose of fine root structures, and consequently, many devices and technologies have been developed to nondestructively capture and measure root systems in situ in 2D or 3D, such as WinRhizo, minirhizotrons, X-ray computed tomography, and magnetic resonance imaging.

9.1 Introduction

The root system is an essential part of plants and is important not only because it is the main organ that absorbs water and nutrients, but also because it provides anchorage against uprooting forces (Barber 1995; Bailey et al. 2002). The spatial configuration of the entire root system in its growth medium is known as root system architecture (RSA) (Lynch 1995) and can significantly impact many physiological functions, such as water and nutrients acquisition (Bowman et al. 1998; Doussan et al. 1998; Liao et al. 2001; Lambers et al. 2006; Skaggs and Shouse 2008), carbon distribution (Bidel et al. 2000; Coll et al. 2008), and the adaptability to environmental stresses (Miller 1986; Al-Ghazi et al. 2003). Furthermore, RSA plays an important role in protecting against root-borne diseases and pests in some

S. Fang • H. Liao (✉)
Root Biology Center, South China Agricultural University, Guangzhou, 510642, China
e-mail: hliao@scau.edu.cn

R. Clark
US Plant, Soil and Nutrition Laboratory, USDA-ARS, Cornell University, Ithaca, NY 14853, USA

S. Mancuso (ed.), *Measuring Roots*,
DOI 10.1007/978-3-642-22067-8_9, © Springer-Verlag Berlin Heidelberg 2012

leguminous plants (Bailey et al. 2002; Snapp et al. 2003; Cichy et al. 2007). The three-dimensional (3D) root structure of an individual plant includes both the topological arrangement of its root components and the coarse geometric characteristics of its entire root system (Danjon and Reubens 2008). Topology focuses on the physical connections between plant components, while geometry includes the shape, size, orientation, and spatial location of the root components (Godin 2000; Gregory 2006; Reubens et al. 2007). Geometry is mainly involved in plant–environment exchanges, anchorage, and resource capture, while topology can be used to build up the biological sequences embedded in axes or can be considered as the basis for internal fluxes of energy, mass, and information (Godin et al. 1999). Root architecture has traditionally been described by multiple root parameters, such as total root length, branching angle, root depth, root width, ratio of root depth to root width, and distribution of root length or volume in various layers of the growth medium (Fitter and Stickland 1991; Fitter et al. 1991; Zhu et al. 2006; Fang et al. 2009).

Due to the underground growth habitat and complexity of root systems, their measurements can often be difficult. Depending on the objective of the study, data collection on root systems can be performed by a variety of techniques and at different levels of details (Danjon and Reubens 2008). It can be used for detailed architectural traits description, including topology (Fitter 1991), position of the root volume (Drexhage et al. 1999), or both topology and geometry (Danjon et al. 1999a; Oppelt et al. 2000). Qualitative study of plant architecture was first developed by Hallé and Oldeman (1970) on tropical tree crowns and named “architectural analysis” (Barthelemy and Caraglio 2007). The methodology was adapted to other root systems by Atger and Edelin (1994, 1995). Although some early methods are still useful both for fine and coarse roots, in general, proper choice of methodology also depends on the dimensions of the roots to be measured. Detailed quantitative 3D plant architecture assessment was first initiated by de Reffye (1979) and Fisher and Honda (1979).

During traditional approaches for soil-based root measurements, the topology of the root system is often destroyed by the excavation or washed soil cores (Iyer-Pascuzzi et al. 2010). This process limits the types of root traits that can be measured, where typically measured traits include root depth, root width, total root length, and total number of fine roots. During the excavation process, a trench is usually dug around the plant, exposing the extremities of the root system and allowing the width and depth of the root system to be measured by hand. The soil surrounding the root system is then carefully removed, allowing the entire root system to be pulled from the ground, rinsed more thoroughly with water, and measured for total root length and number of fine roots. This approach can be extremely time consuming and labor intensive and it has been suggested that the finest roots are not adequately sampled by destructive soil coring methods because of the difficulty of extracting them from the cores (Guo et al. 2004). Missing the smallest ($\leq$0.3 mm) and most ephemeral roots could bias measurements of root traits. To measure roots automatically and nondestructively, two problems need to be solved. The first is to improve the ability to capture the high quality root images and the second is to accurately quantify the RSA. In recent years, many analysis systems have been developed to image and measure plant roots in situ and nondestructively. These systems are usually classified into two categories: 2D quantification systems and 3D quantification systems.

9.2 2D Quantification Systems

Many 2D quantification systems are available to measure RSA. Current approaches use commercial or freely available software such as ImageJ (Abramoff et al. 2004), OPTIMAS analysis software (Media Cybernetics, http://www.mediacy.com), or WinRHIZO (Regents Instruments, Quebec City, Canada; Arsenault et al. 1995) to assess total root length, lateral root length, and number, etc. WinRHIZO image analysis system is widely used to quantify root morphological traits including total root length, root diameter, root surface area, and root volume. It is comprised of a computer program and a scanner, and is relatively inexpensive and suited for large- and small-scale experiments. During imaging, the root system is spread out on the scanner surface to minimize root overlap, however, changes of the RSA are inevitable, thus making it impossible to measure the original architectural traits such as root depth, root width, and root angles. Bonser et al. (1996) developed a paper pouch system in which all the roots grew along the germination paper surface. This method contains the root system to 2D, but allows for the accurate, temporal measurements of the root system without changing root architectural traits.

Armengaud et al. (2009) developed a free, semiautomatic software EZ-RHIZO for the fast and accurate measurements of RSA. EZ-RHIZO was designed to detect and quantify multiple 2D RSA parameters, including main root length and angle, number of lateral roots, from plants growing on a solid support agar medium. The method is noninvasive, enabling the user to track RSA development over time. French et al. (2009) developed RootTrace, a high-throughput, automatic method to trace Arabidopsis (*Arabidopsis thaliana*) seedling roots grown on agar plates. From the seedling traces, additional software can be used to quantify length, curvature, and stimulus response parameters such as onset of gravitropism. These two tools, however, could only be used to efficiently quantify simpler dicot root system of Arabidopsis. To address this limitation, Iyer-Pascuzzi et al. (2010) developed an automatic gel-based image analysis platform for plant RSA where plants were grown in transparent gel system and root images were taken by a digital camera every 18°. Using this platform, one plant could be imaged, regardless of size, in approximately 10 min. This system was used to measure more complicated monocot root systems of rice and maize and could calculate up to 16 different RSA traits including number of roots, average root radius, specific root length, total surface area, total root length, depth, volume, length distribution, maximum horizontal width, and width-to-depth ratio.

9.3 3D Quantification Systems

The main difficulties for quantifying the 3D RSA include the opaque growth environment, which makes it difficult to observe roots, and the complexity of overlapping roots, which limit the accuracy of the measurements (Fang et al. 2009). To measure the 3D root systems nondestructively, 3D root images need to be captured. In recent years, several advanced instruments and technologies have been employed to capture and analyze 3D root images. The remainder of this chapter will focus on seven commonly used methods.

9.3.1 Minirhizotrons

Minirhizotrons are clear glass or plastic tubes that are installed in the soil under plants. Using a specialized video camera, researchers can take pictures of roots growing along the outside walls of the tubes (Fig. 9.1, Johnson et al. 2001). By taking repeated time series images, the progress of the roots can be followed as they appear, mature, and eventually senesce and decay. Minirhizotrons provide a nondestructive, in situ method for directly viewing and studying fine roots with diameters <2.0 mm (Johnson et al. 2001; Strand et al. 2008). They are one of the best tools available for directly studying roots in field. Minirhizotrons are not limited by root size and have been installed in forests (Hendrick and Pregitzer 1993; Joslin and Wolfe 1999),

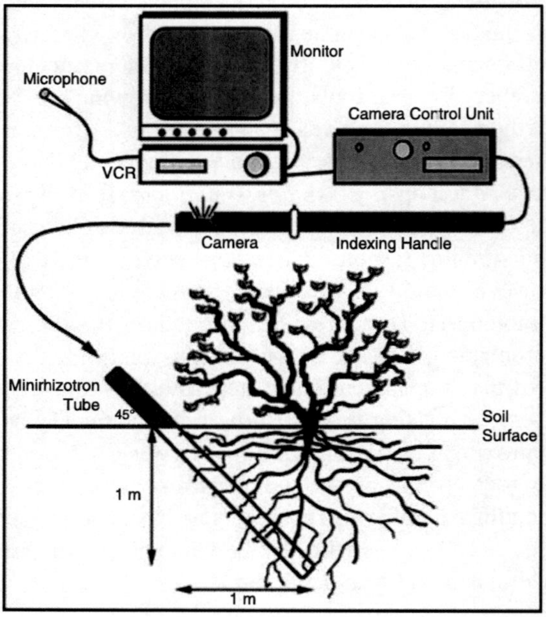

Fig. 9.1 General setup of the minirhizotron camera system and minirhizotron tube (Johnson et al. 2001)

orchards (Kosola et al. 1995; Wells and Eissenstat 2001), grasslands (van Noordwijk et al. 1985), deserts (Reynolds et al. 1999; Phillips et al. 2000), agricultural systems (Samson and Sinclair 1994), controlled environment chambers (Tingey et al. 1996), and greenhouse lysimeters (Fitter et al. 1999).

Minirhizotron techniques have improved significantly since they were first proposed (Bates 1937) and are widely utilized to study the fine roots in agricultural and natural plant communities (Hendrick and Pregitzer 1996; Guo et al. 2008; Ao et al. 2010). However, tube installation is critical and steps must be taken to insure good soil/tube contact without compacting the soil. Ideally, soil adjacent to minirhizotrons will mimic bulk soil. Tube installation causes some degree of soil disturbance and has the potential to create artifacts in subsequent root data and analysis. Johnson et al. (2001) recommended a waiting period between tube installation and image collection of 6–12 months to allow roots to recolonize the space around the tubes and nutrient levels to return to predisturbance levels. This waiting period, however, is too long and is not suitable for imaging seedling roots. Installing minirhizotrons before planting or when root biomass is at an annual low will reduce the amount of injury to the root system and subsequent wound responses. Furthermore, when scratches, water, or other circumstances that obstruct the view arise, there are few options to remedy the situation other than removing the minirhizotron tube and cleaning or replacing it which is time consuming and labor intensive. Removing minirhizotron tubes during the plant growth period prior to harvest can cause root damage and is likely to make precise repositioning impossible which is needed for repeated observation of individual roots (van Noordwijk et al. 1985).

Establishing minirhizotron data quality objectives a priori will guide experimental design and the methods used to collect the data. A well-written standard operating procedure should guide the collection of images and extraction of minirhizotron data (Johnson et al. 2001). Currently, several softwares are available for tracing and measuring the growth of roots in minirhizotron images such as RooTracker (http://www.biology.duke.edu/rootracker), Rootfly (http://www.ces.clemson.edu/~stb/rootfly), and RootView (http://www.mv.helsinki.fi/aphalo/RootView.html). Users can use the software to measure the length, diameter, and color of roots, as well as the birth and death rates.

9.3.2 *X-ray Computed Tomography (CT)*

X-ray CT can capture the undisturbed 3D root images from plant grown in soils (Gregory et al. 2003; Di Iorio et al. 2005; Perret et al. 2007; Tracy et al. 2010). CT uses X-rays to acquire cross-sectional image slices of an object, which contain information regarding the attenuation of the X-rays as a function of the density of the sample material (Mahesh 2002). X-ray CT overcomes some of the limitations associated with previous methodologies for studying roots by providing a noninvasive way to capture 3D images (Heeraman et al. 1997; Pierret et al. 2002; Lontoc-Roy et al. 2006). The use of X-ray CT to visualize roots in undisturbed soil has, until now, been limited to species with coarse roots based on images with a resolution >100 μm (Kaestner et al. 2006). However, current, state-of-the-art, X-ray CT scanners can now achieve a spatial resolution of <500 nm, enabling fine roots, such as those of Arabidopsis plants, to be visualized in situ. RootViz, root tracing software, can be used to determine whether specific pixels within images represent root material. Avizo® software can be used to extract and provide 3D images of root distribution within undisturbed soil columns (Tracy et al. 2010).

X-ray CT offers major benefits, including the nondestructive measurements and 3D visualization of root systems. However, a current limitation of X-ray CT systems for probing root architecture is the attenuation coefficient, which is typically similar for root and other nonspecific soil organic matter. This attenuation problem can also be exacerbated if the soil water content is high. Drying soil sample prior to scanning has been used to help address this issue, but can alter soil structure and root architecture. As an alternative strategy, substrates containing little organic matter (e.g., homogeneous sand and loamy sand) have been used as growth media (Lontoc-Roy et al. 2006). Confounding upon technical issues, X-ray tomography is also very expensive when studying larger root systems (Gregory et al. 2003).

9.3.3 Magnetic Resonance Imaging

MRI techniques use the phenomenon of nuclear magnetic resonance (NMR) to image protons of water to deliver structural information in a nondestructive manner (Jahnke et al. 2009). Over the last two decades, they have received substantial attention with respect to plant studies (MacFall and Johnson 1996; MacFall, 1997; van der Weerd et al. 2001) and have been used to measure 3D structures (e.g., Köckenberger et al. 2004; Jahnke et al. 2009).

MRI provides spatial resolution of up to 30 μm^3 per voxel (pixel element of a 3D image) (Jahnke et al. 2009); however, MRI method is very sensitive to water content in the soil. Using MRI to image roots is more challenging than imaging above-ground parts, because the root structures are surrounded by water filled pores and soil structures that often contain para- or ferromagnetic impurities that can exacerbate image distortion. Zhou and Luo (2009) reported that in a 3D MRI root image of a 6-week tomato root system with 88.5% of water content, the diameter of the root that could be shown clearly was above 1 mm. To solve this problem, an opposite method, positron emission tomography (PET), can be used as a complementary method to MRI. PET is a powerful technology for detecting positron-emitting radionuclides such as ^{11}C, and has the capacity to measure (and image) the transport and distribution of ^{11}C-labeled photo-assimilates in plants in 3D. MRI–PET has been used to image and analyze 3D maize root architecture (Jahnke et al. 2009), however, the extraction and analysis of MRI–PET data is still a challenge under continuous development, particularly for medical diagnostics (Doi 2006; Cho et al. 2008).

9.3.4 3D Digitizing

Three-dimensional digitizing in plants has been primarily used for the spatial and architectural analysis of the aerial parts of trees (Sinoquet et al. 1997), where both the localization of branch growth units and topology (links between branches) is determined (Danjon et al. 1999b). This method is a contact measurement in which the device needs to touch some points of plants to record the 3D spatial coordinates.

The first 3D digitizer for plants was described by Lang (1973), who used articulated arms with rotational angles recorded from high-precision potentiometer resistance values. A commonly used contact measurement device within the plant community is the FASTRAK® magnetic 3D digitizer (Polhemus, Colchester, VT, USA) (Sinoquet and Rivet 1997; Rakocevic et al. 2000; Evers et al. 2005). It includes a magnetic signal receiver and pointer and allows the user to record the 3D spatial coordinates of the pointer within a hemisphere that is 3 m in diameter around receiver. Individual plants are digitally reconstructed by recording a series of point coordinates and the relevant connectivity between the points. Due to its principle of creating a magnetic field, the magnetic digitizer can be used in outdoor environments with relative ease; however, in greenhouse environments the surrounding iron frames can disturb measurements (Van der Heijden et al. 2007). Another commonly used technique is the sonic digitizer GTCO Freepoint 3D (Watanabe et al. 2005). It consists of a handheld probe with two or more sonic emitters and a triangular detector array with three microphones. Calibration is necessary to correct for differences in temperature and humidity of the air. Godin et al. (2005) recommend that sonic digitizers can only be used in the laboratory setting due to their sensitivity to wind fluctuations in the field. A large advantage of contact point measurements is that the points to be measured can be annotated directly, especially when recording is done in a structured way using software like Floradig (CSIRO, Australia) or AMAPmod (AMAP, France).

9.3.5 Ground-Penetrating Radar

In forest environments, root excavation is limited because of the disturbance it causes and the time it requires. Additionally, it is nearly impossible to excavate the entire, large root systems, thus the number of excavation studies has been limited (Danjon et al. 2005). Ground-Penetrating Radar (GPR) has been primarily used to identify compacted soil horizons, stone lines, bedrock, and water tables in soil surveys (Doolittle 1987), but has also been introduced as a nondestructive method for root sampling and measurements. GPR relies on portable radar technology that projects high-frequency impulses of energy into the soil, which are reflected back when they strike an interface between large soil particles. This reflected energy is then recorded on a portable computer for further analysis (Stokes et al. 2002). GPR is most commonly used for measuring large and coarse tree roots in forest environments. Additionally, GPR has also been used to detect tree roots, map root distributions (Hruska et al. 1999; Cermak et al. 2000; Cox et al. 2005; Hagrey 2007), and estimate root biomass (Barton and Montagu 2004).

Although mapping tree root systems with GPR has been successfully used and tested in forest and woodland environments, where soil is relatively homogenous (Hruska et al. 1999; Sustek et al. 1999), doubts still exist about the accuracy of this technique in an urban environment. Miller et al. (2004) noted that GPR results may vary with changes in the soil and other environmental factors. Butnor et al. (2001) found that GPR is ineffective in soils with high clay or water content and at sites with rough terrain. Differences in water content between the soil and roots can be a crucial factor affecting root detection by GPR. When the water content of the soil was higher than that of roots, the roots were not detected by GPR. To establish GRP as a reliable tool for root studies, basic information is still needed and optimal conditions for root detection must be established for various tree species and site conditions (Dannoura et al. 2008).

9.3.6 3D Laser Scanning

According to Danjon and Reubens (2008), laser scanning is described as the best available technique for measuring the surface and shape of roots. Laser scanners, which have mostly been used in Geosciences to generate digital terrain models (Petzold et al. 1999), permit the measurement of the surface area of objects in 3D with high resolution. Consequently, 3D replications of the scanned objects can be generated using specialized software. Gärtner and Denier (2006) reported on the use of a 3D laser scanning device to map the root surfaces of a large uprooted root system. However, since the laser cannot penetrate opaque growth mediums, it requires the excavation of the root system in order to expose the target root system. Additionally, since only the visible root surfaces that are correctly positioned in front of the scanner can be measured, time-intensive scanning from multiple positions must be performed to obtain data on the entire root system. The scanning times also depends on the size of root system or capture resolution, though capturing smaller root systems generally requires less time. Since 3D laser scanner is a destructive technique where the root systems are manually excavated and washed, smaller roots (approx. 1 mm) can be damaged or lost (Gärtner et al. 2009), which can impact overall measurements. The resulting model displays the time of uprooting and is thus unable to give additional information regarding the spatiotemporal development of the root system (Wagner et al. 2010).

Fang et al. (2009) used a transparent 3D root growth system combined with 3D laser scanner to nondestructively study crop RSA in situ. This method was found to be effective for capturing in situ 3D soybean and rice root architecture images dynamically without any contact or perturbation of the root system or the growth medium. Based on the 3D data for rice and soybean root architectures obtained by this nondestructive approach, 3D root architecture parameters, including total root length, root depth, root width, root angle, ratio of root width to root depth, and percentage distribution of root length in various layers, were measured.

9.3.7 *Digital Cameras*

Digital cameras are ubiquitous and can be used effectively to capture and measure root systems; however, they cannot penetrate the soil and thus the root systems must first be removed before imaging and measuring. Zhu et al. (2006) developed a unique stratified, 3D growth system that enabled soybean root systems to be dried and retain their architecture during harvest, allowing root architecture to be recorded by a digital imaging system. The 3D root architecture imaging system included two synchronously connected CCD cameras, two rotating stages, a computerized numerical control (CNC) machine, a computer with two imaging cards, and a white backboard for removing background noise (Fig. 9.2)(Zhu et al. 2006). This 3D imaging system allowed the root systems to be precisely imaged, reconstructed, skeletonized, and measured for 3D architectural traits. From original projective images at different viewing angles (Zhu et al. 2005), the method of visual hull creation (Matusik et al. 2000) combined with the voting principles of the broad-sense Hough Transform (Seitz and dyer 1995) was used to create a 3D image model. Due to the linear feature and the naturally continuous variation of the root system, root architecture could be described by each root axis or skeleton. Man–machine interaction method is used to look for several skeleton nodes for

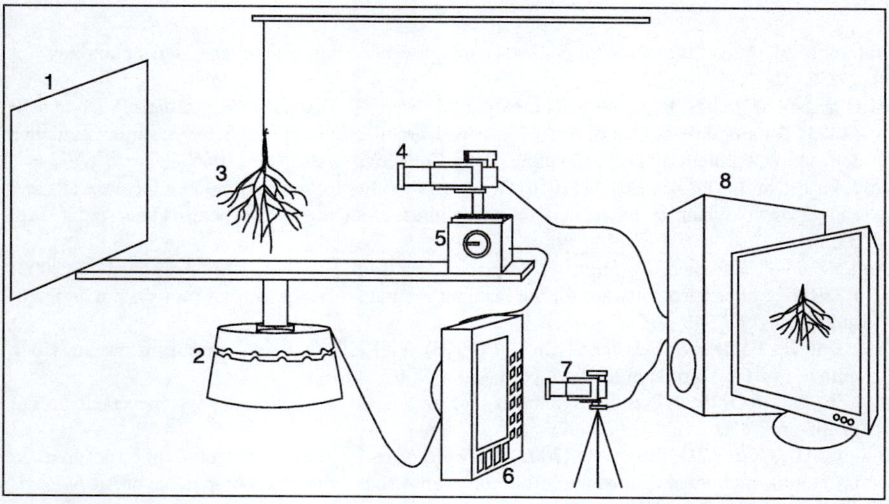

Fig. 9.2 The 3D root architecture image reconstruction system. 1, white backboard; 2, large rotating stage; 3, plant root system and suspension device; 4, synchronously connected CCD camera 1; 5, small rotating stage; 6, CNC machine; 7, synchronously connected CCD camera 2; 8, computer

each tap root and lateral root. These nodes are considered to be the data points, which are connected and fitted by nonuniform B-Spline curve (Zhu 2000); thus, the 3D root skeleton is eventually extracted, based on which the 3D root architecture parameters could be calculated.

All of the above approaches have both advantages and specific limitations that make them suitable for measuring different plant root types. For example, in experiments which only require total root length measurements from small root system, WinRhizo is a good choice because of its fast and easy operation. Minirhizotrons are good for in situ field studies, but are not suitable for imaging small root systems like the Arabidopsis root system. Magnetic resonance imaging and X-ray CT can both provide in situ root images without contact and alteration of the root system, but can be quite expensive and their image precision can be hindered by soil type, heterogeneity, and water content. GPR has low image resolution which makes it only useful for studying the root architecture of large woody trees. 3D digitizers also have low resolution and require the time-consuming excavation of large root systems that can require weeks to obtain architectural measurements on a single plant (Danjon et al. 2008). Digital camera and 3D laser scanner methods also require excavation of the soil around roots and digital cameras also have problems with regard to the calibration accuracy during image capture, and the 3D laser scanner needs about half an hour to capture crop root systems which limits utility during large experiments (Gärtner and Denier 2006).

References

Abramoff M, Magelhaes P, Ram S (2004) Image processing with image. J Biophotonics Int 11:36–42

Al-Ghazi Y, Muller B, Pinloche S, Tranbarger TJ, Nacry P, Rossignol M, Tardieu F, Doumas P (2003) Temporal responses of Arabidopsis root architecture to phosphate starvation: evidence for the involvement of auxin signaling. Plant Cell Environ 26:1053–1066

Ao J, Fu J, Tian J, Yan XL, Liao H (2010) Genetic variability for root morph-architecture traits and root growth dynamics as related to phosphorus efficiency in soybean. Funct Plant Biol 37:304–312

Armengaud P, Zambaux K, Hills A, Sulpice R, Pattison JR, Blatt RM, Amtmann A (2009) EZ-Rhizo: integrated software for the fast and accurate measurement of root system architecture. Plant J 57:945–956

Arsenault JL, Pouleur S, Messier C, Guay R (1995) WinRHIZ O, a root-measuring system with a unique overlap correction method. Hort Sci 30:906

Atger C, Edelin C (1994) Premières données sur l'architecture comparée des systèmes racinaires et caulinaires. Can J Bot 72:963–975

Bailey PHJ, Currey JD, Fitter AH (2002) The role of root system architecture and root hairs in promoting anchorage against uprooting forces in *Allium cepa* and root mutants of *Arabidopsis thaliana*. J Exp Bot 53:333–340

Barber SA (1995) Soil nutrient bioavailability: a mechanistic approach. Wiley-Interscience, New York

Barthelemy D, Caraglio Y (2007) Plant architecture: a dynamic, multilevel and comprehensive approach to plant form, structure and ontogeny. Ann Bot 99:375–407

Barton CVM, Montagu KD (2004) Detection of tree roots and determination of root diameter by ground penetrating radar under optimal conditions. Tree Physiol 24:1323–1331

Bates GH (1937) A device for the observation of root growth in the soil. Nature 139:966–967

Bidel LPR, Pages L, Riviere LM, Pelloux G, Lorendeau JY (2000) MassFlowDyn I: A carbon transport and partitioning model for root system architecture. Ann Bot 85:869–886

Bonser AM, Lynch JP, Snapp S (1996) Effect of phosphorus deficiency on growth angle of basal roots of *Phaseolus vulgaris* L. New Phytol 132:281–288

Bowman DC, Devitt DA, Engelke MC, Rufty TW (1998) Root architecture affects nitrate leaching from bentgrass turf. Crop Sci 38:1633–1639

Butnor JR, Doolittle JA, Kress L, Cohen S, Johnsen KH (2001) Use of ground-penetrating radar to study tree roots in the southeastern United States. Tree Physiol 21:1269–1278

Cermak J, Hruska J, Martinkova M, Prax A (2000) City tree roots and survival near houses analyzed using sap flow and ground penetrating radar technique. Plant Soil 219:103–116

Cho ZH, Son YD, Kim HK, Kim KN, Oh SH, Han JY, Kim YB (2008) A fusion PET–MRI system with high-resolution research tomograph-PET and ultra-high field 7.0T-MRI for the molecular-genetic imaging of the brain. Proteomics 8:1302–1323

Cichy KA, Snapp SS, Kirk WW (2007) Fusarium root rot incidence and root system architecture in grafted common bean lines. Plant Soil 300:233–244

Coll L, Potvin C, Messier C, Delagrange S (2008) Root architecture and allocation patterns of eight native tropical species with different successional status used in open-grown mixed plantations in Panama. Trees Struct Funct 22:585–596

Cox KD, Scherm H, Serman N (2005) Ground-penetrating radar to detect and quantify residual root fragments following peach orchard cleaning. Hort Technol 15:600–607

Danjon F, Reubens B (2008) Assessing and analyzing 3D architecture of woody root systems, a review of methods and applications in tree and soil stability, resource acquisition and allocation. Plant Soil 303:1–34

Danjon F, Bert D, Godin C, Trichet P (1999a) Structural root architecture of 5-year-old *Pinus pinaster* measured by 3D digitizing and analyzed with AMAPmod. Plant Soil 217:49–63

Danjon F, Sinoquet H, Godin C, Colin F, Drexhage M (1999b) Characterization of structural tree root architecture using 3D digitizing and AMAPmod software. Plant Soil 211:241–258

Danjon F, Fourcaud T, Bert D (2005) Root architecture and windfirmness of mature *Pinus pinaster*. New Phytol 168:387–400

Danjon F, Barker DH, Drexhage M, Stokes A (2008) Using three-dimensional plant root architecture in models of shallow-slope stability. Ann Bot 101:1281–1293

Dannoura M, Hirano Y, Igarashi T, Ishii M, Aono K, Yamase K, Kanazawa Y (2008) Detection of cryptomeria japonica roots with ground penetrating radar. plant biosystems – an international journal dealing with all aspects of plant biology. Official J Soc Bot Ital 142:375–380

de Reffye P (1979) Modélisation de l'architecture des arbres par des processus stochastiques. Simulation spatiale des modèles tropicaux sous l'effet de la pesanteur. Application au Coffea robusta. Thèse de doctorat d'Etat. Université de Paris-Sud Orsay, pp 194

Di Iorio A, Lasserre B, Scippa GS, Chiatante D (2005) Root system architecture of *Quercus pubescens* trees growing on different sloping conditions. Ann Bot 95:351–361

Doi K (2006) Diagnostic imaging over the last 50 years: research and development in medical imaging science and technology. Phys Med Biol 51:5–27

Doolittle JA (1987) Using ground-penetrating radar to increase the quality and efficiency of soil surveys. In: Reybold WU, Peterson GW (eds) Soil survey techniques, SSSA special publication No. 20. Soil Science Society of America, Madison, WI

Doussan C, Pages L, Vercambre G (1998) Modelling of the hydraulic architecture of root systems: an integrated approach to water absorption – model description. Ann Bot 81:213–223

Drexhage M, Chauvière M, Colin F, Nielsen CNN (1999) Development of structural root architecture and allometry of *Quercus petraea*. Can J For Res 29:600–608

Evers JB, Vos J, Fournier C, et al (2005) Towards a generic architectural model of tillering in Gramineae, as exemplified by spring wheat (Triticum aestivum). New Phytologist 166(3):801–812

Fang S, Yan X, Liao H (2009) 3D reconstruction and dynamic modeling of root architecture in situ and its application to crop phosphorus research. Plant J 60:1096–1108

Fisher JB, Honda H (1979) Branch geometry and effective leaf area: a study of Terminalia branching pattern. 2 Survey of real trees. Am J Bot 66:645–655

Fitter AH (1991) Characteristics and functions of root systems. In: Wisel Y, Eshel A, and Kafkafi U (eds) Plant roots: The Hidden Half. Marcel Dekker, Inc, New York, Basel, Hongkong, pp 3–24

Fitter AH, Stickland TR (1991) Architectural analysis of plant-root systems 2. Influence of nutrient supply on architecture in contrasting plant-species. New Phytol 118:383–389

Fitter AH, Stickland TR, Harvey ML, Wilson GW (1991) Architectural analysis of plant-root systems 1. Architectural correlates of exploitation efficiency. New Phytol 118:375–382

Fitter AH, Self GK, Brown TK, Bogie DS, Graves JD, Benham D, Ineson P (1999) Root production and turnover in an upland grassland subjected to artificial soil warming respond to radiation flux and nutrients, not temperature. Oecologia 120:575–581

French A, Ubeda-Tomas S, Holman TJ, Bennett MJ, Pridmore T (2009) High-throughput quantification of root growth using a novel image-analysis tool. Plant Physiol 150:1784–1795

Gärtner H, Denier C (2006) Application of a 3D Laser scanning device to acquire the structure of whole root systems- A pilot study. In: Heinrich I, Gärtner H, Monbaron M, Schleser G (eds) TRACE – tree rings in archaeology. Climatology and Ecology Association for Tree-Ring Research, Tervuren, pp 288–294

Gärtner H, Wagner B, Heinrich I, Denier C (2009) 3D laser scanning-a new methodology to analyze coarse tree root systems. For Snow Landsc Res 82:95–106

Godin C (2000) Representing and encoding plant architecture: a review. Ann For Sci 57:413–438

Godin C, Costes E, Sinoquet H (1999) A method for describing plant architecture which integrates topology and geometry. Ann Bot 84:343–357

Godin C, Costes E, Sinoquet H (2005) Plant architecture modelling: virtual plants and complex systems. In: Turnbull CGN (ed) Plant architecture and its manipulation. Blackwell, Oxford, pp 238–287

Gregory PJ (2006) Plant roots: growth, activity and interaction with soils. Blackwell, Oxford

Gregory PJ, Hutchison DJ, Read DB, Jenneson PM, Gilboy WB, Morton EJ (2003) Non-invasive imaging of roots with high resolution X-ray micro-tomography. Plant Soil 255:351–359

Guo D, Mitchell R, Hendricks J (2004) Fine root branch orders respond differentially to carbon source-sink manipulations in a longleaf pine forest. Oecologia 140:450–457

Guo D, Li H, Mitchell RJ, Han W, Hendricks JJ, Fahey TJ, Hendrick RL (2008) Fine root heterogeneity by branch order: exploring the discrepancy in root turnover estimates between minirhizotron and carbon isotopic methods. New Phytol 177:443–456

Hagrey SA (2007) Geophysical imaging of root-zone, trunk, and moisture heterogeneity. J Exp Bot 58:839–854

Hallé F, Oldemann RAA (1970) Essai sur l'architecture et la dynamique de croissance des arbres tropicaux. Masson, Paris, p 178

Heeraman DA, Hopmans JW, Clausnitzer V (1997) Three dimensional imaging of plant roots in situ with X-ray computed tomography. Plant Soil 189:167–179

Hendrick RL, Pregitzer KS (1993) Patterns of fine root mortality in two sugar maple forests. Nature 361:59–61

Hendrick RL, Pregitzer KS (1996) Temporal and depth-related patters of fine root dynamics in northern hardwood forests. J Ecol 84:167–176

Hruska J, Cermak J, Sustek S (1999) Mapping of tree root systems by means of the ground penetrating radar. Tree Physiol 19:125–130

Iyer-Pascuzzi AS, Symonova O, Mileyko Y, Hao Y, Belcher H, Harer J, Weitz JS, Benfey PN (2010) Imaging and analysis platform for automatic phenotyping and trait ranking of plant root systems. Plant Physiol 152:1148–1157

Jahnke S, Menzel MI, Van Dusschoten D, Roeb GW, Bühler J, Minwuyelet S, Blümler P, Temperton VM, Hombach T, Streun M (2009) Combined MRI–PET dissects dynamic changes in plant structures and functions. Plant J 59:634–644

Johnson MG, Tingey DT, Phillips DL, Storm MJ (2001) Advancing fine root research with minirhizotrons. Environ Exp Bot 45:263–289

Joslin JD, Wolfe MH (1999) Disturbances during minirhizotron installation can affect root observation data. Soil Sci Soc Am J 63:218–221

Kaestner A, Schneebeli M, Graf F (2006) Visualizing three dimensional root networks using computed tomography. Geoderma 136:459–469

Köckenberger W, De Panfilis C, Santoro D, Dahiya P, Rawsthorne S (2004) High resolution NMR microscopy of plants and fungi. J Microsc 214:182–189

Kosola KR, Eissenstat DM, Grahm JH (1995) Root demography of mature citrus trees: the influence of *Phytophthora nicotianae*. Plant Soil 171:283–288

Lambers H, Shane MW, Cramer MD, Pearse SJ, Veneklaas EJ (2006) Root structure and functioning for efficient acquisition of phosphorus: matching morphological and physiological traits. Ann Bot 98:693–713

Lang ARG (1973) Leaf orientation of a cotton plant. Agric Meteorol 11:37–51

Liao H, Ge Z, Yan XL (2001) Ideal root architecture for phosphorus acquisition of plants under water and phosphorus coupled stresses: from simulation to application. Chin Sci Bull 46:1346–1351

Lontoc-Roy M, Dutilleul P, Prasher SO, Han LW, Brouillet T, Smith DL (2006) Advances in the acquisition and analysis of CT scan data to isolate a crop root system from the soil medium and quantify root system complexity in 3-D space. Geoderma 137:231–241

Lynch J (1995) Root architecture and plant productivity. Plant Physiol 109:7–13

MacFall JS, Johnson GA (1996) Plants, seeds and roots as applications for magnetic resonance imaging. In: Grant DM, Harris RK (eds) Encyclopedia of nuclear magnetic resonance. Wiley, London, pp 3633–3640

MacFall JS (1997) Visualization of root growth and development through magnetic resonance imaging. In: Flores HE, Lynch JP, Eissenstat D (eds) Radical Biology: Advances and Perspectives on the Function of Plant Roots. American Society of Plant Physiologists, Rockville, MD, pp 57–80

Mahesh M (2002) The AAPM/RSNA physics tutorial for residents: search for isotropic resolution in CT from conventional through multiple-row detector. RadioGraphics 22:949–962

Matusik W, Buehler C, Raskar R (2000) Image-based visual hulls. Sig-Graph, New Orleans, LA, pp 369–374

Miller DE (1986) Root systems in relation to stress tolerance. Hortscience 21:963–970

Miller TW, Hendrickx JMH, Borchers B (2004) Radar detection of buried landmines in field soils. Vod Zone J 3:1116–1127

Oppelt AL, Kurth W, Dzierzon H, Jentschke G, Godbold DL (2000) Structure and fractal dimensions of root systems of four co-occurring fruit tree species from Botswana. Ann For Sci 57:463–475

Perret JS, Al-Belushi ME, Deadman M (2007) Non-destructive visualization and quantification of roots using computed tomography. Soil Biol Biochem 39:391–399

Petzold B, Reiss P, Stössel W (1999) Laser scanning surveying and mapping agencies are using a new technique for the derivation of digital terrain models. ISPRS J Photogramm Remote Sens 54:95–104

Phillips DL, Johnson MG, Tingey DT, Biggart C, Nowak RS, Newsom JC (2000) Minirhizotron installation in sandy, rocky soils with minimal soil disturbance. Soil Sci Soc Am J 64:761–764

Pierret A, Capowiez Y, Belzunces L, Moran CJ (2002) 3D reconstruction and quantification of macropores using X-ray computed tomography and image analysis. Geoderma 106:247–271

Rakocevic M, Sinoquet H, Christophe A, et al (2000) Assessing the geometric structure of a white clover (*Trifolium repens L.*) canopy using 3-D digitising. Annals of Botany 86(3):519–526

Reubens B, Poesen J, Danjon F, Geudens G, Muys B (2007) The role of fine and coarse roots in shallow slope stability and soil erosion control with a focus on root system architecture: a review. Trees 21:385–402

Reynolds JF, Virginia RA, Kemp PR, De Soyza AG, Tremmel DC (1999) Impact of drought on desert shrubs: effects of seasonality and degree of resource island development. Ecol Monogr 69:69–106

Samson BK, Sinclair TR (1994) Soil core and minirhizotron comparison for the determination of root length density. Plant Soil 161:225–232

Seitz SM, Dyer CR (1995) Complete scene structure from four point correspondences. In: IEEE (ed) Proceedings of the fifth international conference on computer vision. IEEE Computer Society Press, Cambridge, pp 330–337

Sinoquet H, Rivet P (1997) Measurement and visualization of the architecture of an adult tree based on a three-dimensional digitising device. Trees Struct Funct 11:265–270

Sinoquet H, Rivet P, Godin C (1997) Assessment of the three-dimensional architecture of walnut trees using digitising. Silva Fenn 31:265–273

Skaggs TH, Shouse PJ (2008) Roots and root function: introduction. Vadose Zone J 7:1008–1009

Snapp S, Kirk W, Roman-Aviles B, Kelly J (2003) Root traits play a role in integrated management of Fusarium root rot in snap beans. Hortscience 38:187–191

Stokes A, Fourcaud T, Hruska J, Cermak J, Nadyezdhina N, Nadyezhdin V, Praus L (2002) An evaluation of different methods to investigate root system architecture of urban trees in situ: 1. Ground-penetrating radar. J Arboric 28:2–10

Strand AE, Pritchard SG, McCormack ML, Davis MA, Oren R (2008) Irreconcilable differences: fine-root life spans and soil carbon persistence. Science 319:456–458

Sustek S, Hruska J, Druckmuller M, Michalek T (1999) Root surfaces in the large oak tree estimated by image analysis of the map obtained by the ground penetrating radar. J For Sci 45:139–143

Tingey DT, McVeety BD, Waschmann R, Johnson MG, Phillips DL, Rygiewicz PT, Olszyk DM (1996) A versatile sun-lit controlled-environment facility for studying plant and soil processes. J Environ Qual 25:614–625

Tracy SR, Roberts JA, Black CR, McNeill A, Davidson R, Mooney SJ (2010) The X-factor: visualizing undisturbed root architecture in soils using X-ray computed tomography. J Exp Bot 61:311–313

van der Heijden GWAM, de Visser PHB, Heuvelink E (2007) Measurements for functional-structural plant models. In: Vos J, Marcelis LFM, de Visser PHB, Struik PC, Evers JB (eds) Functional-structural plant modeling in crop production, vol 22, Wageningen UR Frontis Series. Springer, Berlin, pp 13–25

van der Weerd L, Claessens MM, Ruttink T, Vergeldt FJ, Schaafsma TJ, Van As H (2001) Quantitative NMR microscopy of osmotic stress responses in maize and pearl millet. J Exp Bot 52:2333–2343

van Noordwijk M, de Jager A, Floris J (1985) A new dimension to observations in minirhizotrons: a stereoscopic view on root photographs. Plant Soil 86:447–453

Wagner B, Gärtner H, Ingensand H, Santini S (2010) Incorporating 2D tree-ring data in 3D laser scans of coarse-root systems. Plant Soil 334:175–187

Watanabe T, Hanan JS, Room PM, Hasegawa T, Nakagawa H, Takahashi W (2005) Rice morphogenesis and plant architecture: measurement, specification and the reconstruction of structural development by 3D architectural modeling. Ann Bot 95:1131–1143

Wells CE, Eissenstat DM (2001) Marked differences in survivorship among apple roots of different diameters. Ecology 82:882–892

Zhou XC, Luo XW (2009) Advances in Non-Destructive Measurement and 3D Visualization Methods for Plant Root Based on Machine Vision. In Proceedings of the 2nd International Conference on BioMedical Engineering and Informatics. Tianjin, BMEI'09, pp 1–5

Zhu XX (2000) Free curves and surfaces modeling technology (in Chinese). Science Press, Beijing, pp 138–169

Zhu TL, Fang SQ, Li ZY, Yan X (2005) A generalized Hough transform template and its applications in computer vision. J Comput Inform Syst 1:601–607

Zhu TL, Fang SQ, Li ZY, Liu Y, Liao H, Yan X (2006) Quantitative analysis of 3-dimensional root architecture based on image reconstruction and its application to research on phosphorus uptake in soybean. Chin Sci Bull 51:2351–2361

Part II
Field Methods

Chapter 10
Geophysical Imaging Techniques

Said Attia al Hagrey

Abstract Soils are the biggest terrestrial carbon store. The contribution of plant roots to soil and atmospheric carbon is significant and difficult to survey accurately. Worldwide interest in reducing greenhouse gases has led to research activities for quantifying the root biomass and evaluating their critical role in space and time. Geophysical methods image the medium under study in 2D and 3D, and monitor changes and processes in 4D. They offer good parametrical and spatiotemporal resolution combined with a minimum-invasive character. The high-resolution techniques of electric resistivity imaging (ERI) and ground penetrating radar (GPR) have extended newly their applications into the less known in vivo investigation of biogeophysical targets of living plants and trees. This includes mapping roots and trunk structures, diagnosing wood decay, analysing fluid content and physiological processes of water redistribution, sap uptake, etc. (e.g., Hagrey and Michaelsen, Eur J Environ Eng Geophys 7:75–93, 2002; Hagrey and Michaelsen, Geophysics 64:746–753, 1999; Hagrey et al., Geophys J Int 138:643–654, 1999; Hagrey et al. Proceedings, Meeting of Engineering Geology, Kiel, Germany, 2003; Hanafy and Hagrey, Geophysics 71:k9–k18, 2006; Hubbard et al., The Leading Edge 21:552–559, 2002). Applications of ERI and GPR techniques to image roots and root-zones are based on the electrical and electromagnetic contrast of these targets relative to surrounding soils (e.g., Hruška et al. Tree Physiol 19:125–130, 1999; Butnor et al., Tree Physiol 21:1269–1278, 2001; Butnor et al., Am J Soil Sci Soc 67:1607–1615, 2003; Hagrey et al. Geophys J Int 138:643–654, 2004, Hagrey 2007; Werban et al. J Plant Nutr Soil Sci 171:927–935, 2007; Amato et al. Tree Physiol 28:1441–1448, 2008; Amato et al. Eur J Agron 31:213–222, 2009; Petersen and Hagrey, The Leading Edge 28(10):1220–1224, 2009; Doolittle and Butnor,

S.A. al Hagrey (✉)
Department of Geophysics, University of Kiel, Otto-Hahn-Platz 1, 24118 Kiel, Germany
e-mail: sattia@geophysik.uni-kiel.de

S. Mancuso (ed.), *Measuring Roots*,
DOI 10.1007/978-3-642-22067-8_10,

Soils, peatlands and biomonitoring. In: Jol HM (Ed) Ground penetrating radar: theory and applications. Elsevier, Amsterdam, 177–192, 2009; Hagrey and Petersen, Geophysics 76(2), G25–G35, 2011). In this chapter, we present principles of ERI and GPR techniques for root studies.

10.1 Electrical Resistivity Imaging Technique

10.1.1 Introduction

Terms of electrical resistivity imaging (ERI) and tomography (ERT) refer to minimally destructive, electrical studies of the internal structure (in 2D and 3D) of a medium and changes (in 4D) with multi-measurements conducted from outer surfaces (e.g., ground surface). Historically, Schlumberger in 1912 began pioneering studies to understand the merits of utilizing electrical resistivity methods for exploring the subsurface (e.g., Dobrin 1960). The progress in applying electrical methods to groundwater exploration began during World War II by French, Russian and German geophysicists (Breusse 1963). In soil science, the first known equipotential map was compiled by Malamphy in 1938 at the site of Williamsburg in USA (Bevan 2000). Since the end of the twentieth century, ERI techniques have experienced a renaissance due to the development of the multi-electrode acquisition systems (e.g., Griffiths and Turnbull 1985; Griffiths et al. 1990) and efficient 2D and 4D inversion schemes (e.g., Loke and Barker 1996; Günther et al. 2006). First published biogeophysical applications of ERI techniques on trees and roots were reported by Hagrey and Michaelsen (1999, 2002).

ERI techniques are now the first choice for many near-surface investigations. Their high-resolution applications are currently of steady increase in the prospecting of heterogeneous soils, roots, their hydrology and (physiological) processes. Unlike georadar, ERI techniques require galvanic contact with the soil and need longer survey time. However, recent advances allow multichannel measurements (i.e., simultaneous multi-readings) and almost continuous (mobile) survey.

Conventional electrical resistivity methods for subsurface exploration are covered in many textbooks, e.g., Keller and Frischknecht (1966), Koefoed (1979), Telford et al. (1990), Lowrie (1997) and Reynolds (1997). More advanced development of this method and high-resolution applications for the vadose soil and root-zone can be found in recent published review articles, e.g., Samouëlian et al. (2003), Binley and Kemna (2005) and Hagrey (2007). We present the resistivity tools available for the investigation of root-zones and hosting vadose soils. In the next, we review the basic principles of ERI methods, electrical properties of soils and woods, electrode configurations and sensitivities, data acquisition, processing and inversion, and technique limitations.

10.1.2 Principles of Geoelectrical Resistivity Technique

According to Ohm's law, the electric resistance (R in Ohm, Ω) of a simple elongated body (e.g., cylinder) against current flow is directly proportional to its length, ℓ and to the inverse of its cross-sectional area, A, i.e.,

$$R = \frac{\rho \ell}{A}, \tag{10.1}$$

where ρ is the proportionality constant of the specific electrical resistivity. It is characteristic of the material and is independent of its shape or size. The resistivity of a material is defined as being numerically equal to the resistance of a specimen of the material of unit dimensions, i.e., in the unit of Ωm. According to Ohm's law, the resistance is given by:

$$R = \frac{\Delta U}{I} \tag{10.2}$$

where ΔU is the potential difference and I strength of electric current. From (10.1) and (10.2) we get

$$\rho = \frac{A \Delta U}{(\ell I)} \tag{10.3}$$

Equation (10.3) may be used to determine the resistivity (ρ) of homogeneous and isotropic materials in the form of regular geometric shapes, e.g., cylinders, and cubes.

Geoelectrical resistivity surveys usually use 4-electrode measurements, two for current injection (C_1, C_2) and two for potential difference or voltage measurements (P_1, P_2). A point source electrode (C_1) is realized by placing a remote electrode C_2 at infinity, i.e., sufficiently far from C_1 (Fig. 10.1). A current flow from C_1 located at the surface of a homogeneous isotropic semi-infinite (half-space) medium will follow (e.g., Keller and Frischknecht 1966):

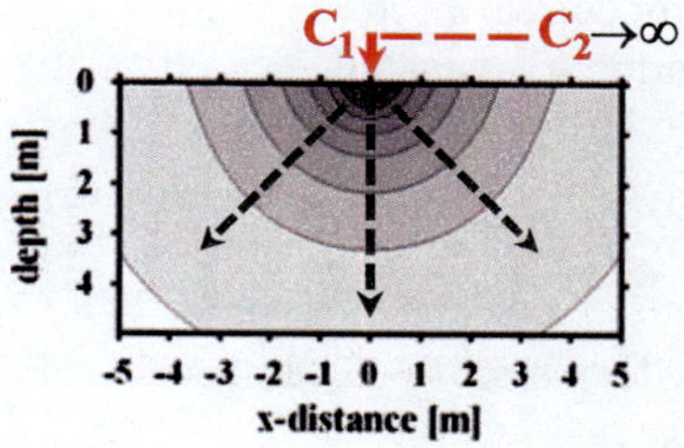

Fig.10.1 Distribution of current flows (*dashed lines*) and equipotential lines (*hemicircles*) in a homogeneous half-space soil resulting from a point source of current electrode, $C1$. Electrode $C2$ is placed very far from $C1$ (remote electrode)

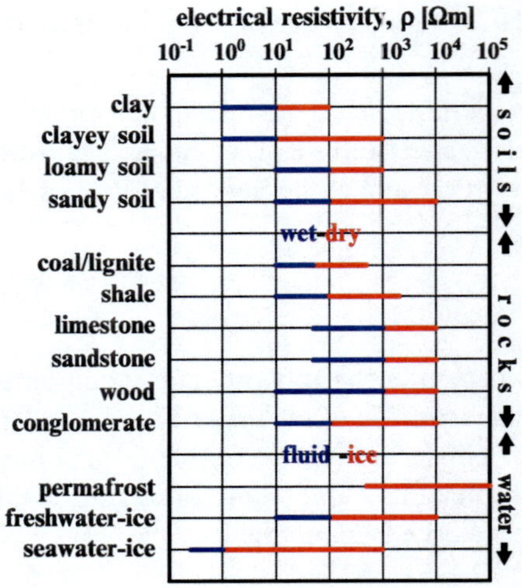

Fig. 10.2 Electrical resistivity ρ of soils and woods (Keller and Frischknecht 1966; Skaar 1988)

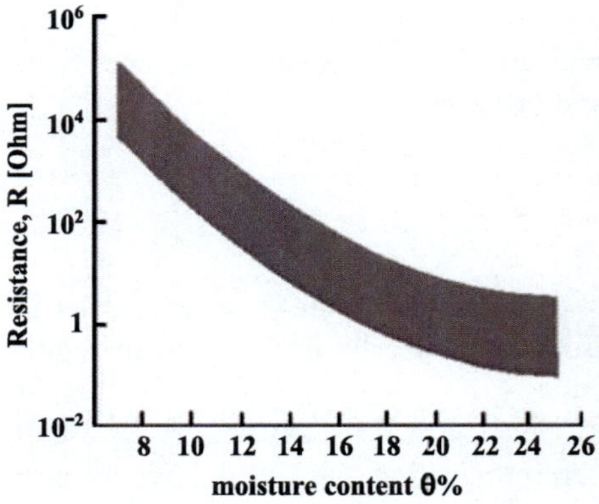

Fig. 10.3 Resistance versus θ of wood for many tree species (Simpson and TenWolde 1999)

$$U(r) = \frac{I\rho}{(4\pi r)} \tag{10.4}$$

where U is the potential of a single current of point source, r distance and ρ medium resistivity.

In this half-space, electrical equipotentials are hemispherical (Fig. 10.1). Now for current electrode pair (C_1, C_2) over a hemispherical surface, the potential at any arbitrary potential electrode P_1 is:

$$U_{\mathrm{P1}} = \frac{I\rho}{2\pi}\left(\frac{1}{r_1} - \frac{1}{r_2}\right) \tag{10.5}$$

where $r_1 = P_1\text{–}C_1$ distance and $r_2 = P_1 - C_2$ distance.

A pair of potential electrodes measures the difference (ΔU) by subtracting the potential at P_2 from that at P_1, i.e., $\Delta U(=U_{P1}-U_{P2})$ or:

$$\Delta U = \frac{I\rho}{2\pi}\left(\frac{1}{r_1}-\frac{1}{r_2}-\frac{1}{r_3}+\frac{1}{r_4}\right) \tag{10.6}$$

where $r_3 = P_2{-}C_1$ distance and $r_4 = P_2{-}C_2$ distance.

Introducing the geometric factor (k)

$$k = 2\pi\left(\frac{1}{r_1}-\frac{1}{r_2}-\frac{1}{r_3}+\frac{1}{r_4}\right)^{-1} \tag{10.7}$$

we achieve

$$\rho = \frac{k\Delta U}{I} \tag{10.8}$$

The geometric factor (k) depends on the electrode arrangement. For a uniform, homogeneous and isotropic half-space, (10.8) yields a true resistivity (ρ), which is constant and independent of both the electrode spacing and surface location. For an inhomogeneous layered subsurface, this equation will produce the so-called apparent resistivity (ρ_a). The ρ_a value is a function of the relative position of the electrodes (i.e., their geometrical arrangement or the electrode configuration) and layer characteristics (e.g., true ρ, thicknesses, dip angle, anisotropy, Zohdy 1989). It enables a qualitative estimation of the electrical parameters of the medium but does not give its true ρ distribution.

10.1.3 Electrical Resistivity of Vadose Soils and Root-Zones

The electrical resistivity (ρ) of soil root-zones depends mainly on the nature (quantity and distribution) of their constituents of soil grains/organic tissues, water and air content and dissolved electrolytes. This includes the porosity/hydraulic conductivity, clay content of soils, wood species, anisotropy (particularly of woods) and temperature. Most soils (clay free) and woods have a very high resistivity matrix and conduct electricity mainly via the electrolytes (ion mobility) of the interstitial water or tissue sap (e.g., Keller and Frischknecht 1966; Skaar 1988). The resistivity (ρ) within soils and roots decreases with increasing pore or cell water content (θ) and salinity or total dissolved solids (TDS). The fluid resistivity is inversely proportional to the dissolved ion concentration and ion mobility (Langwig 1971; McNeill 1980). The mobility of an ion is mainly determined by its molecular size. Small monovalent ions (e.g., Na, K) have a high mobility (Skaar 1988). The ionic content varies greatly according to the wood species. The high ion concentration inside the cell relative to its outside produces an electric potential controlling the movement of electrolytes to and from the cell.

Clay minerals have a negative charged surface (from isomorphic substitution) and adsorb positive cations (in an exchangeable form) around grains in the "double layer". Thus strong clay conductivity is provided by the cation exchange capacity (CEC) and surface conductance. Fine clay grains or soil colloids (<2 μm) hold a high adsorptive water capacity causing a decline of resistivity.

The electrical resistivity (ρ) and/or resistance (R) of natural soils, rocks and woods display the widest range of all other physical properties (Figs. 10.2 and 10.3), (Schön 1997; Simpson and TenWolde 1999). Resistivity values for unconsolidated sediments commonly range from <1 Ωm for clays or soils saturated with saline water, to several thousands Ωm for rocks, dry sand and gravel. The resistivity of sand and gravel saturated with fresh water ranges from about 15 to 600 Ωm. Resistivity measurements of wood cores yield $\rho \approx 10^8$ Ωm for the dead dry cell wall (mainly of cellulose compounds), 100–250 Ωm for healthy wood and 1,000 Ωm for dry heartwood (Venkateswaran 1972).

Wood cells show generally a well-developed anisotropic structure; they possess a greater extension in the axial (longitudinal) direction than in the transverse (radial and tangential) directions. Electrical measurements display a pronounced wood anisotropy of the ρ ratio: $\rho_{\text{longitudinal}}:\rho_{\text{radial}}:\rho_{\text{tangential}} = 1:1.8:2$ (Skaar 1988; Simpson and TenWolde 1999). In addition, a temperature rise reduces the fluid viscosity and increases the degree of water dissociation, ion mobility and electrical conductivity ($\sigma = 1/\rho$).

10.1.4 Measurements and Classic Electrode Arrays

Geoelectric surveys from the ground surface and in boreholes are carried out using modern multielectrode and multichannel data acquisition systems. The measurements are accomplished by four-point electrodes, mainly of stainless steel (Fig. 10.4). An electric current (I) is injected into the heterogeneous subsurface material via a pair of current electrodes (C_1 and C_2) and the resulting voltage (U) of the inhomogeneous electric field is measured between a second pair of potential electrodes (P_1 and P_2). The apparent specific electrical resistivity ρ_a

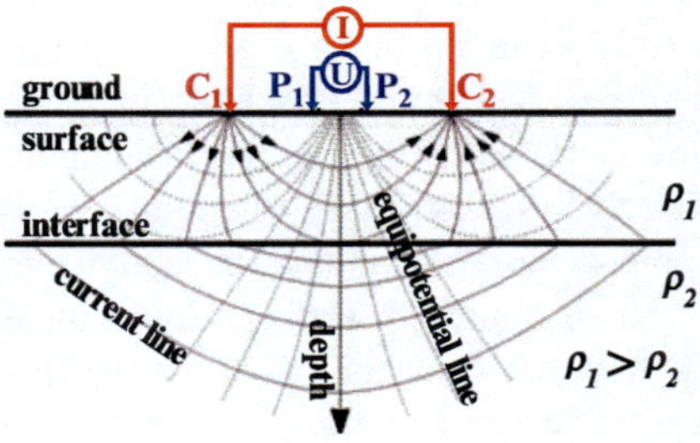

Fig. 10.4 Four-point electrode configuration in two-layer soil model of resistivities ρ_1 and ρ_2. I = current intensity, U = voltage, C = current electrode, P = potential electrode

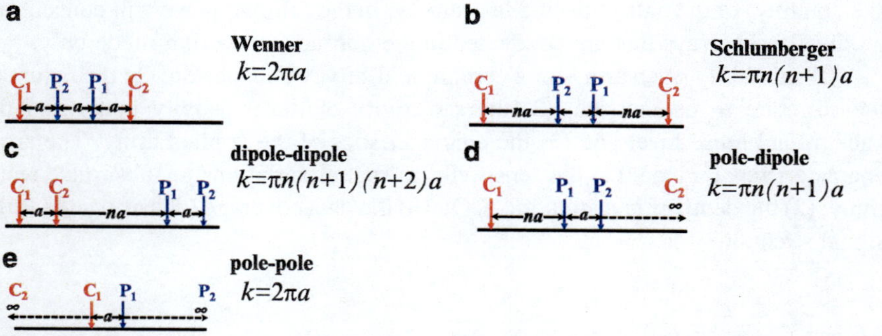

Fig. 10.5 Common electrode configurations with geometric factors (k) applied in geoelectrical surveys, a = electrode spacing

(in Ωm) over this semi-infinite, heterogeneous medium is calculated by (10.8) (e.g., Koefoed 1979; Parasnis 1997). These surveys are usually conducted in diverse electrode configurations or arrays depending on the goal of the survey, the resolution of results, the depth of investigation (DOI), etc. These arrays are the ways of placing current and potential electrode pairs relative to each other. The most common arrays used in resistivity surveys (also for soil and root-zone studies) together with their geometric factors are shown in Fig. 10.5. They include:

1. Wenner array
2. Schlumberger array
3. Dipole–dipole array
4. Pole–dipole array
5. Pole–pole array

Moreover, combining Wenner and Schlumberger arrays together yields the hybrid Wenner–Schlumberger array (Pazdirek and Blaha 1996). It includes the sum of their single data and should show their single advantages. All these arrays are symmetrical around the midpoint except the pole–dipole array. All arrays use four active electrodes that are placed at finite distance from each other except the pole–dipole and pole–pole arrays with one and two remote electrodes, respectively. Remote electrodes are placed sufficiently far from the other electrodes (more than five times the applied largest electrode spacing).

Generally electrical resistivity surveys can also be conducted in the tripotential measurements of α (CPPC), β (CCPP) and γ (CPCP), (Carpenter and Habberjam 1956). Each array can be conducted in two directions of forward (or normal) and reverse (or reciprocal) configurations. In reciprocal measurements, the current and the potential electrodes are interchanged compared with the forward measurements. For example, the configurations CCPP and PPCC represent a forward and reciprocal dipole–dipole array, respectively. Ideally, normal and reciprocal measurements should be identical since their electrode geometry remains the same. Any deviation between the normal and reciprocal measurement is a measure for the error (i.e., quality) of the data collected in a survey. In the following, we will concentrate on the classic arrays that are conducted in the normal (forward) α mode only.

The suitability of an array for a certain field survey depends on: (1) the nature of the structure to be mapped, (2) the sensitivity of the resistivity meter and the background noise level and (3) the characteristics of the applied array. The array characteristics include: (1) the sensitivity to spatial variations in subsurface resistivity, (2) the depth of investigation, DOI, (3) the data coverage (density) and (4) the signal strength.

10.1.5 Array Sensitivity: Depth of Investigation

Sensitivity analyses determine the influence that small spatial changes in the electrical resistivity (ρ) of a subsurface have on the potential measured by an electrode array (Geselowitz 1971; McGillivray and Oldenburg 1990). They determine the change in the potential as measured by the array on the surface if the resistivity of a thin horizontal layer is changed. The higher the sensitivity value, the greater is the influence of the subsurface region on the measured apparent resistivity (ρ_a). Positive and negative sensitivities indicate regions where an increase in ρ leads to an increase and decrease in ρ_a, respectively. The 1D function value defines the median depth of investigation (DOI) characteristic of the various arrays (e.g., Roy and Apparao 1971; Edwards 1977, Table 10.1). The penetration depth depends on the applied electrode array, maximum electrode spacing and the data density, and resistivity structures and its contrast in the subsurface. In resistivity sounding surveys, the increase of the separation between the electrodes increases the depth of sounding to deeper layers but decreases the signal strength. Principally electrical current flows in soil follow paths with least resistivity (highest conductivity). A conductive surface layer causes a decrease in the penetration depth since the current primarily would flow in this layer. If a resistivity decreases downwards the penetration depth may increase as the current tends to flow in the deeper conductive layers.

Results of the 1D (or DOI) and 2D sensitivity functions are listed in Table 10.1 and shown in Fig. 10.6 for the different conventional electrode configurations. Together with the geometric factor (k), DOI is shown relative to the electrode spacing (a) and total array length (L) for these configurations. Generally the signal strength is inversely proportional to k, since the larger the k factor the larger the electrode spacing and thus the lower the signal strength. Moreover, a comparative summary of the characteristics of the different conventional arrays is given in Table 10.2. All these results will be discussed together for the single classic electrode arrays in the next part.

1. *Wenner array*: This well-known array was first proposed by Wenner (1916). Its four electrodes are placed at the surface along a straight line at an equal spacing (a) from each other (Fig. 10.5a). Starting with the least a value, a is increased continuously by the same amount to sound for deeper layers. This array is

Table 10.1 The depth of investigation (z) and geometric factor (k) for different conventional arrays (Edwards 1977)

Array ($\alpha = 1$ m)		z/α	z/L	k
Wenner		0.519	0.173	6.283
Schlumberger	$n = 1$	0.519	0.173	6.283
	$n = 8$	3.247	0.191	226.19
Dipole–dipole	$n = 1$	0.416	0.139	18.850
	$n = 6$	1.730	0.216	1,055.60
Pole–dipole	$n = 1$	0.519	0.296	12.566
	$n = 8$	3.247	0.361	452.39
Pole–Pole		0.867		6.283

L = total length of the array (remote electrodes are not considered), α = electrode spacing (*s.* Fig. 10.5). k is calculated for $\alpha = 1$ m

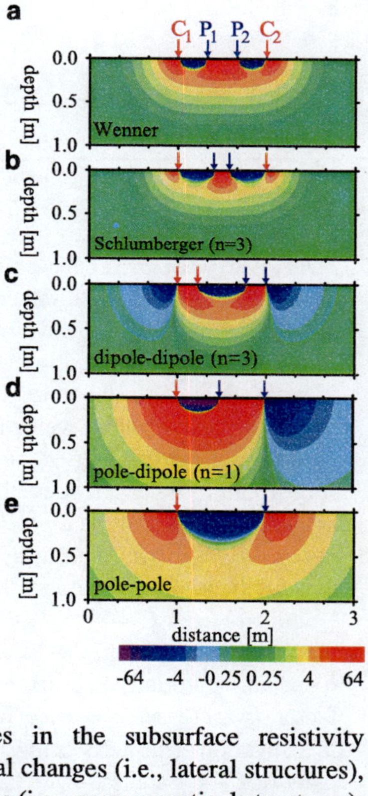

Fig. 10.6 2D sensitivity sections of conventional Wenner (**a**), Schlumberger (**b**), dipole–dipole (**c**), Pole–dipole (**d**), Pole–pole (**e**) electrode arrays

sensitive to vertical (less lateral) changes in the subsurface resistivity (Fig. 10.6a). It is effective in resolving vertical changes (i.e., lateral structures), but relatively poor in detecting lateral changes (i.e., narrow vertical structures). Compared to other arrays, the Wenner array (α, CPPC) has a moderate DOI ($\approx 0.5\alpha$, Tables 10.1 and 10.2). Among the common arrays, the Wenner (like pole–pole) array has the least geometric factor and thus the strongest signal strength. Accordingly it is robust in field surveys of areas with high background noise. For 2D surveys, this array has the disadvantage of the relatively poor horizontal coverage as the electrode spacing is increased.

Table 10.2 Characteristics of different 2D arrays configurations types

Parameter	Wenner	Schlum-berger	Dipole–dipole	Pole–dipole	Pole–pole
Sensitivity for lateral structures	***	***	**	***	**
Sensitivity for vertical structures	*	**	****	**	**
lateral data coverage (data density)	*	**	***	***	****
Lateral resolution (decrease with depth)	**	***	****	***	*
Depth of investigation (DOI)	**	***	*	***	****
Signal strength, directly proportional to $1/k$ or to →	**** $(1/\alpha)$	*** $(1/n^2)$	* $(1/n^3)$	*** $(1/n^2)$	*** $(1/\alpha)$
Suitability for multichannel acquisition	*	**	****	****	**

The parameter quality is classified form least (*) to highest (****), $k =$ geometry factor

2. *Schlumberger array*: This is called after Schlumlberger who defined resistivity in terms of the electric field rather than the potential difference (as in the Wenner array), Fig. 10.5b. This symmetric array has a focused pattern of positive sensitivities beneath the array centre, i.e., any resistivity change here will result in large variations of the measured potential (Fig. 10.6b). This directional characteristic makes this array well suited for vertical sounding. The signal strength is inversely proportional to n^2, $n =$ distance ratio of C_1–P_1 or C_2–P_2 to P_1–P_2 (Fig. 10.5b). The signal strength is weaker than that for the Wenner array, but it is higher than that of the dipole–dipole and pole–dipole arrays (Table 10.2).
3. *Dipole–dipole array*: This array becomes common in electrical prospecting after development of the necessary theory by Al'pin (1950). This array has the dipole length α ($\alpha = C_1$–$C_2 = P_1$-P_2 distance) and factor n (the ratio of the dipole–dipole spacing, C_2–P_1 to α, Fig. 10.5c). For surveys with this array, the α spacing is initially kept fixed at the smallest unit electrode spacing and the n factor is increased stepwise from 1 to 6 in order to increase the depth of investigation. This array is most sensitive to resistivity changes below the dipole pairs of C_1–C_2 and P_1–P_2. The sensitivity distribution is almost vertical for n values greater than 2 (Fig. 10.6c). Thus the array is very sensitive to horizontal (less vertical) changes in resistivity. It is good in mapping vertical (not horizontal) structures, e.g., root-zones and cavities. Depending on both α and n values, this array generally has a shallower depth of investigation compared to the Wenner array (Table 10.1). For 2D surveys, this array has better horizontal data coverage than the Wenner array. A common disadvantage of this array is

the very low signal strength for large n values, where the voltage is inversely proportional to n^3 (Table 10.2). To overcome this problem the n factor should not exceed the value of 6.

4. *Pole–dipole array*: In this array, a remote electrode, C_2, is placed sufficiently far from the survey site (Fig. 10.5d). The greatest sensitivity beneath P_1–P_2 dipole pair is increasing vertically with increasing n factor (Fig. 10.6d). Thus, similar to the dipole–dipole array, this array is more sensitive to vertical structures. The signal strength decreases with n^2 (Table 10.2). It is intermediate between the higher one of the Wenner array and the lower one of the dipole–dipole array. It is recommended to use n values of less than 8–10. Beyond this, the α spacing between the P_1–P_2 dipole pair should be increased to obtain stronger signal strength. This array has a good horizontal coverage and therefore is suited for areas with hard accessibility (Table 10.2). Unlike the other common arrays, this array is asymmetric and produces asymmetric apparent resistivity anomalies over symmetrical structures (Fig. 10.5d). To eliminate this asymmetry effect, one may repeat the measurements with the electrodes arranged in the reverse or reciprocal mode. The depth of investigation of this array is comparable to that of Schlumberger which is higher than that of dipole–dipole and Wenner but lower than that of pole–pole (Table 10.1).
5. *Pole–pole array*: This pole–pole array is less common. It consists of two near electrodes (C_1 and P_1) and two remote electrodes (C_2 and P_2, Fig. 10.5e). There might be practical problems in finding suitable remote locations for the C_2 and P_2 electrodes. This array shows a wider sensitivity curve around the maximum, i.e., a poorer vertical resolution, compared with the other arrays (Fig. 10.6d). Another disadvantage of this array is that the large P_1–P_2 distance may lead to include telluric noise in the measurements and thus yield data with low signal strength (Table 10.2). It is suitable for 3D surveys with the multichannel system and for some applications such as archaeological surveys where small electrode spacings are needed (Li and Oldenburg 1992). This array has the widest horizontal coverage and the largest depth of investigation (0.867α, α = pole–pole spacing). However, it has the poorest resolution (Tables 10.1 and 10.2).

10.1.6 Advantages and Disadvantages of Classic Electrode Arrays

Table 10.2 summarizes and compares the characteristics of the different conventional 2D arrays regarding the: (1) array sensitivity to spatial (vertical and lateral) heterogeneities, (2) depth of investigation, DOI, (3) data coverage and (4) signal strength. A numerical study shows that the dipole–dipole and pole–dipole arrays have high resolution but are most sensible to noise compared to the other conventional arrays (Dahlin and Zhou 2004). Of all arrays the Wenner is least sensible to noise. The pole–dipole, dipole–dipole and Schlumberger arrays are recommended for 2D resistivity surveying because of their recorded high data density and thus resolution. Generally, the use of multiple arrays (e.g., Wenner–Schlumberger) in the same survey enhances the data density and quality which improve the mapping resolution of subsurface targets and leads to a better interpretation (Hesse et al. 1986), see next sections.

10.1.7 Field Survey and Data Acquisition

Here we will concentrate on the modern 2D and 3D surveys only and neglect the 1D survey based on the almost unrealistic assumption of 1D subsurface model of laterally homogeneous subsurface. The 2D/3D surveys better suit the strong heterogeneous media of vadose soils and root-zones. Typical surveyed data set approaches 10–20 readings for 1D resistivity sounding, 100–1,000 for 2D imaging surveys and several thousand measurements for a 3D survey. Starting with the 2D survey, two methods are described: (1) lateral electrical profiling (LEP) and (2) continuous vertical electrical sounding (CVES) of electrical resistivity imaging (ERI).

10.1.7.1 Lateral Electrical Profiling (LEP)

LEP describes lateral variations of the apparent electrical resistivity (ρ_a) at a constant depth. Profiling is typically carried out at the soil surface along transects (profiles) using a particular electrode array type. Quadripole measurements are made at different positions (or stations) with constant electrode spacing depending on the target depth. Each station provides an integrated ρ_a value and is presented at the geometric centre of the 4-electrode array. The lateral ρ_a-variability depends on the array type and the unknown resistivity distribution of the soil. Profiling is useful for relatively rapid lateral coverage for the purpose of lateral mapping of soil targets at a single depth, e.g., single coarse roots, root-zones and archaeological structures.

Technological improvements allow rapid resolution both in space and time. Different mobile platforms are developed for a rapid and accurate, lateral mapping of electrical soil properties at large field areas (up to 10 ha/day). They can be towed by a vehicle (tractor) and allow an automatic positioning of the data points using GPS systems. Two examples of these mobile platforms are shown in Fig. 10.7.

1. The system MUCEP (Multi-Continuous Electrical Profiling) uses eight rolling electrodes (one pair for the current injection and the other three pairs for voltage measurements) corresponding to three depths of investigation (Fig. 10.7a, Panissod et al. 1997).
2. The PACEP (Pulled Array Continuous Electrical Profiling, Sørensen 1996) and the OhmMapper (GSSI Inc.) use a streamer of multi-electrode cable. One electrode pair is used for transmitting current while other electrode pairs are used for measuring the potentials at different spacings and thus depths (Fig. 10.7b). Unlike

Fig. 10.7 Mobile systems for geoelectrical surveys: (a) Multi- Continuous Electrical Profiling (MUCEP) and (b) sketch of pulled array system (Sørensen 1996). C_1/C_2 = current electrodes

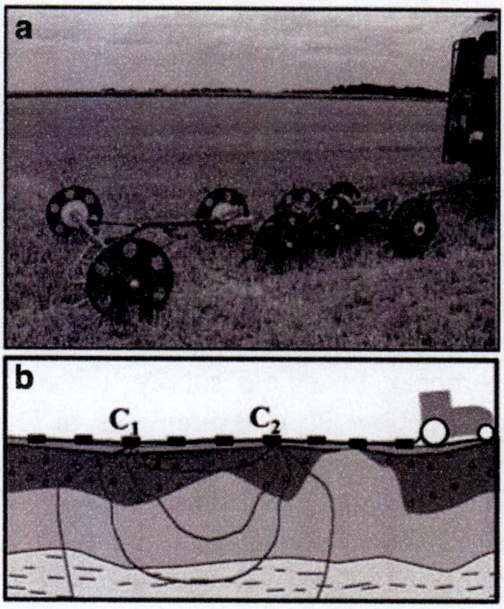

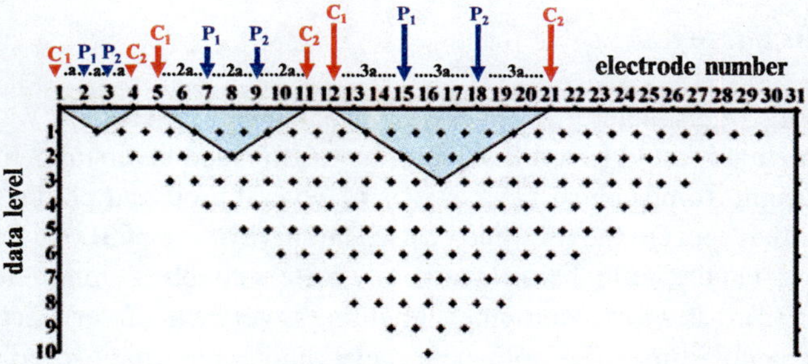

Fig. 10.8 Acquisition of a 2D apparent resistivity pseudosection using the Wenner array ($C1P1$ $P2C2$). C = current electrode; P = potential electrode; α = electrode spacing

the other systems, they measure ρ_a of soils without cumbersome ground stakes used in traditional resistivity surveys.

10.1.7.2 ρ_a Pseudosection of Continuous Vertical Electrical Sounding

The 2D data acquisition is necessary to get an accurate 2D image of the subsurface. Figure 10.8 shows an example of a possible sequence of the so-called "pseudosection" of 2D measurements for ρ_a using the Wenner array. This survey uses a linear system with N (here 31) fixed electrodes placed at a constant spacing, α. All possible quadripole measurements (of the order of $C_1P_1P_2C_2$) are conducted in all possible data levels of electrode spacings (1α–10α). For level 1 with 1α electrode spacing, the measurements start using electrodes number 1, 2, 3 and 4 for the first reading then 2, 3, 4 and 5 for the second reading. This is repeated until electrodes 28, 29, 30 and 31 are used for the last measurement in level 1. For this system with

$N = 31$ electrodes, there are 28 ($N-3$) possible measurements in level 1 (1a spacing) for the Wenner array. After completing the sequence of measurements with level 1, the next data sequence with level 2 (2a electrode spacing) is measured. In level 2, electrodes 1, 3, 5 and 7 are used for the first measurement and electrodes 2, 4, 6 and 8 for the second measurement. This process is also repeated until electrodes 25, 27, 29 and 31 are used for the last measurement with level 2. For this system with $N = 31$ electrodes, there are 25 ($N-2 \times 3$) possible measurements with 2a spacing. The same process is repeated for measurements with spacings of 3a (level 3), 4a (level 4), ..., and 10a (level 10 = $N/3$ as integer number). In a field survey, the best data coverage and results are achieved by conducting the measurements in all possible electrode spacings and data levels which result in a total of $[(N-1)(N-2)/6]$, i.e., 145 data points for the Wenner array (Dahlin and Loke 1998). The survey procedure with the other arrays is similar to that used for the Wenner array to yield 2D pseudosections of apparent resistivity (Hallof 1957; van Overmeeren and Ritsema 1988; Dahlin 1996).

10.1.7.3 3D Survey

Almost all soil targets including root-zones are 3D in nature. The best way to handle these 3D structures would be a full 3D data acquisition and inversion. However, the data set resulting from such survey consists of several thousand points and is very expensive with respect to the time and costs. The survey time of 3D studies is highly reduced using: (1) the multichannel resistivity meters enabling simultaneous multi-readings and (2) faster microcomputers enabling inversions of very large data sets within a reasonable time. The pole–pole, pole–dipole and dipole–dipole arrays as well as unconventional (optimized) arrays are frequently used for 3D surveys because of their suitability for multichannel acquisition (Table 10.2). In addition, they have good data coverage near the edges of the survey grid. The electrodes are usually arranged in a square or rectangular grid depending on the target nature and accessibility of the study site. To get a uniform resolution, the spacing between the inline electrodes and line offset should be equal. One method of building a 3D electrical tomogram consists of the reconstruction of a 2D network of parallel pseudosections (e.g., Hagrey et al. 1999; Zhou et al. 2001; Chambers et al. 2002). An accurate 3D electrical image can be achieved if the inline electrodes are perpendicular to the anomaly strike. Therefore, electrical surveys of a heterogeneous subsurface (e.g., root-zone) should include measurements using linear arrays oriented in many possible directions (Fig. 10.9). Such multidirectional resistivity survey is also important to study anisotropic media. Square and circular electrode grids that centred at the tree stem may be convenient for the 3D mapping of the root-zone in various directions (e.g., Lazzari 2007). For an in vivo root study using a square grid, the data are commonly collected along electrode grids of 5–20 cm spacings and 5–15 m side length depending on the expected size and depth of the root-zone. This yields a penetration depth of about 1–5 m depending on the soil conductivity. If the target is a small plant with shallower roots, the fixed electrode

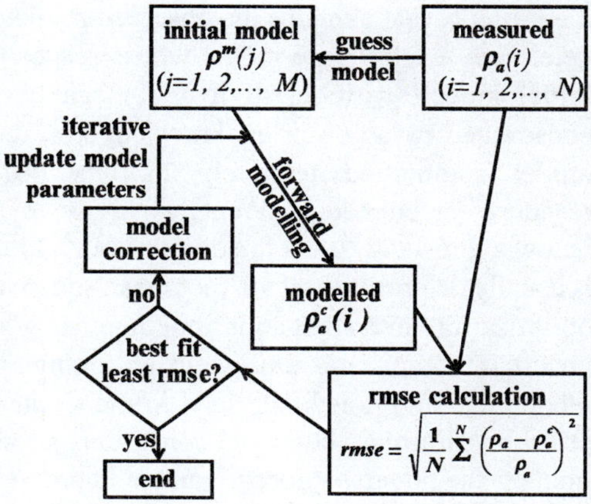

Fig. 10.9 A flow chart showing the individual steps of the iterative inversion of geoelctric data. ρ_a, ρ_a^c = measured and calculated apparent resistivity, respectively; M = number of mesh cells; N = number of data points; rmse = root mean square error

spacing can be decreased down to 1 cm (Petersen and Hagrey 2009). In this case fine electrodes (about 0.2 mm diameter) are used and carefully installed to ensure good contact with the soil material.

10.1.8 Data Processing

Geoelectrical data are normally processed by simply removing data influenced by noise. The removal is done manually or using a filter. Plots of the raw data are commonly used to pick out these outliers (with a "jiggered" appearance) in the data. This filter commonly applies a maximum threshold value for the geometric factor (k) value and removes all data of higher k values. The data noise increases with increasing k of the data (Coscia et al. 2008; Hagrey 2009). The noisy data results from current and potential leakages during the acquisition. This increase in data noises decreases the reliability of the subsurface model resulting from the inversion of the measured apparent resistivities.

10.1.9 Inversion

The ultimate goal of geoelectrical methods applied to a soil medium is to determine the "true" resistivity (ρ) distribution inside the medium from measurements conducted at its boundaries. The problem of inverting measured apparent resistivity (ρ_a) data in a true ρ-model is nonlinear and inherits a non-unique solution. Therefore the solution is carried out using a regularized optimization process. It includes: (1) a forward problem solution to calculate ρ_a for a given electric soil model (with the true ρ distribution) using a specific electrode array (e.g., Dey and Morrison 1979; Silvester and Ferrari 1990) and (2) an inverse problem solution to calculate the ρ-model that explains the observed ρ_a distribution. All inversion methods try to determine a subsurface model whose response agrees with the observed data. An initial model (with resistivity, ρ^m) can be guessed by the observed data and constrained by any a priori knowledge on the subsurface (Fig. 10.9). This initial model is modified iteratively until the difference (misfit) between the model response (calculated apparent resistivity, ρ_a^c) and observed data (ρ_a) is reduced below a threshold (predefined) value, e.g., the average error of the data. This misfit is usually determined by the root mean square error, rmse (e.g., Zohdy 1989). In this optimization inversion, an infinite number of models fit the data set within a given uncertainty limit and thus yield an ambiguity in the interpretation. In addition, incomplete and noisy data lead to more interpretation ambiguity. However, an a priori constraining the model constraints strongly reduce the non-uniqueness solution of the inverse process, and a more reliable model can be obtained. The mathematical treatment of these optimized inversion algorithms are presented in many publications, e.g., Oldenburg and Li (1994), Loke and Dahlin (2002), Günther et al. (2006).

10.1.10 Examples

Figure 10.10 shows synthetic examples of ρ_a pseudosections of the Wenner, and Schlumberger arrays and their inverted ρ models in the resistive root-zone within host soils. The distribution and polarity of the ρ_a anomalies of the Wenner and Schlumberger pseudosections are in good agreement with their corresponding sensitivity patterns (Fig. 10.6). Comparing the models with their corresponding pseudosections shows the qualitative character of ρ_a anomalies in the interpretation. Pseudosections also reflect that the data coverage is least for the Wenner array, moderate for the Schlumberger and highest for the Wenner–Schlumberger. All

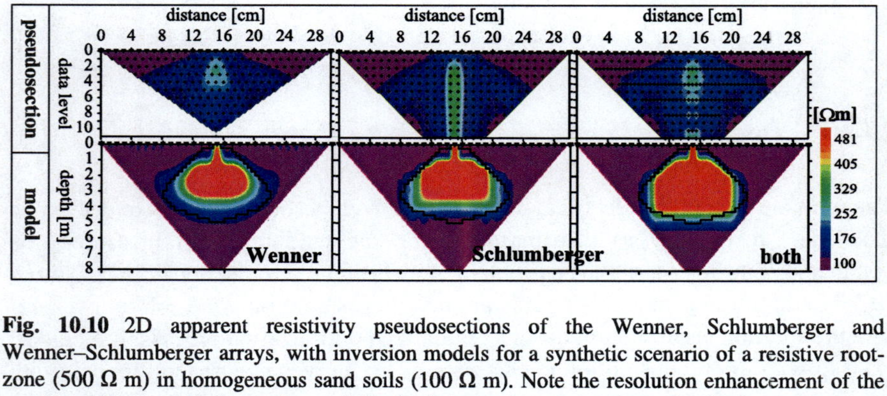

Fig. 10.10 2D apparent resistivity pseudosections of the Wenner, Schlumberger and Wenner–Schlumberger arrays, with inversion models for a synthetic scenario of a resistive root-zone (500 Ω m) in homogeneous sand soils (100 Ω m). Note the resolution enhancement of the combined data set relative to these of the single ones. *Dots, triangles and solid lines* show data, electrodes and root-zone positions, respectively

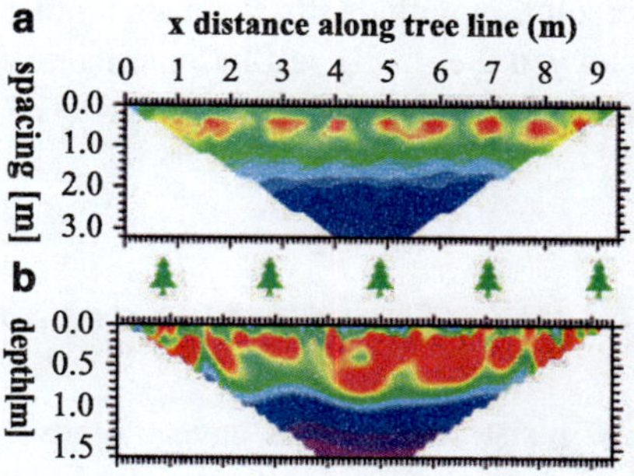

Fig. 10.11 Measured apparent resistivity pseudosection using Wenner array (**a**) and 2D inverted subsurface model below a row of peach trees (**b**)

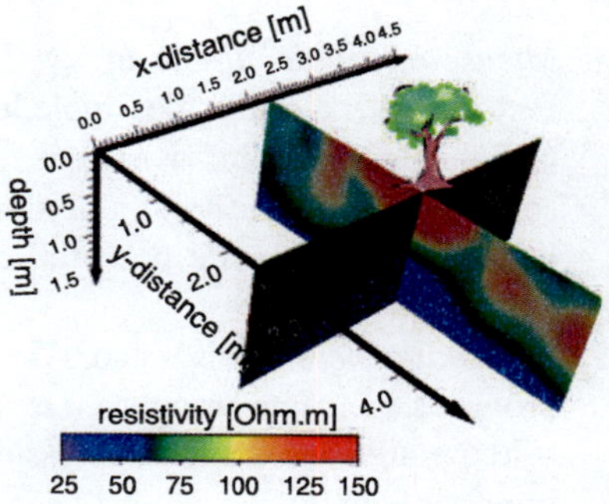

Fig. 10.12 3D resistivity inversion model for data collected along squares around a tree using dipole–dipole and Wenner configurations. It reflects a high-resistivity anomaly in the root-zone (Lazzari 2007)

inversion models clearly reflect the oval resistivity high of the root-zone with nearly similar magnitude at the centre (500 Ωm). We can also see the advantageous resolution enhancement of the mapped root-zone (better distribution and magnitude of the anomaly) by combining the data of Wenner and Sclumberger arrays in the inversion.

Figures 10.11 and 10.12 show field applications for 2D and 3D inversion tomograms for Wenner and dipole–dipole measurements carried out along linear and square array of electrodes, respectively, for peach trees (Hargrey and Michaelsen 2002; Lazzari 2007).

10.1.11 Resolution

It is difficult to give a quantitative measure for the resolution. It depends on the number and the type of the used electrode arrays, and the magnitude and contrast of the resistivity structures. Dahlin and Zhou (2004) show how different electrode arrays resolve five different 2D models. By a rule of thumb, a horizontal resistivity structure should have a thickness of about 1.5–2 times the layers above it to be well modelled in a surface survey. Thinner resistivity structures may be detected in the resistivity model, but it may suffer from the equivalence and suppression principles (see next section). Buried electrodes, as in borehole arrays, strongly enhance the resolution with depth. Petersen and Hagrey (2009) and Hagrey and Petersen (2011) showed the strong resolution enhancement of mapping root-zones by adding borehole electrodes to the surface arrays.

10.1.12 Limitations, Error Sources and Pitfalls

2D resistivity surveys have the advantage of mapping many complex structures, but caution must still be exercised in interpreting the results from the data. Some of the common pitfalls are as follows:

1. *Misuse of dipole–dipole array*: This includes the monotonic increase of the n factor, while keeping the dipole length a fixed, in an effort to increase the depth of investigation. This usually results in very noisy data for n greater than 6. To solve this problem, the n value should not exceed 6. Instead one may increase the a value to reduce the potential drop and continue the measurements at the n levels of 1–6.
2. *Non-uniqueness*: This occurs in the inversion where more than one model produces the same response, i.e., an ambiguity in the interpretation. It is related to the equivalence and suppression principles. This equivalence states that it is impossible to distinguish between two highly resistive beds of different thickness (h) and resistivity (ρ) values if the resistance ($h \times \rho$) is the same, or between two highly conductive beds if the conductance (h/ρ) is the same. The suppression principle states that if a bed is very thin compared to those above and below, its effect on the sounding curve is insignificant unless its resistivity is extremely high or low. Equivalences occur, where a thin resistive layer is sandwiched by two conductive layers or vice versa. For a resistive or conductive thin layer, it may only be possible to determine the layer resistance or conductance, respectively. A layer (either resistive or conductive) should have a thickness more than 1.5–2 times the overburden before its ρ and h can be resolved independently. These problems restrict the capability of ERI techniques for mapping thin soil horizons and root-zones. This ambiguity can be strongly reduced by constraining the inversion with a priori information from other techniques.
3. *Cultural noise sources*: To avoid any (electromagnetic) noise, one should avoid locating profiles parallel to high-voltage power lines. In addition, elongated (conductive or resistive) bodies may distort the data if they are parallel to the survey lines. Metallic objects cause current channelling. A safety distance of about 1.5–2 times the expected penetration depth should ensure a good data quality.
4. *Poor electrode coupling*: Poor contacts between the electrodes and the ground arise in stony or dry soils. It is difficult to obtain good data quality or even to collect data. Poor contacts result in a relative high contact resistance with insufficient current injection into the ground. Wet soils guarantee good contacts with high current injections into the ground and accordingly high data quality.

5. *Poor current penetration*: This may happen in extremely resistive or conductive soils. In dry resistive soils it might be very difficult to get enough current to flow through the ground. In well conductive clayey soils (with brackish porewater) the electric current might be trapped in the top soil horizon.
6. *Polarization of current electrodes*: This happens in saturated soils where charges tend to build up around the current electrodes. After stopping the current injection, the electrodes need enough time (minutes) for the depolarization. If the same electrode is used as a potential electrode immediately after it has been used as a current electrode, this could result in an erroneous reading (Dahlin 2000; Merriam 2005).
7. *2D assumption for real 3D soil structure*: Vadose soils and root-zones are typically heterogeneous 3D structures. In this case, 2D models are strongly affected by heterogeneities perpendicular to the survey line causing a distortion in the model obtained. These 2D models should be treated with some reservations and needs confirmation from excavation results.
8. *Optimization and inversion*: Most non-linear inversion algorithms of geo-electrical data carry out an optimization (i.e., not a direct one-to-one inversion with a unique solution). In the optimization one tries to reduce any misfit (estimated by rmse) between the calculated and measured apparent resistivity values. Data acquisition should not have an effect on the results if there is an infinite, noise-free data and a perfect fit (i.e., rmse approaches 0). However, with real, noisy and limited data, the art of acquisition has an effect on the resulting model. For instance, two different datasets collected from the same place may yield two different models even with the same rmse values.

10.2 Ground Penetrating Radar GPR

10.2.1 Introduction

RADAR is an acronym for *ra*dio *d*etection *a*nd *r*anging (Buderi 1996). The geophysical technique is also called *ground penetrating (probing) radar* (GPR), georadar, subsurface radar or earth sounding radar. The history may be traced to an initial German patent by Hülsmeyer in 1904. The first GPR survey was performed in Austria in 1929 to determine a glacier's depth (Stern 1929, 1930). GPR instrumentation was developed during World War II to implement effective ground-to-air systems for aircraft detection. The technology received more attention starting the 1950s when US Air Force radars were penetrated by ice as planes while trying to land in Greenland, misread the altitude and crashed into the ice (Barringer 1965). In 1967, a radar system was flown on Apollo 17 to the moon for the surface electrical properties experiment (Simmons et al. 1972). In 1972 commercial GPR systems became available and an explosion of applications started (Morey 1974). Neal (2004) displayed the exponential increase of the research publications in the period 1980–2001.

GPR is an electromagnetic (EM) reflection technique for exploration, imaging, characterization and monitoring of the subsurface. GPR utilizes the transmission and reflection of high frequencies, f (10–2,500 MHz) within the near surface zone. GPR mapping capability is largely determined by the contrast of the EM properties (between the target and its surrounding soil material). GPR techniques have the advantages of: (1) being a fast, non-invasive and non-destructive geophysical technique, (2) using compact equipment and fast data acquisition, (3) having the highest resolution (down to cm range) of any geophysical exploration technique and (4) being highly sensitive to water content in soil. GPR may be performed from the ground surface, from a borehole or between boreholes (crosshole). Therefore, GPR is very suitable for a wide range of applications in fields of geology, geotechniques, hydrology, environment, archaeology and agroforestry. This includes the resolution and 2D and 3D mapping of fine soil horizons and objects, including roots. Descriptions of the fundamental principles can be found in publications by Davis and Annan (1989), Bristow (2003), Annan (2004), Daniels (2004), Neal (2004), Jol (2009), etc.

In the next, we review the EM wave propagation in soil and root-zone media, then summarize the data acquisition and processing and finalize by the limitations and pitfalls.

10.2.2 Theoretical GPR Background

Maxwell's equations for EM waves describe a coupled set of electric and magnetic fields when the fields vary with time. Changing electric fields creates magnetic fields which in turn induces electric fields. Depending on the energy loss, this continuing succession of one field inducing the other results in fields which may diffuse or propagate through the medium. The EM wave propagation of radar signal in an isotropic medium is expressed by the time harmonic solution from Maxwell's equation:

$$\nabla^2 E = \mu\sigma \frac{\partial E}{\partial t} + \mu\varepsilon \frac{\partial^2 E}{\partial t^2} \tag{10.9}$$

with E = electric field amplitude (V/m), ε = dielectric permittivity (F/m), μ = magnetic permeability (H/m) and σ = electric conductivity (S/m).

On the right-hand side of (10.9), the first and second terms represent energy dissipation (conduction current) and energy storage (displacement current), respectively. In highly conductive medium (very high σ) the energy dissipation dominates over the energy storage and the equation reduces to the diffusion term, i.e., EM waves do not propagate in this medium. In low loss medium (very low σ) and high frequency, f, the energy storage dominates over the energy dissipation and EM waves propagate. The propagation of GPR signals in a homogeneous and isotropic medium can be described by the plane wave solution of the Helmholz equation, which is obtained from (10.9) (e.g., Ward and Hohmann 1988; Hayt 1989). The solution for one component of the electric field intensity vector, E travelling in the z direction is:

$$E = E_0\, e^{-\alpha z}\, e^{j(\omega t - \beta z)} \tag{10.10}$$

where E_0 = complex initial amplitude, α = attenuation constant [Np/m], β = phase constant [Rad/m], $\omega = 2\pi f$ = angular frequency (s^{-1}) and $j = \sqrt{-1}$.

The phase velocity (v) and attenuation (α) are the main parameters describing the GPR wave propagation. Both depend on the dielectric and conductivity properties of the material. The skin depth (δ) in a medium is defined by the distance through which an EM wave is attenuated by a factor of e^{-1} (=0.368) of its initial amplitude. To obtain δ and v, α and β can be expressed separately as real quantities by (e.g., Hagrey and Müller 2000):

$$\alpha = \frac{1}{\delta} = \omega\sqrt{\frac{\mu\varepsilon'}{2}\left(\sqrt{1+\tan^2\vartheta} - 1\right)} \tag{10.11}$$

$$\beta = \omega\sqrt{\frac{\mu\varepsilon'}{2}\left(\sqrt{1+\tan^2\vartheta} + 1\right)} \tag{10.12}$$

with ε' = the real part of the complex dielectric permittivity ε, ϑ = loss angle and $\tan\vartheta$ is a measure for dielectric loss.

At low conductivities ($\sigma < 100$ mS/m, $\tan\vartheta << 1$) and high frequencies (f = 10–2,500 MHz), the radar velocity (v) remains essentially constant in low-loss materials (e.g., Hagrey and Müller 2000). For non-magnetic soils, μ is almost equal to μ_0 (μ-vacuum) and μ_r ($=\mu/\mu_0$) is close to unity. This simplifies v and δ or α of a wave propagating in the z direction into:

$$v \approx \frac{c}{\sqrt{\varepsilon'_r}} \approx \frac{c}{\sqrt{\varepsilon_r}} \tag{10.13}$$

$$\alpha = \frac{1}{\delta} = \frac{\sigma}{2}\sqrt{\frac{\mu}{\varepsilon}} = \frac{60\pi\sigma}{\sqrt{\varepsilon_r}} \tag{10.14}$$

with c = EM wave (light) velocity in vacuum = 0.3 [m/ns], σ in [mS/m], $\mu_0 = 1.25 \times 10^{-6}$ H/m and $\varepsilon_0 = 8.89 \times 10^{-12}$ Faraday/m.

A plane EM wave incident on an interface (far larger than the wave length, λ) separating two homogeneous, isotropic media is partly transmitted or refracted and partly reflected. For normal incidence (i.e., $\theta_i = \theta_r = 90°$), the reflection coefficient, R is given by:

$$R = \frac{Z_2 - Z_1}{Z_2 + Z_1} \tag{10.15}$$

where Z is the intrinsic impedance that can be reduced in low loss media of negligible σ, μ and dispersion into:

$$Z = \sqrt{\frac{j\omega\mu}{\sigma + j\omega\varepsilon'}} = \sqrt{\frac{\mu}{\varepsilon}} = \frac{Z_0}{\sqrt{\varepsilon_r}} \tag{10.16}$$

with $Z_0 = \sqrt{\frac{\mu_0}{\varepsilon_0}} = 377$ Ohm = impedance in vacuum (Annan 2005). Applying this to (10.15) results in the simple equation:

$$R = \frac{\sqrt{\varepsilon_{r1}} - \sqrt{\varepsilon_{r2}}}{\sqrt{\varepsilon_{r1}} + \sqrt{\varepsilon_{r2}}} = \frac{v_2 - v_1}{v_2 + v_1} \tag{10.17}$$

Obviously the propagation of GPR signals is dominated by ε_r and σ of the non-magnetic media of soils and roots. Since ε_r is 81 for water and below 10 for most soil and wood materials (see next section), the effect of water content (θ) on ε_r and R is dominant. The soil conductivity (σ) is inversely proportional to δ and Z ((10.14) and (10.16), Wensink 1993). The contrast in Z (e.g., at interfaces to roots and root-zones) is a potential target for GPR mapping. Normally, R lies in the range $0 \leq |R| \leq 1$. At interfaces with EM contrast R has either positive or negative value as follows:

$R > 0$	For $Z_2 > Z_1 \rightarrow \varepsilon_1 > \varepsilon_2 \rightarrow v_2 > v_1$, subscript 1 is for overburden soils and 2 is for target. For void target $\varepsilon_2 = 1$, air has the highest Z value
$R < 0$	For $Z_2 < Z_1 \rightarrow \varepsilon_1 < \varepsilon_2 \rightarrow v_2 < v_1$, e.g., contact to wet root-zones
R = -1	For metallic target ($Z_2 \approx 0 \rightarrow \varepsilon_2 \approx \infty$), metal has the least Z value

10.2.3 EM Properties of Soil Root-Zone Media

The propagation properties of GPR waves (velocity, v, attenuation, α, and intrinsic impedance, Z) are controlled mainly by ε and σ of the soil media consisting mainly of non-magnetic materials ($\mu_r = 1$). EM properties are directly related to the amount, distribution, chemistry and phase (liquid, solid, gas) of interstitial water of roots and soils, etc. Tables 10.3–10.5 show the typical values for the EM parameters characterizing the different non-magnetic media (soils, rocks, water and roots). They display the effects of the water content (θ), water salinity or total dissolved solids (TDS) and temperature as well as the wood density. These tables clearly reflect the inverse relationship between ε_r and v on one side and σ and α on the other. Roots consist mainly of wood (cellulose of cell wall) and water. An increase in θ in soils and roots strongly increases ε_r, σ and α, and decreases v. This dominance of water effect is related to ε_r values, which is highest for water ($\varepsilon_r = 81$)

Table 10.3 Electromagnetic (EM) characteristics of materials at 100 MHz (Torgovnikov 1993; Daniels 1996; Asprion 1998). α = attenuation, ε_r = relative permittivity, σ = electric conductivity, v = GPR velocity

Material	ε_r [-]		σ [mS/m]		v [m/ns]		α [dB/m]	
	Dry	Wet	Dry	Wet	Dry	Wet	Dry	Wet
Sand	3–5	20–30	0.01	0.1–1.0	0.06	0.15–0.16	0.01	0.03–0.3
Gravel	3.5	16.5	0.01	0.7–9	0.06	0.1–0.13	0.01–0.1	0.03–0.5
Silt	3–5	22–30	1–100	100	0.05–0.07	0.1–0.12	1–100	1–300
Clay	4–7	8–40	1–100	100–1,000	0.05–0.1	0.09–0.12	1–100	10–300
Till	7–21	24–34	2.5–10	2–5	0.1–0.12	0.1–0.12	10–100	–
Sandy soil	4–6	15–30	0.1–10	10–100	0.05–0.08	0.12–0.15	0.1–2	1–5
Clayey soil	5–6	10–15	0.1–100	100–1,000	0.08–0.09	0.11–0.15	0.3–3	5–30
Loamy soil	4–6	10–20	0.1–1	10–100	0.07–0.09	0.12–0.15	0.5–3	1–6
Sandstone	4–6	5–10	10^{-6}–10^{-3}	10^{-3}–10^{-2}	0.11	0.12–0.15	0.3–0.5	10–20
Limestone	7	8	10^{-6}–10^{-3}	10–100	0.11	0.13	0.4–1	10–25
Coal	4–5	8–25	10–2	100–1,000	0.11	0.16	1–10	2–20
Cell wall/ cellulose	4.5	22	0.24	4	0.141	0.064	0.187	1.35

Table 10.4 Effect of freezing temperature (T) and water salinity on electromagnetic properties

Material	ε_r [-]		σ [mS/m]		v [m/ns]		α [dB/m]	
	$T \approx 20°C$	$\leq 0°C$	$\approx 20°C$	$\leq 0°C$	$\approx 20°C$	$\leq 0°C$	$\approx 20°C$	$\leq 0°C$
Peat/permafrost	57–81	4–8	<40	0.01–10	0.03–0.06	0.11–0.15	0.3	0.1–5
Freshwater/ice	81	3–4	0.1–10	0.01	0.034	0.16	0.1	0.01
Seawater/ice, 33 g/ 1	81	4–3	30,000	1	0.034	0.15	1,000	10–30
Air, vacuum	1		0		0.30		0	
Metal	∞		∞		–		∞	

Table 10.5 Effect of wood density on electromagnetic parameters

Wood density [g/cm^3]	ε_r [-]	σ [mS/m]	v [m/ns]	α [dB/m]
0.4	16	2.27	0.075	0.95
0.6	23	4.89	0.062	1.66
0.7	26	6.45	0.058	2.05

Wood parameters were measured at 100% water saturation

and below 10 for most soil and wood substances. In soil root-zones, σ ($1/\rho$, ρ = resistivity) increases with increasing θ, TDS and/or clay content (McNeill 1980).

Freezing temperatures decrease ε_r, α and σ, and increase v (Table 10.3). The wood density has a similar effect as θ (Table 10.5).

Table 10.6 Factors affecting electric and electromagnetic properties

Method		ERI	GPR	References, e.g.
Controlling physical property		ρ	σ, ε_r, μ_r (≈ 1)	Archie (1942), Keller, and Frischknecht (1966), Topp et al. (1980), Torgovnikov (1993), Sandoz (1996), Simpson and TenWolde (1999), Martinis (2002), Bucur (2003), Nicolotti et al. (2003)
Output attribute		ρ	v, α	
Controlling factor	↑proportional to↑			
• Water content, θ		ρ^{-1}	v^{-1}, α (ε_r, σ)	
• Total dissolved solids, *TDS*		ρ^{-1}	v^{-1}, α (ε_r, σ)	
• Clay/organic matter content		ρ^{-1}	v^{-1}, α (ε_r, σ)	
• Wood density, *d*		ρ^{-1}	v^{-1}, $\alpha(\varepsilon_r)$	
• Frequency, *f*		–	v, α (ε_r)	
• Temperature, *T*		ρ^{-1}	v^{-1}, α (ε_r, σ)	
• Wood anisotropy		$\rho_l << \rho_d < \rho_t$	$\varepsilon_l >> \varepsilon_d > \varepsilon_t$	

α = energy attenuation, ε_r = relative dielectric permittivity, μ_r = relative magnetic permeability ρ = electrical resistivity, σ = electrical conductivity, v = radar velocity, subscripts l, d, t = longitudinal, radial, and transversal component, respectively. *ERI* = electrical resistivity imaging, *GPR* = ground penetrating radar

10.2.4 Factors Affecting Electrical/EM Properties

Backgrounds of ERI and GPR techniques show that the electric resistivity field and radar wave propagation in soil and wood media are strongly governed by: (1) fluid content, θ, (2) salinity or total dissolved solids TDS, (3) density, *d*, (4) energy decay, α, (5) anisotropy, (6) temperature, *T* and (7) applied GPR wave frequency, *f*, causing physical property dispersion. Table 10.6 summarizes the effect of these factors on applied geophysical attributes.

10.2.5 GPR Antenna and Data Acquisition

Like soil interfaces, single roots and whole root-zone envelops possess an EM contrast with regard to hosting vadose soils. These objects constitute potential targets for GPR reflections. Commercial radar systems consist of either one single transmitting (Tx) and receiving (Rx) antenna (monostatic), or two, separate, Tx and Rx antennas (bistatic). In most bistatic systems, both antennas are housed in the same shielded unit at a fixed spacing that allows radiation from one side only. This minimizes undesired radiation interferences from outside the study medium.

The data acquisition in reflection profiling of GPR is displayed in Fig. 10.13. The transmitter antenna (Tx) radiates a short pulsed EM wave or "wavelet" into the medium. The wavelet contains a number of frequencies, but is usually referred to by the centre frequency (f_c) of the antennas. As the EM wave propagates downwards it experiences materials of differing EM properties (ε, σ, μ), which alter its velocity. At interfaces of objects and layers, the wave energy is partly reflected, diffracted and scattered back to the surface and partly transmitted downward. The reflected signal is recorded by the receiver antenna (Rx). The time for the path from the transmitter to the reflector and back to the receiver is referred to as two-way travel time (TWT) and is measured in nanoseconds (1 ns $= 10^{-9}$ s). Recorded TWT from a reflector is a function of its depth, the antenna spacing and the average GPR velocity in the overlying soils.

Recorded data during a survey consists of a sequence of reflection traces (amplitude versus time) that build up a radar reflection profile (Fig. 10.14a). Each trace consists of the direct pulse followed by reflected pulses. A scan is a trace with coloured-scaled amplitude. A radar data record/section, or simply radargram results from moving antenna along the survey line; it consists of series of traces/scans collected at discrete points to display the subsurface. The data can either be displayed in a wiggle or coloured scan mode (Fig. 10.14b, c). In a wiggle mode the amplitude variation of each trace is displayed as a curve. In a coloured scan mode the amplitudes are colour coded.

The antennas used in a certain survey are selected on the basis of the depth and size of the study target as well as the EM properties (mainly ε and σ) of the medium. The centre frequency (f_c) of the antenna is proportional to the resolution but to the inverse of the penetration depth.

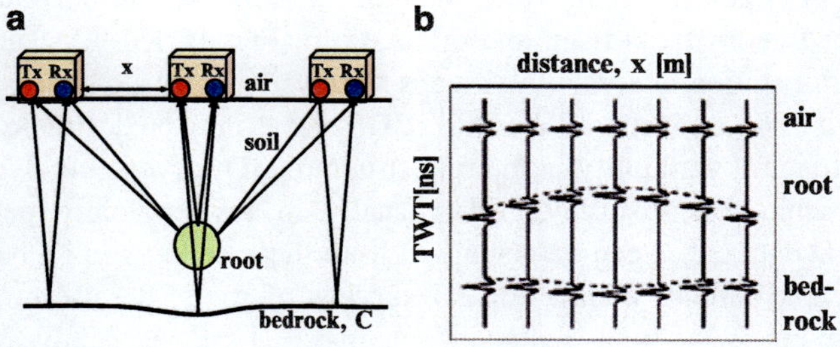

Fig. 10.13 (a) Radar reflection profiling with a fixed transmitter, Tx – receiver, Rx offset over an anomalous root and bedrock, (b) Recorded radargram with air waves and reflections from the root and bedrock. TWT = 2-way travel time (Hagrey 2007)

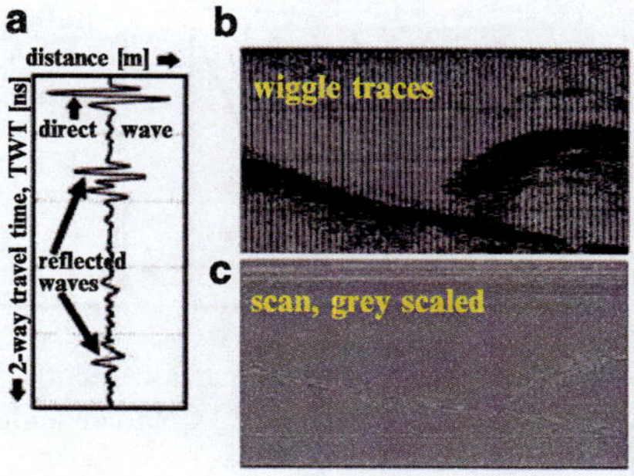

Fig. 10.14 Bistatic GPR antenna and signal display as a single trace (amplitude-time) or radargram of wiggle traces (**b**) or scan (**c**) with colour-scaled amplitude. *Tx* transmitter, *Rx* receiver

10.2.6 GPR Operation Mode

GPR field survey and data acquisition are carried out in two modes: the reflection profiling and the transmission tomography. One measures the TWT for the reflection mode and the one-way travel time (for the direct path from the transmitter to the receiver) for the transmission mode. The reflection profiling is acquired in the following three main configurations (Fig. 10.15):

1. Common-offset reflection mode
2. Multi-offset reflection of common midpoint (CMP) and wide angle reflection and refraction (WARR)
3. Common-source or common-receiver mode

These survey modes as well as the transmission mode are described in the followings in detail.

10.2.6.1 Common-Offset Reflection Survey

Common (or single)-offset reflection mode is the most frequently used in GPR studies to characterize the subsurface. It uses bistatic systems consisting of a single transmitting (Tx) and receiving (Rx) antenna (often housed in the same shielded unit) at a fixed geometry (offset, and orientation) (Fig. 10.15). Measurements are made at regular station intervals. An area is surveyed mostly along parallel lines or rectangle grid of lines. Pseudo-3D or truly 3D surveys may be desirable when there is significant lateral variability in internal structure. 3D surveys involve collecting data on closely spaced survey grids, usually in two mutually perpendicular directions. 3D data cubes can be generated from these surveys (e.g., Nitsche et al. 2002; Heinz and Aigner 2003). For 3D surveys of roots, we use normally grid spacing in cm range and high-frequency antennas (≥1 GB) especially in dry sandy soils (Fig. 10.16). An accurate, fast position and elevation of 3D data points for mobile field surveys (using a platform and towing vehicle) is determined using a

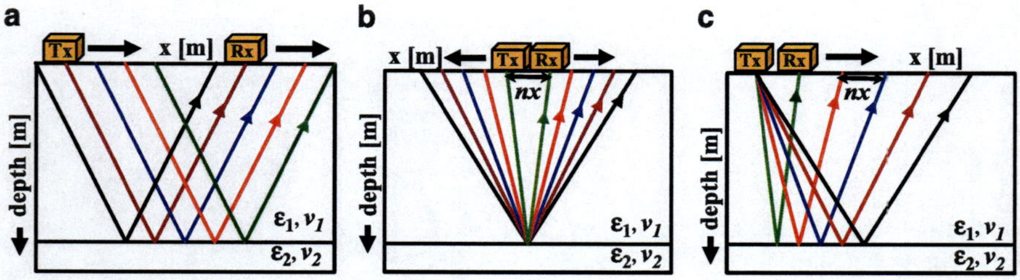

Fig. 10.15 Main types of radar reflection survey in a non-magnetic two-layer soil medium. **(a)** Common Tx–Rx offset, **(b)** Common middle point (CMP), **(c)** Common source. *Tx* transmitter, *Rx* receiver, *nx* Nyquist interval

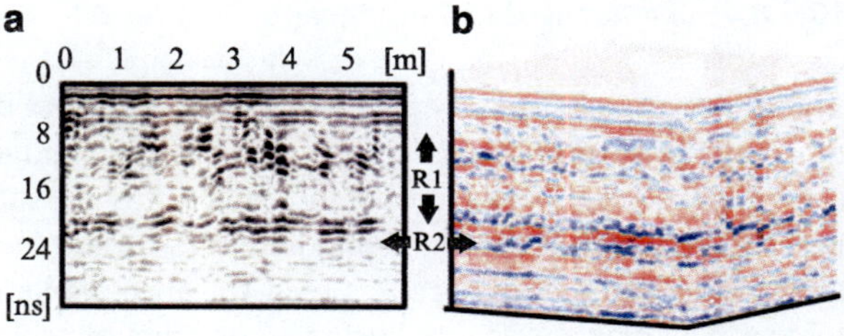

Fig. 10.16 Visualization of GPR reflection data in 2D radargram (**a**) and 3D data cube (**b**) for a survey around a poplar tree. R1, single roots; R2, subsurface interface of groundwater

semi-automated self-tracking laser theodolite (Lehmann and Green 1999) or GPS system (e.g., Urbini et al. 2001).

10.2.6.2 Multi-Offset Reflection, CMP/WARR Mode

Multi-offset reflection of a common midpoint (or depth, CMP) and wide angle reflection and refraction (WARR) soundings are used to determine the velocity (v) versus depth in the medium under study. Surveys are performed using two separate Tx and Rx antennas. A planer subsurface reflector should be identified on the radargram from a GPR reflection survey conducted prior to this sounding. Antennas are positioned on a surface point over this reflector with an initial (minimal) Tx–Rx offset, nx (Nyquist interval, Fig. 10.15b, c). Antennas are moved apart sequentially at fixed intervals. Two-way travel time (TWT) data to the reflector are then collected by increasing Tx–Rx offsets as integer multiples of nx. In the CMP mode, both antennas are moved from the centre point in steps of nx/2 (Fig. 10.15b). In the WARR mode, one antenna is held fixed while the other is moved out in steps of nx (equivalent to common-source mode of Fig. 10.15c). The CMP mode is the standard mode; the reflected signal comes from a fixed point rather than the moving one of the WARR sounding. Depending on the ground attenuation, the maximum Tx–Rx offset (x) is about 1–2 times the reflector depth. Resulting reflection events (TWT versus x) are used to calculate average radar-wave velocities to a given reflection using semblance analyses (Annan and Davis 1976; Reynolds 1997).

10.2.6.3 Common-Source/Receiver Surveys

Both common-receiver and common-source surveys are multi-fold surveys to enhance the reflections (Fig. 10.15c). Common-source with multi-receivers is equivalent to the WARR. Compared with the common-offset (single fold) data, such surveys have the advantages of improved signal-to-noise ratio, reduction of coherent noise, increased penetration depth, greater reflection continuity, more accurate spatial positioning, etc. (e.g., Greaves et al. 1996; Pipan et al. 1999).

10.2.6.4 Transmission Tomography Surveys

These measurements are conducted where Tx and Rx can be put on opposite sides of a medium so as to image it. The survey can be conducted between two boreholes (crossholes) and/or between boreholes and ground surface. The data coverage is best for a combined crosshole and borehole-surface survey. Direct one-way travel times and/or amplitudes are recorded for all possible sets of Tx–Rx pairs. These tomography data are inverted using reconstruction techniques to image the volume between the measurement points. The reconstructed tomograms reflect the velocity (v for travel time data) anomalies or the attenuation (α for amplitude data) anomalies. These surveys provide accurate mapping of the bulk root system, particularly of a tap root with limited accessibility from surface. Hanafy and Hagrey (2006) used the travel time tomography to reconstruct the whole root-zone (including coarse and fine root branches) of a poplar tree as a negative velocity anomaly (Fig. 10.17). This negative anomaly is due to the relative high water content within the root-zone, which results in high dielectric permittivity and thus lower velocity relative to that of hosting soils.

10.2.7 GPR Wave Types

Figure 10.18 illustrates the CMP propagation paths of GPR waves with the diagram of antenna offset-TWT and an example of a data record. This displays the following wave types:

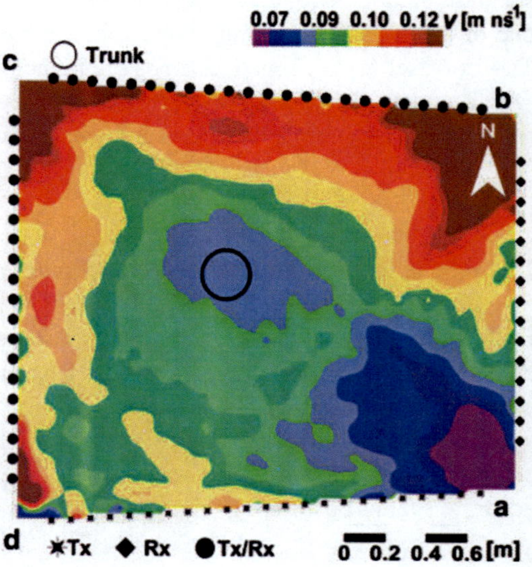

Fig. 10.17 2D tomogram of radar velocity around a young popular tree (*plan view*), Kiel, Germany. It resulted from the inversion of travel time data collected with 500 MHz antennas in the trenches ABCD (0.5 m depth). The central and southwester negative anomalies are caused by root and soil heterogeneities excavated before the survey, respectively (Hanafy and Hagrey 2006)

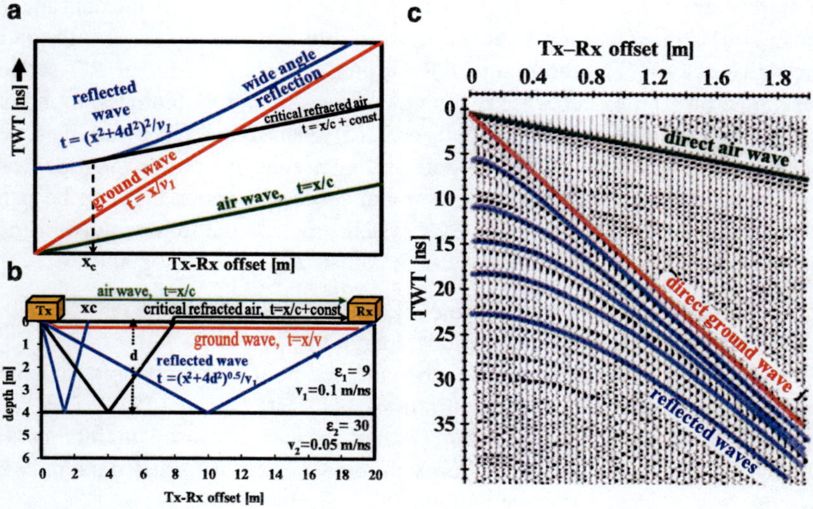

Fig. 10.18 Illustration of CMP propagation paths of radar waves in layered soils of contrasting dielectric permittivity (ε_1, ε_2,...) and velocities (v_1, v_1,...), (**a**) with event arrival-time versus antenna offset (**b**) and an example of a field record. *Tx* transmitter, *Rx* receiver antenna, TWT = 2-way travel time

1. Air wave: This is the first recorded pulse which travels directly from Tx antenna to Rx antenna in the air at the light speed (0.3 m/ns).
2. Ground wave: It is the second arrival, which travels directly through the ground medium between the Tx and Rx antennas at the ground velocity. Events of each of the air and ground waves lie on a straight line passing through the origin. These events mask any primary reflections in the upper part of a radar reflection profile.
3. Critically refracted air waves or lateral waves: They result from shallow reflections that approach the surface at the appropriate critical angle and are subsequently refracted along the air–ground interface (Clough 1976).
4. Vertical reflected wave and wide angle reflected wave: Here the angle of incidence is equal to angle of reflection.

10.2.8 Data Processing

Post-acquisition processing of radar data is accomplished to reduce clutter, minimize the effects of multiple hyperbolic reflections and enhance the signal-to-noise ratio (Daniels 1996). Signal processing serves to transform the image to be representative of the actual dimensions of the target that can be more readily interpreted. Digital processing and display of GPR data can be performed now on minicomputers.

GPR applicants use many of the highly developed techniques of seismic data analysis (Yilmaz 2001). Transforming GPR information into soil and root-zone information follows two ways: (1) presenting GPR response measurements in 2D sections (radargrams) or 3D data cubes for mapping purposes (e.g., root biomass) to indicate spatial change, texture, target location, etc., (2) quantifying GPR variables, e.g., velocity, attenuation, etc. to derive soil and root-zone quantities such as water content, porosity, and salinity. Some essential processing steps that can be partly conducted in the field using modern GPR systems include the following (for further explanations see e.g., Jol and Bristow 2003; Annan 2009; Cassidy 2009):

1) *Data editing*: This includes the removal and correction of bad data, update data header, file sorting, reverse directions, merge files, etc.
2) *Dewow* (or de-wowing): It removes the inherent, nonlinear wow noise of a low f signal trend and low DC (direct current) bias in data (Fig. 10.19). WOW noise arises from the close proximity of Rx to Tx and is caused mainly by EM induction and/or the swamping or saturation of the recorded data by early arrivals, e.g., ground/air wave (Annan 1973; Gerlitz et al. 1993).
3) *Correction of time zero*: The start time that may not be detected precisely by the system is corrected here to match with surface position and ensure correct depths in the profile.
4) *Filtering*: 1D and 2D filtering to enhance signal-to-noise ratio and visual quality. It includes band pass filtering to remove high f noise and a spatial low pass filter to reduce noise, enhances (nearly) flat reflections and suppresses effects of rapid changing features, e.g., diffraction hyperbolas and steeply dipping reflections.
5) *Display/time gains*: They include selection of appropriate gains for data display and interpretation. Radar signals of high f are very rapidly attenuated as they propagate into the ground. A simultaneous display of strong and weak signals from shallow and large depths, respectively, requires preconditioning for visual analysis display (Fig. 10.20). An automatic gain control (AGC) compensates the

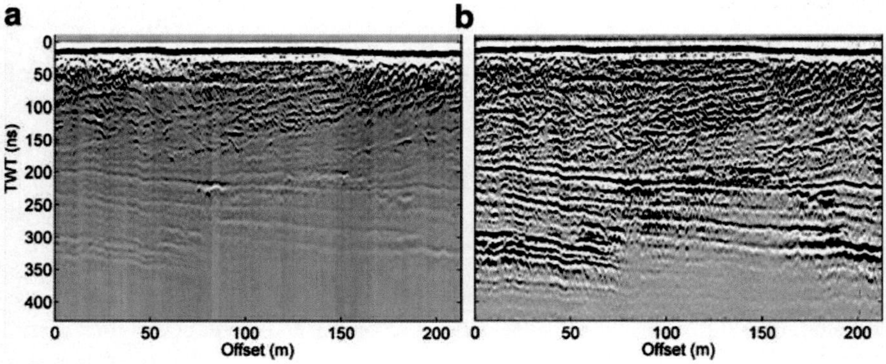

Fig. 10.19 Raw radargram with wow noise (bright white energy near TWT = 0 and gray colour at TWT>> 0, (**a**) and after dewing (**b**). Wow biases the response of the source to positive values (Battista et al. 2009)

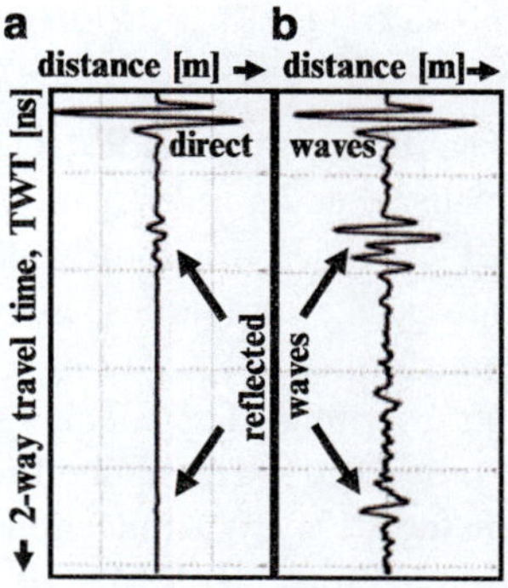

Fig. 10.20 Single GPR trace before (**a**) and after (**b**) applying AGC time gain

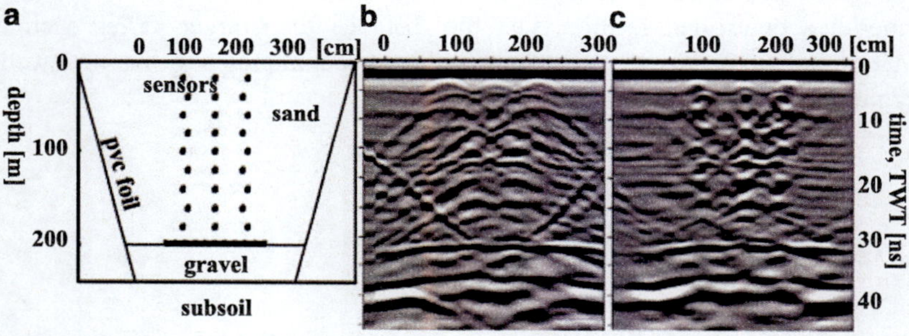

Fig. 10.21 (**a**) Vertical section of a soil model with buried sensors (PVC, radius = 1.2 cm), (**b**) Raw radargram (500 MHz antenna) with strong interfering diffractions from sensors, (**c**) Migrated section with a layered velocity model (0.16–0.14 m/ns) showing a good focusing of diffractions (Hagrey and Müller 2000)

effect of attenuation and spherical EM wave spreading of the signal in the GPR data.

6) *Migration*: It aims to reconstruct the distribution of reflections with their correct geometries and amplitudes (i.e., actual size and shape). It enhances the display of the reflections significantly since diffraction hyperbolas are collapsed and dipping reflections are moved to the true geometrical position. Migration requires knowledge of the velocity distribution that should be determined before in the field (e.g., CMP-method, see next section). Multiple hyperbolas of root reflections are migrated using Kirchhoff or Hilbert transform that use reflection geometry or return signal magnitude to guide the decomposition to a representative and compact size. However, Kirchhoff migration may be confused by multi-oriented roots and Hilbert transform may be impacted by the soil moisture (Doolittle and Butnor 2009). Figure 10.21 shows vertical section within the GeoModel of Kiel University with 27 buried sensors and radargrams before and after applying a layered velocity model (Hagrey et al. 2003). The migration clearly focuses and separates the individual sensor reflections.

10.2.9 GPR Velocity (v) Determination

The full utility of GPR data requires knowledge of how fast the signals travel in the material under investigation. GPR velocity (v) values are important to quantify: (1) GPR images in depth values (i.e., to convert the time axis into depth) and, (2) subsurface properties into dielectric pedo-hydrologic information (i.e., v is transformed into ε_r using equation (8) and ε_r is converted in water content and porosity (e.g., Topp et al. 1980; McCann et al. 1988; Rust et al. 1999; McCann and Forde 2001). The applied techniques for calculating v include: (1) Ground wave method, (2) TWT reflections from known depth, (3) Hyperbolic fitting and diffraction tail matching, (4) CMP (common mid-point) or WARR (wide angle reflection and refraction), and (5) Transmission tomography of travel time. All of these techniques require GPR measurements along a traverse where the geometry is varying in a controlled fashion, i.e., the distance to a target varies such that estimations of velocity can be extracted. These techniques may be reviewed in many publications (e.g., Annan 2004).

10.2.10 GPR Penetration Depth

The penetration depth of GPR technique is controlled by the antenna frequency (f), the electrical conductivity (σ) and the attenuation (α) of the soil materials. In low-loss (i.e., electrically resistive) soils the penetration depth expressed by the skin depth is inversely proportional to f and σ (10.14). An increase in σ strongly increases α and decreases the penetration depth. GPR investigations in sediments report penetration depths of 5–15 m for 200 MHz, few metres for 500–1,000 MHz and only few dm-cm for 1,500–2,500 MHz. The maximum penetration depth in soils is obtained in dry clean sand and gravel (e.g., Jol et al. 2002; Bakker 2004).

10.2.11 GPR Resolution (Horizontal, Vertical)

GPR has the highest resolution of any geophysical method for imaging the subsurface. The resolution is the ability to distinguish between two closely spaced features (signals) from each other (Fig. 10.22a). Resolution is controlled mainly by the wavelength (λ) of the EM wave propagating in the ground, the contrast in EM properties, the geometry (size, shape, and orientation) of the target and the field noise. The wavelength is determined by the GPR frequency, f and velocity, v of the ground material as $\lambda = v/f$. An increase in f (decrease in λ) enhances the resolution, but decreases the depth of investigation. In most materials, resolution decreases with increasing depth as frequency-dependent properties.

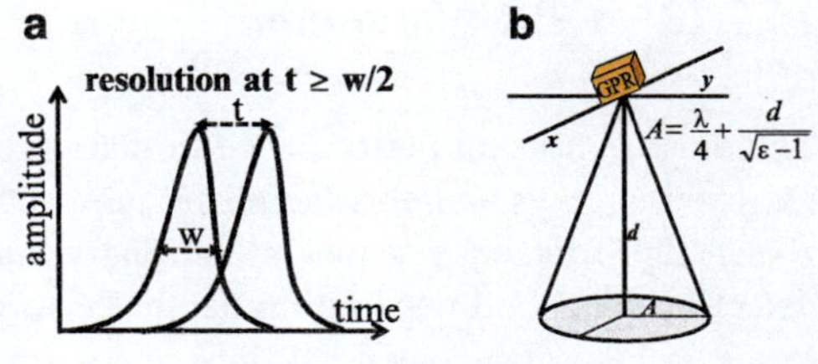

Fig. 10.22 GPR resolution (**a**) and footprint controlling lateral or spatial resolution (**b**). Pulse is clearly separable at $t \geq w/2$ (w = half width, i.e., pulse width at half amplitude) only. A = footprint axis, d = depth, λ = wavelength, ε = dielectric permittivity

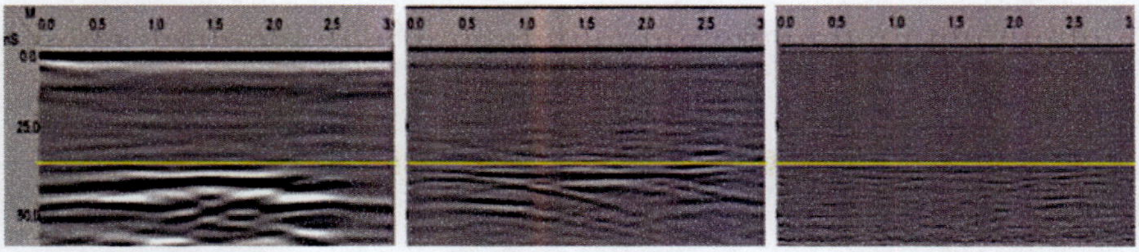

f = 200 MHz f = 400 MHz f = 900 MHz

Fig. 10.23 Radagrams over a fine sandy soil with a reflector at 34 ns (2 m depth, yellow line) using GPR antennas of different frequencies, f. Note, the resolution increases with increasing f

For the vertical resolution, the distance between two reflectors should at least be $\frac{1}{8}\lambda$–$\frac{1}{2}\lambda$ to be resolved theoretically (Sheriff and Geldart 1995; Møller and Vosgerau 2006). Using $\frac{1}{4}\lambda$ in practice, the vertical resolution in low saturated sand with a velocity of 0.1 m/ns is about 0.25–0.01 m for a 100–2,500 MHz centre frequency (f_c). Within this vertical distance any reflections will interfere in a constructive manner and result in a single, observed reflection. In order to increase vertical resolution, the high-frequency content of the data must be enhanced. Figure 10.23 displays GPR profiles that are acquired with f_c = 200, 400 and 900 MHz. This figure clearly illustrates that the vertical resolution is enhanced by increasing f_c. Vertical resolution of a radar reflection profile has important implications for the root interpretation, as it will determine whether root-sets can be imaged and the scale of root structure that can be observed. Sets of roots (with individual thicknesses down to <0.02 m) are most likely to be resolved in low-loss media such as sand and gravels with high-frequency antennas (e.g., ≥1 GHz) whose maximum vertical resolution is ≤0.02 m.

The lateral or spatial resolution is determined by the area illuminated by a GPR antenna, the antenna footprint (Fig. 10.22b). The footprint is elliptical in shape and its size increases with depth. The radar footprint is estimated by the width of its long axis (A): $A = \lambda/4 + d/\sqrt{\varepsilon - 1}$, d = depth to the reflector. The decrease in lateral resolution with depth has important implications for interpretation of root-zone. A further consideration with respect to lateral resolution is the horizontal spacing between traces on GPR profiles. Neal (2004) discusses in detail the different aspects that have to be taken into account in the evaluation of the lateral resolution.

10.2.12 GPR Limitations

The most significant performance limitation of GPR is in highly conductive media enriched in clays and/or mineralized porewater. In wet, conductive soils and root-zones high-frequency waves are strongly attenuated that limit the resolution for detecting roots and penetration depth. Performance is also limited by signal scattering in heterogeneous conditions, e.g., stony soils. Attenuation describes either real intrinsic losses by transforming EM energy to another energy type, e.g., heat, or apparent losses through geometric effects. Apparent losses are the result of geometric spreading, surface and volume scattering, waveguides and multipathing. In the apparent losses, the energy follows a path that is no longer observable. Attenuation always results in frequency-dependent properties (dispersion).

References

Al'pin LM (1950) The theory of dipole sounding: Moscow, Gcetoptekhizdat, 89 p. [Translation in Dipole methods for measuring earth conductivity]: Plenum Press, New York, 1966, pp 1–60

Amato M, Basso B, Cellano G, Bitella G, Morelli G, Rossi R (2008) In situ detection of tree root distribution and biomass by multielectrode resistivity imaging. Tree Physiol 28:1441–1448

Amato M, Bitella G, Rossi R, Gomez JA, Lovelli S, Gomes JF (2009) Multi-electrode 3D resistivity imaging of alfalfa root zone. Eur J Agron 31:213–222

Annan AP (1973) Radio interferometry depth sounding: Part I – theoretical discussion. Geophysics 38:557–580

Annan AP (2004) Ground penetrating radar, principles, procedures & applications. Sensors & Software, Mississauga, Canada, p 303

Annan AP (2005) GPR methods for hydrogeological studies, chapter 7. In: Rubin Y, Hubbard S (eds) Hydrogeophysics. Springer, Heidelberg, p 523

Annan AP (2009) Electromagnetic principles of ground penetrating radar. In: Jol HM (ed) Ground penetrating radar: theory and applications. Elsevier, Amsterdam, pp 3–37

Annan AP, Davis JL (1976) Impulse radar sounding in permafrost. Radio Sci 11:383–394

Archie GE (1942) The electrical resistivity log as an aid in determining some reservoir characteristics. Trans Am Inst Min Eng 146:54–62

Asprion U (1998) Ground-penetrating radar (GPR) analysis in aquifer-sedimentology: case studies, with an emphasis on glacial systems of SW Germany. Tübinger Geowissenschaftliche Arbeiten, Reihe A, B and 43

Bakker MAJ (2004) The internal structure of Pleistocene push moraines. A multidisciplinary approach with emphasis on ground penetrating radar: PhD thesis, Queen Mary, University of London, p 177

Barringer AR (1965) Research directed to the determination of subsurface terrain properties and ice thickness by pulsed VHF propagation methods. Air Force Cambridge Research Laboratories contribution AF 19(628):2998

Battista BM, Addison AD, Knapp CC (2009) Empirical mode decomposition operator for dewowing GPR data. J Environ Eng Geophys 14:163–169

Bevan BW (2000) An early geophysical survey at Williamsburg, USA. Archaea Prospec 7:51–58

Binley A, Kemna A (2005) Electrical methods. In: Rubin Y, Hubbard S (eds) Hydrogeophysics. Springer, Heidelberg, pp 129–156

Breusse JJ (1963) Modern geophysical methods for subsurface water exploration. Geophysics 27 (4):633–657

Bristow CS (2003) Ground penetrating radar in sediments. Geol Soc, London

Bucur V (2003) Nondestructive characterization and imaging of wood. New York, Springer, p 324

Buderi R (1996) Invention that changed the World. New York, NY, p 575

Butnor JR, Doolittle JA, Kress L, Cohen S, Johnsen KH (2001) Use of ground-penetrating radar to study tree roots in the southeastern United States. Tree Physiol 21:1269–1278

Butnor JR, Doolittle JA, Johnsen KH, Samuelson L, Stokes T, Kress L (2003) Utility of ground-penetrating radar as a root biomass survey tool in forest systems. Am J Soil Sci Soc 67:1607–1615

Campbell JE (1990) Dielectric properties and influence of ionic conductivity in soils at one to fifty megahertz. Soil Sci Soc Am J 54:332–341

Carpenter EW, Habberjam GM (1956) A tripotential method for resistivity prospecting. Geophysics 21:455–469

Cassidy NJ (2009) Ground penetrating radar data processing, modelling and analysis. In: Jol HM (ed) Ground penetrating radar: theory and applications. Elsevier, Amsterdam, pp 141–172

Chambers JE, Oglivy RD, Kuras O, Cripps JC, Meldrum PI (2002) 3D electrical imaging of known targets at a controlled environmental test site. Environ Geol 41:690–704

Clough JW (1976) Electromagnetic lateral waves observed by earth-sounding radars. Geophysics 41:1126–1132

Coscia I, Marescot L, Maurer H, Greenhalgh S, Green AG (2008) Experimental design for cross hole electrical resistivity tomography datasets. EAGE Near Surface Geophysics, Cracow, p 5

Dahlin T (1996) 2D resistivity surveying for environmental and engineering applications. First Break 14:275–284

Dahlin T (2000) Short note on electrode charge-up effects in DC resistivity data acquisition using multi-electrode arrays. Geophys Prospec 48:181–187

Dahlin T, Loke MH (1998) Resolution of 2D Wenner resistivity imaging as assessed by numerical modelling. J Appl Geophys 38:237–249

Dahlin T, Zhou B (2004) A numerical comparison of 2D resistivity imaging with ten electrode arrays. Geophys Prospect 52:379–398

Daniels DJ (1996) Subsurface penetrating radar. The Institution of Electrical Engineers, London, p 320

Daniels DJ (2004) Ground-penetrating radar. Institute of Electrical Engineers, London

Davis JL, Annan AP (1989) Ground penetrating radar for high-resolution mapping of soil and rock stratigraphy. Geophys Prospect 37:531–551

Dey A, Morrison HF (1979) Resistivity modelling for arbitrarily shaped three-dimensional structures. Geophysics 44:753–780

Dobrin MB (1960) Geophysical prospecting. McGraw Hill, New York, p 10020

Doolittle JA, Butnor JR (2009) Soils, peatlands and biomonitoring. In: Jol HM (ed) Ground penetrating radar: theory and applications. Elsevier, Amsterdam, pp 177–192

Edwards LS (1977) A modified pseudosection for resistivity and IP. Geophysics 42:1020–1036

Emerson DW (1990) Emerson, notes on mass properties of rocks-density, porosity, permeability. Explor Geophys 21:209–216

Gerlitz K, Knoll MD, Cross GM, Luzitano RD, Knight R (1993) Processing ground penetrating radar data to improve resolution of near-surface targets. In: Proceeding of the symposium on the application of geophysics to engineering and environmental problems. San Diego, California

Geselowitz DB (1971) An application of electrocardiographic lead theory to impedance plethysmography. IEEE Trans Biomed Eng 18:38–41

Greaves RJ, Lesmes DP, Lee JM, Toksöz MN (1996) Velocity variations and water content estimated from multi-offset, ground-penetrating radar. Geophysics 61:683–695

Griffiths DH, Turnbull J (1985) A multi-electrode array for resistivity surveying. First Break 3(7):16–20

Griffiths DH, Turnbull J, Olayinka AI (1990) Two-dimensional resistivity mapping with a computer-controlled array. First Break 8:121–129

Günther T, Rücker C, Spitzer K (2006) 3-d modeling and inversion of DC resistivity data incorporating topography – Part II: inversion. Geophys J Int 166:506–517

Hagrey SA al (2006) Electrical resistivity imaging of wooden tree trunks. Near Surf Geophys 4:177–185

Hagrey SA al (2007) Geophysical imaging of root zone, trunk and moisture heterogeneity. J Exp Bot 58:839–854

Hagrey SA al (2009) 2D Optimisation of electrode arrays for borehole surveys. EAGE Near Surface Geophysics, Dublin, p 5

Hagrey SA al, Michaelsen J (1999) Resistivity and percolation study of preferential flow in vadose zone at Bokhorst, Germany. Geophysics 64:746–753

Hagrey SA al, Michaelsen J (2002) Hydrogeophysical soil study at a drip irrigated orchard, Portugal. Eur J Environ Eng Geophys 7:75–93

Hagrey SA al, Müller C (2000) GPR-study of pore water content and salinity in sand. Geophys Prospect 48:63–85

Hagrey SA al, Schubert-Klempnauer T, Wachsmuth D, Michaelsen J, Meissner R (1999) Preferential flow, first results of a full scale flow model. Geophys J Int 138:643–654

Hagrey SA al, Rabbel W, Meissner R, Werban U (2003) The "GeoModel" at Kiel-A hydrogeophysical full scale model for Engineering Geology to study pore water, contamination and structure of soils. In: Proceedings, meeting of engineering geology. Kiel, Germany, pp 361–362

Hagrey SA al, Meissner R, Werban U, Rabbel W, Ismaeil A (2004) Hydro-, Bio-Geophysics. The Leading Edge 23:670–674

Hagrey SA al, Petersen T (2011) Numerical and experimental mapping of small root zones using optimized surface and borehole resistivity tomography. Geophysics 76(2):G25–G35

Hallof PG (1957) On the interpretation of resistivity and induced polarization measurements, Ph.D Thesis, Massachusetts, p 256

Hanafy SM, Hagrey SA al (2006) Radar tomography for soil moisture heterogeneity. Geophysics 71:k9–k18

Hayt W (1989) Engeneering electromagnetics. McGraw-Hill, New York

Heinz J, Aigner T (2003) Three-dimensional GPR analysis of various quaternary gravel-bed braided river deposits (southwestern Germany). In: Bristow CS, Jol HM (eds) Ground penetrating radar in sediments. Geol Soc London Spec Publ 211:99–110

Hesse A, Jolivet A, Tabbagh A (1986) New prospects in shallow depth electrical surveying for archaeological and pedological applications. Geophysics 51:585–594

Hruška J, Cermák J, Sustek S (1999) Mapping of tree root systems by means of the ground penetrating radar. Tree Physiol 19:125–130

Hubbard SS, Grote K, Rubin Y (2002) Mapping the volumetric soil water content of a California vineyard using high-frequency GPR ground wave data. The Leading Edge 21:552–559

Jol HM (ed) (2009) Ground penetrating radar – theory and applications. Elsevier, Amsterdam

Jol HM, Bristow CS (2003) GPR in sediments: advice on data collection, basic processing and interpretation, a good practice guide. In: Bristow CS, Jol HM (eds) Ground penetrating radar in sediments. Geol Soc London Spec Publ 211:9–27

Jol HM, Lawton DC, Smith DG (2002) Ground penetrating radar: 2-D and 3-D subsurface imaging of a coastal barrier spit, Long Beach, WA, USA. Geomorphology 53:165–181

Keller GV, Frischknecht FC (1966) Electrical methods in geophysical prospecting. Pergamon, Oxford

Koefoed O (1979) Geosounding principles 1: resistivity sounding measurements. Elsevier, Amsterdam

Langwig JE (1971) Trace element-electrical conductivity relationships in wood, PhD Thesis: State University Coll. For. Syracuse, NY, p181

Lazzari L (2007) Study of spatial variability of soil root zone properties using electrical resistivity technique, Ph.D. thesis: University of Basilicata, Italy

Lehmann F, Green AG (1999) Semiautomated georadar data acquisition in three dimensions. Geophysics 64:719–731

Li Y, Oldenburg DW (1992) Approximate inverse mappings in DC resistivity problem. Geophys J Int 109:343–362

Loke MH, Barker RD (1995) Least-squares deconvolution of apparent resistivity pseudosection. Geophysics 60:1682–1690

Loke MH, Barker RD (1996) Rapid least square inversion of apparent resistivity pseudosections using quasi-Newton method. Geophys Prospect 48:131–152

Loke MH, Dahlin T (2002) A comparison of the Gauss-Newton and quasi-Newton methods in resistivity imaging inversion. J Appl Geophys 49:149–162

Loke MH, Acworth I, Dahlin T (2003) A comparison of smooth and blocky inversion methods in 2D electrical imaging surveys. Explor Geophys 34:182–187

Lowrie W (1997) Fundamentals of geophysics. Cambridge University Press, Cambridge

Martinis R (2002) Nondestructive techniques for decay diagnosis on standing trees. PhD thesis: University of Florence, Italy, p300

McCann DM, Forde MC (2001) Review of NDT methods in the assessment of concrete and masonry structures. NDT and E (Independent Nondestructive Testing and Evaluation) International 34(2):71–84

McCann DM, Jackson PD, Fenning PJ (1988) Comparison of the seismic and ground probing radar methods in geological surveying. Proc Inst Elect Eng 135:380–390

McGillivray PR, Oldenburg DW (1990) Methods for calculating Frechet derivatives and sensitivities for the non-linear inverse problem: a comparative study. Geophys Prosp 38:499–524

McNeill JD (1980) Electrical conductivity of soils and rocks: Geonics Ltd. Technical Note TN 5:p22

Merriam JB (2005) Injection electrode overprinting. J Environ Eng Geophys 10:365–370

Møller I, Vosgerau H (2006) Testing ground penetrating radar for resolving facies architecture changes - a radar stratigraphic and sedimentological analysis along a 30 km profile on the Karup Outwash Plain, Denmark. Near Surf Geophys 4:57–68

Morey RM (1974) Continuous subsurface profiling by impulse radar. In: Proceedings of the Engineering Foundation Conference on Subsurface Exploration for Underground Excavation and Heavy Construction, Henniker, NH, August 11–16, 1974, American Society of Civil Engineers, pp 213–232

Neal A (2004) Ground-penetrating radar and its use in sedimentology: principles, problems and progress. Earth-Sci Rev 66:261–330

Nicolotti G, Socco LV, Martinis R, Godio A, Sambuelli L (2003) Application and comparison of three tomographic techniques for detection of decay in trees. J Arboricult 29:66–78

Nitsche FO, Green AG, Horstmeyer H, Büker F (2002) Quaternary depositional history of the Reuss delta, Switzerland: constraints from high-resolution seismic reflection and georadar surveys. J Quat Sci 17:131–143

Oldenburg DW, Li Y (1994) Inversion of induced polarization data. Geophysics 59:1327–1341

Oldenburg DW, Li Y (1999) Estimating depth of investigation in dc resistivity and IP surveys. Geophysics 64:403–416

Panissod C, Dabas M, Jolivet A, Tabbagh A (1997) A novel mobile multipole system (MUCEP) for shallow (0–3 m) geoelectrical investigation: the 'Vol-de-Canards' array. Geophys Prospect 45:983–1002

Parasnis DS (1997) Principles of applied geophysics. Chapman & Hall, London

Pazdirek O, Blaha V (1996) Examples of resistivity imaging using ME-100 resistivity field acquisition system. In: EAGE 58th Conference and Technical Exhibition Extended Abstracts, Amsterdam

Petersen T, Hagrey SA al (2009) Mapping root zones of small plants using surface and borehole resistivity tomography. Leading Edge 28(10):1220–1224

Pipan M, Baradello L, Forte E, Prizzon A, Finetti I (1999) 2-D and 3-D processing and interpretation of multi-fold ground penetrating radar data: a case history from an archaeological site. J Appl Geophys 41:271–292

Reynolds JM (1997) An Introduction to applied and environmental geophysics. Wiley, New York
Roy A, Apparao A (1971) Depth of investigation in direct current methods. Geophysics 36:943–959
Rust AC, Russell JK, Knight RJ (1999) Dielectric constant as a predictor of porosity in dry volcanic rocks. J Volcanol Geotherm Res 91:79–96
Samouëlian A, Cousin I, Richard G, Tabbagh A, Bruand A (2003) Electrical resistivity imaging for detecting soil cracking at the centimetric scale. Soil Sci Soc J Am 67:1319–1326
Sandoz JL (1996) Ultrasonics solid wood evaluation in industrial applications. In: Tenth international symposium on nondestructive testing of wood. Lausanne, Switzerland, pp 147–154
Schön JH (1997) Physical properties of rocks: fundamentals and principles of petrophysics, handbook of geophysical exploration. Elsevier, Amsterdam
Sheriff RE, Geldart LP (1995) Exploration seismology. Cambridge University Press, New York
Silvester PP, Ferrari RL (1990) Finite elements for electrical engineers. Cambridge University Press, Cambridge
Simmons G, Strangway DW, Bannister L, Baker R, Cubley D, La Torraca G, Watts R (1972) The surface electrical properties experiment. In: Z Kopal, DW Strangway, D Reidel (eds) Lunar geophysics. Dordrecht, 258–271
Simpson W, TenWolde A (1999) Physical properties and moisture relations of wood. In: Wood handbook-Wood as an engineering material. Madison. US Department of Agriculture, Forest Service, Forest Products Laboratory, Report FPL–GTR–113, Ch. 3, p 463
Skaar C (1988) Wood-water relations. Springer, Heidelberg, p 279
Sørensen K (1996) Pulled array continuous electrical profiling. First Break 14(3):85–90
Stern W (1929) Versuch einer elektrodynamischen Dickenmessung von Gletschereis. Ger Beitr zur Geophysik 23:292–333
Stern W (1930) Uber Grundlagen, Methodik und bisherige Ergebnisse elektrodynamischer Dickenmessung von Gletschereis. Z Gletscherkunde 15:24–42
Telford WM, Geldart LP, Sheriff RE (1990) Applied geophysics. Cambridge University Press, Cambridge
Topp GC, Davis JL, Annan AP (1980) Electromagnetoc determination of soil water content. Measurements in coaxial transmission lines. Water Resour Res 16:574–582
Torgovnikov G (1993) Dielectric properties of wood and wood-based materials. Springer, New York, p 196
Urbini S, Vittuari L, Gandolfi S (2001) GPR and GPS data integration: examples of application in Antarctica. Annali di Geofisica 44:687–702
van Overmeeren RA, Ritsema IL (1988) Continuous vertical electrical sounding. First Break 6(10):313–324
Venkateswaran A (1972) A comparison of the electric properties of milled wood, milled wood cellulose, and milled wood lignin. Wood Sci 4:248–253
Ward SH, Hohmann GH (1988) Electromagnetic theory for geophysical aplications. In: Nabighian MN (ed) Electro-magnetic methods in applied geophysics, 1, Theory. Society of Exploration Geophysicists, Tulsa, Okla, pp 130–311
Wenner F (1916) A method of measuring earth resistivity: US. Natl Bureau Stand Bull 12(3):469–478
Wensink WA (1993) Dielectric properties of wet soils in the frequency range 1–3000 MHz. Geophys Prospect 41:671–696
Werban U, Hagrey SA al, Rabbel W (2007) Hydrogeophysical monitoring of root zone water content in the lab. J Plant Nutr Soil Sci 171:927–935
Yilmaz Ö (1987) Seismic data processing. Society of Exploration Geophysicists, Tulsa, Oklahoma, p 526
Yilmaz Ö (2001) Seismic Data Analysis. Society of Exploration Geophysicists, Tulsa, Oklahoma
Zhou QY, Shimada J, Sato A (2001) Three-dimensional spatial and temporal monitoring of soil water content using electrical resistivity tomography. Water Resour Res 37:273–285
Zohdy AAR (1989) A new method for the automatic interpretation of Schlumberger and Wenner sounding curves. Geophysics 54:245–253

Chapter 11
Multi-electrode Resistivity Imaging

Mariana Amato, Vincenzo Lapenna, Roberta Rossi, and Giovanni Bitella

Abstract Multi-electrode soil imaging is a promising way to investigate root systems by visualizing the distribution of soil volumes with different root densities, based on relationships between root biomass (RD) and electrical resistivity (ρ) of soils.

Its most distinctive features are spatial coverage, rapidity, and minimum disturbance. Spatial patterns and frequency of ρ match those of RD, but calibration is needed and small RD values may not be clearly discriminated in soils with large ρ or variability in other features. Therefore it has been envisaged as:

A nondestructive method for spatial quantification in two and three-dimensions
A basis for spatially sound sampling
A support for differential soil management

Available data indicate that a definite response is not found for roots <2 mm and data sets are positively skewed. Statistical procedures to handle deviations from normality with the advantage of simplicity are discussed, as well as field of application, advantages, drawbacks, and future needs.

11.1 Electrical Prospecting and Root Studies

Multi-electrode soil imaging has been proposed and tested on field and container systems as a nondestructive technique to unravel the spatial structure of root-related processes and to provide quantitative relationships with root properties of large use

M. Amato (✉) • R. Rossi • G. Bitella
Department of Crop, Forest and Environmental Sciences, University of Basilicata, Via Ateneo Lucano, 10, Potenza 85100, Italy
e-mail: mariana.amato@unibas.it

V. Lapenna
CNR-IMAA Contrada S. Loja - Zona industriale C.P. 27, 85050 Tito Scalo, Potenza, Italy

S. Mancuso (ed.), *Measuring Roots*,
DOI 10.1007/978-3-642-22067-8_11,

in ecology and management such as biomass (Amato et al. 2008; Rossi et al. 2010) and length (Amato et al. 2009). The main advantages of the method include minimal time requirements and disturbance, and the possibility to provide nearly instant depth coverage and intensive spatial data (Tabbagh et al. 2000). This latter issue allows to overcome one of the biggest problems in root research, which is the lack of appropriate sampling strategies because the spatial structure of data is hidden (Amato and Ritchie 2002). Traditional and innovative techniques are often mono-dimensional and imply sampling or exploration at the millimeter or centimeter scale; they therefore provide only local information. In addition, constraints linked to time, cost, and disturbance of the studied system limit the number of samples, and as a consequence the description of root systems is rarely performed at the appropriate scale, and the spatial variability of root systems is often neglected in spite of its importance in soil processes (Amato and Ritchie 2002). Multi-electrode imaging allows to detect and visualize the spatial distribution of properties related to root systems in two and three dimensions and therefore provides a basis for spatial description and spatially sound sampling.

Resistivity imaging is based on 2-D and 3-D applications of electrical resistivity tomography as described in Chap. 11 of this book. The method consists of measurements of electrical resistivity (ρ) in multiple soil volumes through a series of conductors applied at the soil surface, in boreholes, or a combination of the two.

Electrical resistivity (ρ) is the electrical resistance of a uniform body of unit length and unit cross-sectional area and is the reciprocal of conductivity. Measurements are made by applying electrical currents through a set of electrodes (current electrodes), reading the resulting differences in electric potential on a set of separate conductors (potential electrodes), and calculating according to the equation:

$$\rho = K(\Delta V/I)(\text{Ohm m}) \tag{11.1}$$

where:

ρ = electrical resistivity (Ohm m)

ΔV = difference in electrical potential (V)

K = geometrical coefficient, depending on the electrode configuration

I = current (A)

The current is direct or weakly alternate in order to avoid induction and electrodes may be placed in different geometrical configurations, based on a minimum set of four conductors (a “quadrupole”) which yields a single value of ρ in one single soil volume. The quadrupole can be replicated in multiple-electrode arrays placed along lines or grids to give measurements in multiple soil volumes arranged in 2-D or 3-D sections or “tomograms”. The survey is conducted by measuring resistivity on a single quadrupole of the array at a time. All possible quadrupole spacings along the line are used for measurements, from the lowest – corresponding to adjacent electrodes – to maximum spacing, determined by the total array length. Different configurations refer to the relative position of current and potential electrodes, their number and spacing allow to attain different depths of investigation, resolution, and signal-to-noise ratio as treated in Chap. 11.

The distribution in space of voltage differences is a function of the different resistivities of soil volumes. In homogeneous materials the current flow-lines are radial and equi-potential surfaces are hemispherical, while in heterogeneous media the current flow-lines are deformed. Therefore resistivity calculation in soil requires two steps: in a first approximation values are derived from (11.1) in the hypothesis of homogeneous soil and called "apparent resistivity"; their arrangement in space is a "pseudo-section". Further treatment accounts for heterogeneity through numerical modeling in order to calculate the correct values and attribute them to the right position in space; this process is called inversion and yields true resistivity values arranged in a true resistivity section or tomogram (Fig. 11.1).

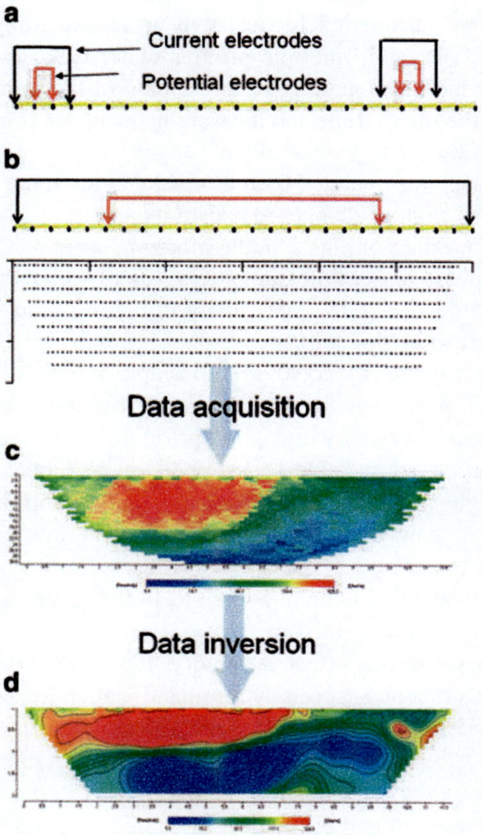

Fig. 11.1 Multi-electrode resistivity imaging data acquisition and processing; (**a**) liner array of electrodes with electrode positions for two quadrupoles at minimum spacing (*top*) and one quadrupole at maximum spacing (*bottom*). *Dots* represent electrodes; (**b**) spatial distribution of soil volumes to which resistivity values are attributed based on the hypothesis of soil homogeneity in 2-D acquisition. *Dots* represent the center of each volume; (**c**) Spatial distribution of soil apparent resistivity based on the hypothesis of soil homogeneity (2-D pseudo-section) obtained after data acquisition; (**d**) 2-D section of true soil resistivity obtained after data inversion with numerical modeling. (From Amato et al. 2010)

11.2 Applications of an Emerging Remote Sensing Technique for Roots

The first evidences of patterns related to the presence of roots within soil images have been found by researchers conducting multi-electrode field studies with other purposes. Electrical resistivity is affected by several permanent and dynamic soil properties – as reviewed in Chap. 11 – therefore it is used in field surveys for the mapping of soil and management-related features (Samouelian et al. 2005). Applications in agronomy and hydrology have shown areas of contrasting ρ below herbaceous crops (Panissod et al. 2001; Michot et al. 2003; Srayeddin and Doussan 2009) and trees (Hagrey and Michaelsen 2002). Hagrey et al. (2004) found a conductive volume surrounded by resistive areas extending to 0.4 m depth beneath a poplar in silt sand; multiple profiles under Oak also showed positive anomalies (areas of high ρ) departing from the base of the trunk and less evident with distance from the plant. Repeated measurements under Cork Oak in Mediterranean climate showed that ρ was always higher under trees, and its values increased as soil water was depleted by root uptake in the top 2 m. Field locations far from trees showed slighter changes in resistivity and lower absolute values at all times (Hagrey 2007). Zones of high ρ in the root-zone were interpreted as volumes of low water content due to water uptake or as volumes where the flux of electrical charges was slowed down due to "electrically insulating dead bark cell layers of root branches" (Werban et al. 2008).

A series of experiments explored relationships of images with actual root features, in order to discriminate the direct effect of roots on ρ from indirect effects due to water content caused by uptake. The main experimental systems and procedures are summarized in Table 11.1. Loperte et al. (2006) and Satriani et al. (2010) detected single woody roots of peach and apricot with high resolution lab measurements in containers with homogeneous soil materials, where 1–3-year-old plants had been just transplanted. The effect of roots was always resistive, causing an increase in ρ beyond 3,000 Ohm m and up to 14,000 Ohm m in sand, and up to about 450 Ohm m in sand–clay mixture.

Since many soil properties affect resistivity, the feasibility of using ρ-based imaging for root detection and especially for quantification in real systems requires that root effects predominate in the soil electrical response, or that the other soil features that affect current can be measured and their effect discounted. Therefore quantitative studies were undertaken comparing resistivity tomograms, root biometrics, and the relevant soil properties. The effect of roots was detectable but confounded by other soil factors such as water content, salinity, and stone fraction in a peach orchard of southern Italy as reported by Loperte et al. (2006) and Lazzari et al. (2008). In a natural soil the effect roots was dominating over that of other factors (Amato et al. 2008): an intense sampling scheme showed that spatial patterns of soil resistivity (Fig. 11.2a) corresponded to spatial distribution of root biomass (RD) (Fig. 11.2b) both in the horizontal and in the vertical dimensions. Multiple regression showed that RD and soil water content were related to soil resistivity,

Table 11.1 Multi-electrode imaging studies on root systems with ground-truth measurements on root properties. ES = spacing between electrodes

Imaging method	Soil–plant system	Effect	Ground-truth and comparison method	Statistical treatment	Range of ρ and RD	References
Surface 2-D ES = 0.02 m	1-year old apricot and 3-year old peach, containers (sand, clay, sand–clay mixture)	Single roots detected as resistive areas on the image	Whole root system extraction and qualitative comparison with image	n.a.	$1<\rho<14000$. $2<\rho<300$	Loperte et al. (2006), Satriani et al. (2010)
Surface 2-D and 3-D ES = 0.25 m L = 11.75 m	12-year old peach, field, sandy loam	Resistive effect of soil volumes with known root density	Soil coring with root (RD) and soil parameters. Statistical analysis	Multiple regression	$20<\rho<300$	Loperte et al. (2006)
Surface 2-D ES = 0.25 m L = 11.75 m	Mature stand of *Alnus glutinosa L.*, Dystri-Andic Cambisol, sandy-loam	Quantitative resistive effect of soil volumes with measured root density	Soil coring with root (RD) and soil parameters. Statistical analysis	Univariate logistic; multiple regression	$7<\rho<700$ $0<RD<329$	Amato et al. (2008)
Surface 2-D and 3-D ES = 0.25 m L = 11.75 m	12 year old peach, field, sandy loam	Quantitative resistive effect of soil volumes with known root density	Soil coring with root (RD, RLD) and soil parameters. Statistical analysis	Multiple regression	$20<\rho<300$	Lazzari et al. (2008)
Surface 2-D L = 3.5 m	*Populus_canadensis* Moench, Regosols, Silty loam	Single roots detected as resistive spots	Root identification on trench wall and qualitative comparison with image	n.a.	$150<\rho<1000$	Zenone et al. (2008)
Surface 3-D L × 1 = 3.0 × 3.8 Multi-temporal analysis with different soil water content	Pine forest *Pinus pinaster* Ait. and *Pinus pinea* L Albic Arenosols, Sand	Resistive effect and/or different percent variation in ρ with time of rooted volumes or single roots	Whole root system extraction with compressed air and laser scanning. Visual comparison with image	n.a.	$50<\rho<3000$	Zenone et al. (2008)
Surface and buried 2-D ES = 0.02 m L = 0.30 and 0.14 m	*Lupinus polyphyllus* LINDL.,	A conductive zone was identified within a larger	Extraction and sketching of total root system and visual	n.a.	$25<\rho<2800$	Werban et al. (2008)

(continued)

Table 11.1 (continued)

Imaging method	Soil–plant system	Effect	Ground-truth and comparison method	Statistical treatment	Range of ρ and RD	References
Multi-temporal analysis during water uptake	container, fine sand	zone of high root content	comparison with image. Measurement of soil water content			
Borehole 3-D ES = 0.01 m D = 0.19 m	*Medicago sativa L.*, container, silt loam	Quantitative resistive effect of soil volumes with measured root density	Soil coring with root (RD, RLD) and soil parameters. Statistical analysis	Univariate linear; multiple regression	$13<\rho<350$ 0<RD< 0.001	Amato et al. (2009)
Surface – borehole 3-D ES = 0.01 m D = 0.12 m	*Hibiscus rosa-sinensis L.*, *container sand*	A conductive zone was identified in volumes where the roots were placed	Extraction of total root system and visual comparison	n.a.	$30<\rho<500$	Petersen and Hagrey (2009)
Surface 2-D ES = 0.25 m L = 11.75 m	*Olea Europaea L, silt loam; Arundo plinii* Turra, *saline sand*	Resistive effect of soil volumes with measured root density	Soil coring with root (RD, RLD) and soil parameters. Statistical analysis	Student *t*-test	$10<\rho<160$ $24<\rho<300$	Amato et al. (2010)
Surface 2-D ES = 0.25 m L = 11.75 m Surface 3-D ES = 0.25 m L × l = 1.75 × 1.25 m	*Citrus spp, Olea europaea L.*, Typic haploxeralf fine, mixed thermic soil, Silt loam	Quantitative resistive effect of soil volumes with measured root density	Soil coring with root (RD, RLD) and soil parameters. Discrimination between fine and coarse roots. Statistical analysis	Univariate glm; multiple regression; anova	$10<\rho<460$ 0<RD< 0.173	Rossi et al. (2010)

L = length of electrode array; D = depth of borehole; ρ = soil electrical resistivity (Ohm m); RD = root biomass per unit soil volume (Mg m^{-3}); RLD = root length per unit soil volume; n.a. = not applied. Soil classification FAO-ISRIC-ISSS 2006. Soil parameters = volumetric water content, coarse fragments, texture, pH, paste electrical conductivity, organic matter, bulk density

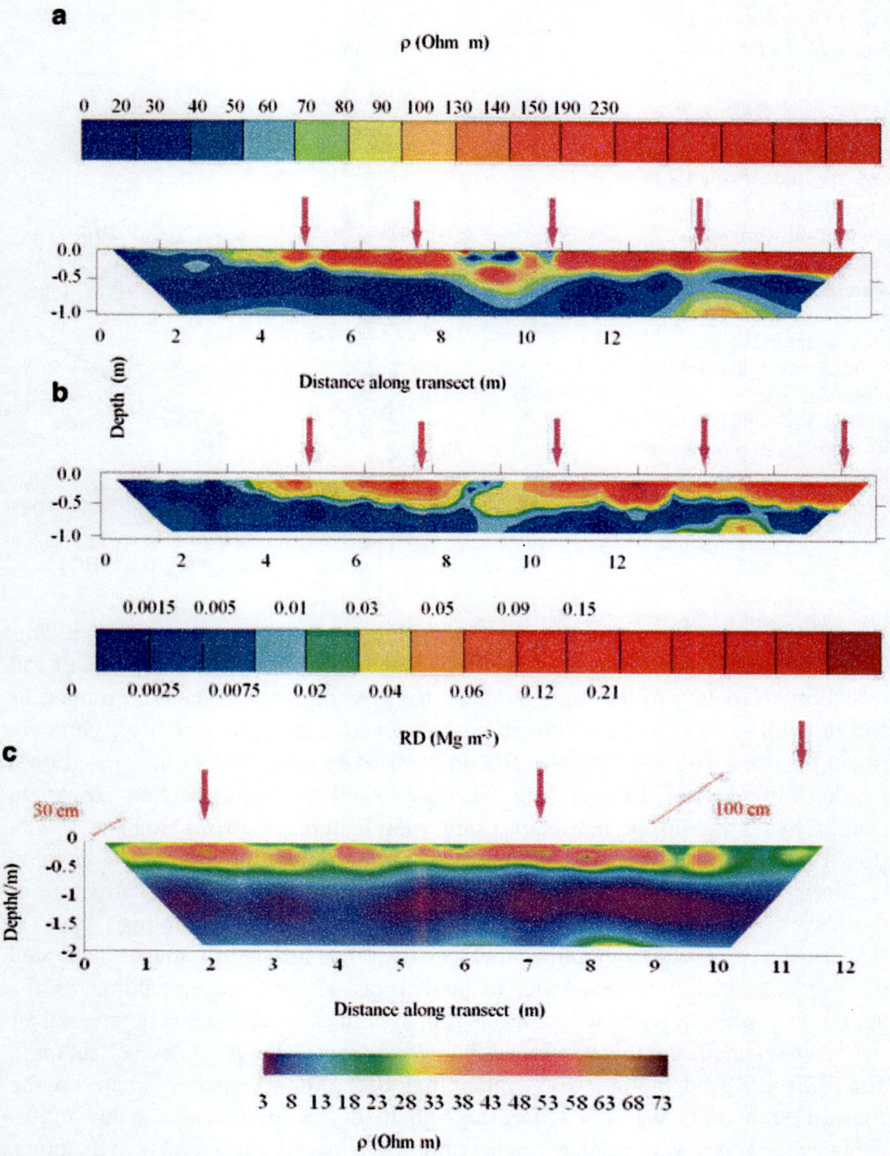

Fig. 11.2 Images from a transect under *Alnus Glutinosa L.* in southern Italy (from Amato et al. 2008). (**a**) soil electrical resistivity (ρ) tomogram and (**b**) root biomass (RD); (**c**) soil electrical resistivity (ρ) tomogram from a transect in an orchard at Ginosa-Italy from Rossi et al. (2010). *Arrows* mark the position of trees

but a very high proportion of the variability in ρ was explained by root biomass alone. The overall response to plant roots was resistive, since the presence and amount of roots were associated to an increase in ρ from tens of Ohm m in un-rooted soil to about 700 Ohm m in soil with high root biomass (Fig. 11.3). Nevertheless, at

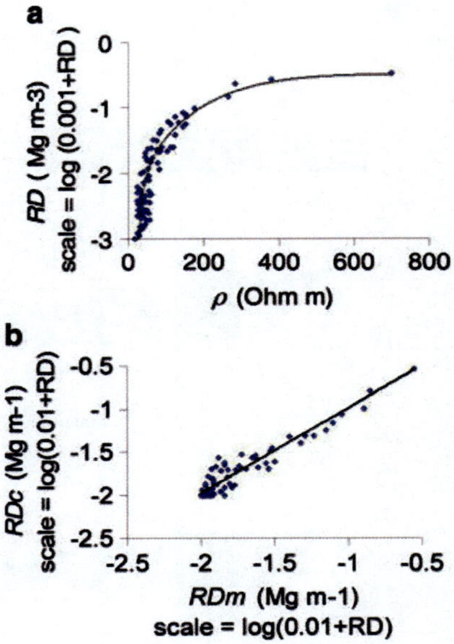

Fig. 11.3 (**a**) RD as a function of ρ for *Alnus glutinosa L.* Symbols: data points. Line: regression model RD = ((0.32-0.0464850) / (1+ exp((ρ-241.01)/-109.48)))-0.046485. $R^2 = 0.93$; (**b**) Root dry mass per unit soil volume calculated with the logistic regression model (RDc) versus measured RD (RDm) on a separate data set. Symbols: data points. Line: regression line RDc = 1.0017*RDm. $R^2 = 0.97$. From Amato et al. (2008)

low RD it was difficult to discern the effect of roots from the background noise. This was commented formulating the hypothesis that at low biomass the response of soil resistivity to roots is of the same order as that of variations in other soil properties and therefore too weak to be discriminated. Overall, though, a highly significant found for the single regression of RD and resistivity (Fig. 11.3a), and tested on a relationship separate data set (Fig. 11.3b) allowed to derive the first univariate quantitative relationship between ρ and root features with the logistic model (Fig. 11.3-caption). Quantitative results were very encouraging as a basis to develop resistivity-based methods for root detection or for bringing the spatial structure of root data to view, and provide a foundation for spatially sound sampling.

Zenone et al. (2008) ascertained that resistive spots in 2-D soil images generated in a poplar plantation corresponded to single roots. Three-dimensional images in a pine forest were compared with a whole root system excavated with compressed air and laser-scanned, and this confirmed that woody roots have a resistive effect on ρ. The soil was a sand, though, with a rather high background resistivity, therefore the contrast with roots was not strong enough to detect all structures, and multi-temporal analysis was needed to discriminate between roots and soil features. Differences in ρ between images acquired at different soil water content provided a good overlap with root positions. Also, roots with different direction and angle provided different resistivity increase due to their different interactions with water infiltration, redistribution, and drainage.

In natural soils, therefore, the effects of woody roots on ρ can be strong enough to give quantitative and qualitative relationships that can be the basis of root-detection methods. Agricultural soils pose a different challenge: management practices – like tillage, traffic, localized irrigation, and fertilization – introduce structured variation in soil properties that have an effect on ρ. This may make roots difficult to discern. Rossi et al. (2010) acquired 2-D and 3-D images under citrus and olive to explore the technique in agricultural soils, using indices of root biomass and length, and discriminating between coarse and fine roots (respectively collected and passing through a 2-mm sieve after soil dispersion). Both 2-D and 3-D images showed strong increase of the order of tens to hundreds of Ohm m getting closer to the base of plants (Figs. 11.2c and 11.4a,b,c). The effect of other soil variables was overshadowed by the presence of roots, therefore no significant multivariate relation was found: variability in ρ was related to RD with a highly significant relationship, due to coarse roots only and to values above 35 Ohm m. The lack of significant relation of ρ with anything else than below-ground biomass allowed the authors to attribute resistive volumes in the images to densely rooted soil, and to reconstruct the root system's spatial distribution in 3-D through a network of electrical resistivity iso-surfaces with $\rho = 35$ Ohm m (Fig. 11.4d, e, f).

Another Mediterranean orchard setting was explored in Amato et al. (2010) who found a significantly higher resistivity and RD close to vegetation than in adjacent

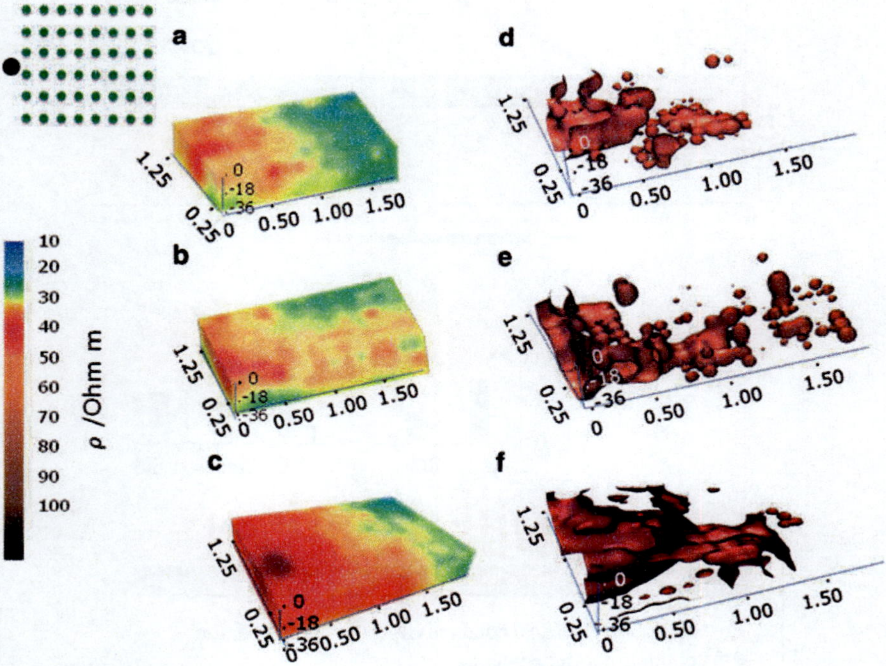

Fig. 11.4 Electrical resistivity 3-D images; (**a–c**) volume tomograms: (**a**) for an olive interrow grid; (**b**) for an olive row grid; (**c**) for a citrus row grid; (**d–f**) volume renders of densely rooted soil volumes identified by iso-surfaces at $\rho > 35$ Ohm m: (**d**) for olive interrow grid; (**e**) for olive row grid; (**f**) for citrus interrow grid. Units of all axes are meters. Top *left* corner: position of the electrodes (*green dots*) and tree trunk (*black circle*). Tree positions correspond to the middle of *y*-axes for all images. From Rossi et al. (2010)

areas with sparse or no vegetation under both olive and *Arundo plinii* Turra (Fig. 11.5).

The orchard experiments show that the response of soil resistivity to roots at low ρ and biomass was of the same order as that of variations in other soil properties and therefore too weak to be discriminated similarly to what recorded by Amato et al. (2008) in a natural soil. Another experiment was then planned with the aim of collecting results on a low biomass system such as a young herbaceous plant. The test was conducted in homogeneous growth medium and controlled temperature chamber, in order to minimize variation of all factors other than root biomass (Amato et al. 2009). A high resolution system was designed, with four electrode arrays buried in boreholes; 3-D images were reconstructed through the processing

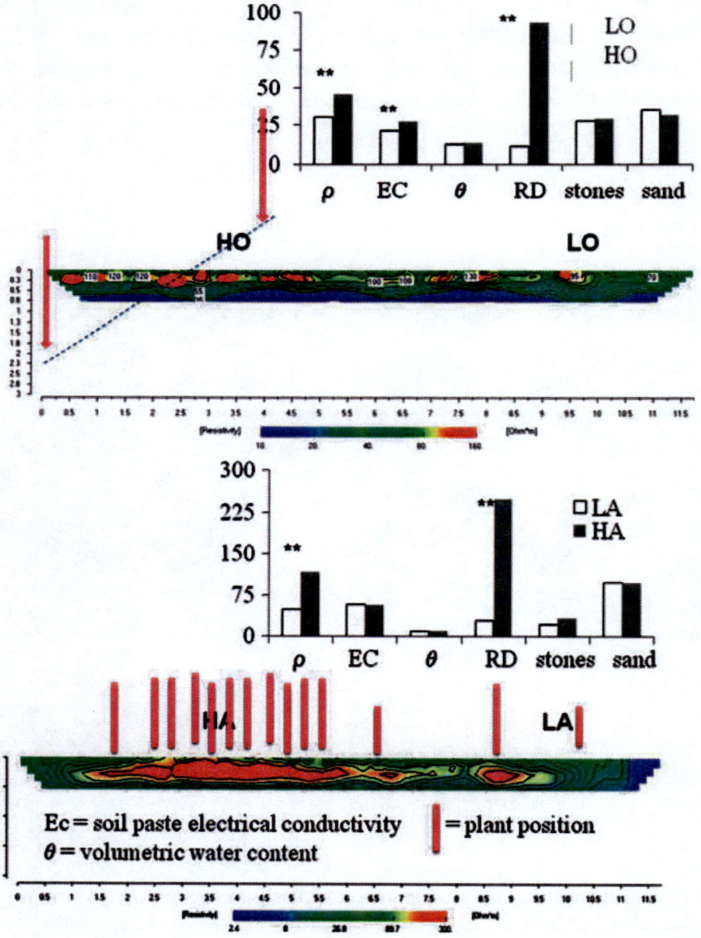

Fig. 11.5 Two-dimensional soil images and average values of soil properties in the canopy projection of an olive row (HO) and outside (LO) (*top*) and in *Arundo plinii Turra* dense vegetation (HA) and sparse vegetation (LA) (*bottom*). Labels indicate values of soil resistivity

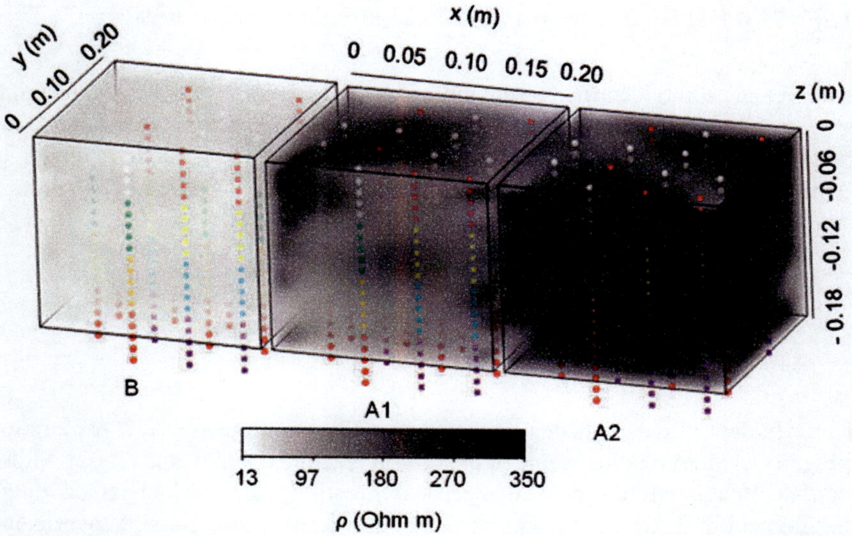

Fig. 11.6 Three-dimensional inverted electrical resistivity images up to 0.18 m of depth in containers equilibrated at the same temperature and water content for: (**b**) bare silt-loam soil; A1: silt-loam soil planted with *Medicago sativa L.*; A2: loam soil planted with *Medicago sativa L.* *Square-dotted lines*: soil moisture and root content sampling points. *Circle-dotted lines*: electrical conductivity sampling points. From Amato et al. (2009)

of data from all arrays. A bare container with a silt loam soil was compared with a treatment where alfalfa (*Medicago sativa* L.) was planted on the same soil and one where it was planted on a loam. Both the presence of roots and a higher content of sand resulted in higher soil resistivity (Fig. 11.6). Across all data roots and soil texture explained a high proportion of variability in soil resistivity. The pattern of ρ matched the spatial distribution of volumetric soil water content in bare soil, but in rooted soil the effect of roots was more important than that of water. Univariate linear relations were found between ρ and both root biomass and length. This allows to conclude that soil resistivity may be quantitatively related to root biomass even in herbaceous plants at low root density (biomass <0.001 Mg^{-3}), but the effect of roots is of the same order of magnitude as the effects of grain size and water content, therefore it is discernable only in case of homogeneous background soil. In field studies other factors may mask the effect of low-density roots, and the effect of variation in other soil properties should be explicitly addressed.

Another young herbaceous plant (*Lupinus polyphyllus* LINDL) was used by Werban et al. (2008) in a study of water content space–time dynamics where a zone of lower resistivity was identified at 4–5 cm depth. Root excavation and visual comparison with 2-D images revealed that the zone of highest root content corresponded to 4–7 cm of depth. Through a composite surface–borehole geometry Petersen and Hagrey (2009) also found a conductive soil volume in the soil region where roots of a young *Hibiscus rosa-sinensis L.* plant had been just transplanted.

11.3 Data Interpretation and Statistical Approaches

Zenone et al. (2008) identified future directions of research on geophysical methods for root studies in two areas: the improvement of standard field procedures and the development of statistical processing tools to relate below-ground biomass to geophysical parameters. This requires the analysis of a series of issues as reported in the following paragraphs.

11.3.1 Discrimination from Background Variation

The usefulness of data depends on their reliability: in all geophysical exploration ambiguity in interpretation is one of the limiting factors (Dahlin et al. 2002). Multi-electrode imaging relies on soil electrical resistivity data, which are a direct measurement of distinct but concurrent natural processes with possible overlaying effects that compromise image interpretation.

Electrical resistivity of earth materials ranges from units of Ohm m or less for saline water to hundred thousands Ohm m for rock minerals (Fig. 11.7), and the mixture of fine-earth materials in soils results in resistivity values of the order of tens of Ohm m (Panissod et al. 2001). Reviewing the relevant literature Amato et al. (2008) report that responses of the order of tens Ohm m are used to detect variations

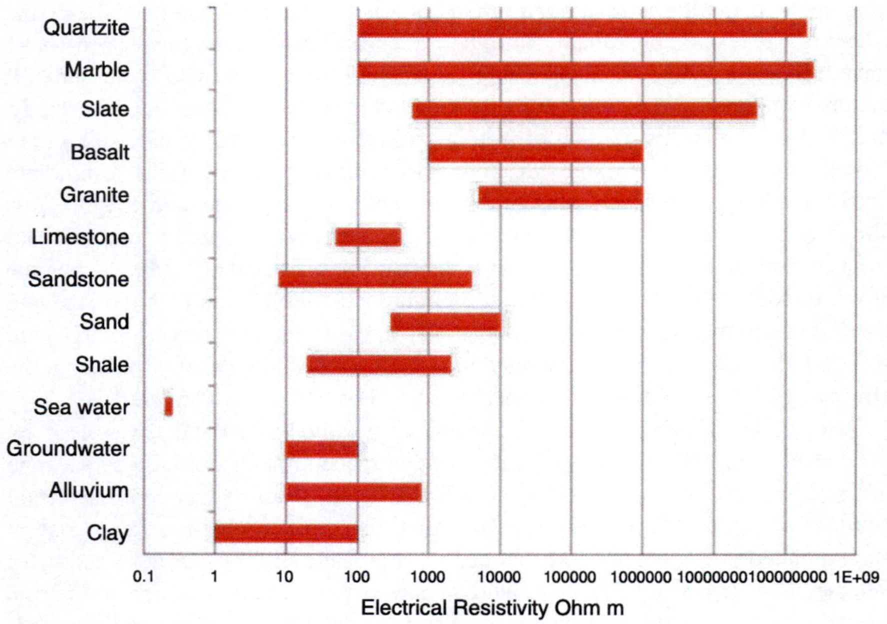

Fig. 11.7 Resistivity ranges for earth materials

in clay content, soil porosity, and other features linked to agriculture. Soil water content gives stronger responses of the order of tens to few hundreds Ohm m. Direct measurements of plant structures show a wide range of ρ values, since water-filled conductive volumes correspond to less than 50 Ohm m, whereas resistive lignified parts to tens of thousands Ohm m (Hagrey 2006, 2007), with an average of hundreds to thousands Ohm m. Research reviewed in Sect. 12.3 shows a highly resistive behavior of soil volumes affected by woody root colonization, with increases of the order of tens to several hundreds Ohm m, well compatible with reported ρ values for plant materials. Therefore the ability to detect roots is based on a strong response of resistivity compared to that of other soil features (Amato et al. 2008). The same electrical response, though, can be due to the possible overlap between synergic or opposite effects, therefore ground truth data should always be acquired: stone content and sand exhibit ρ values higher than those of average soil components and in the same range of values as roots; therefore a resistive soil volume may be due to a high concentration of roots but also to local increases in sand or stone content.

Imaging is not based on absolute values of resistivity but rather on variation of ρ in a given soil volume. Hence, no matter how important the effect of roots on ρ, it will give no readable image if the variability of other properties affecting ρ in the same range is larger. Although roots strongly affect electrical resistivity, therefore, many permanent and transient soil conditions may have an effect on the electrical behavior of soil, and this is a limitation of electrical imaging methods: in some field studies, the effect of soil water, stones, or texture was of the same order or stronger than that of roots (Lazzari et al. 2008; Zenone et al. 2008; Amato et al. 2009).

Success of electrical root imaging therefore requires that the effect of other soil properties is minimized, discounted, or included in models, and in each instance we ensure that ρ is a reliable predictor of root features.

11.3.2 *Choiche of Models of* ρ*: RD Relationships for Skewed Data*

Roots are branched hierarchical systems and as a consequence the frequency distribution of spatial root data is usually positively skewed (Pielou 1969) and heteroskedastic. The same behavior has been documented for ρ in rooted soil (Amato et al. 2008; Rossi et al. 2010) and has been commented as further evidence that roots may dominate the electrical behavior of soils in the rhizosphere. Deviations from normality and heteroskedasticity require an appropriate selection of statistical models for describing relationships of ρ and root biomass.

Where data have been gathered on a limited range of biomass values, the shape of the univariate ρ-RD relationship has been found to be linear (Amato et al. 2009), but if a wider range is explored the shape is a sigmoid (e.g., Figs. 11.4a and 11.8. Positively skewed data are shown in log scale). A logistic model was first chosen by Amato et al. (2008) to describe it: soil electrical resistivity was correlated with RMD, with a highly significant positive correlation. The regression model was

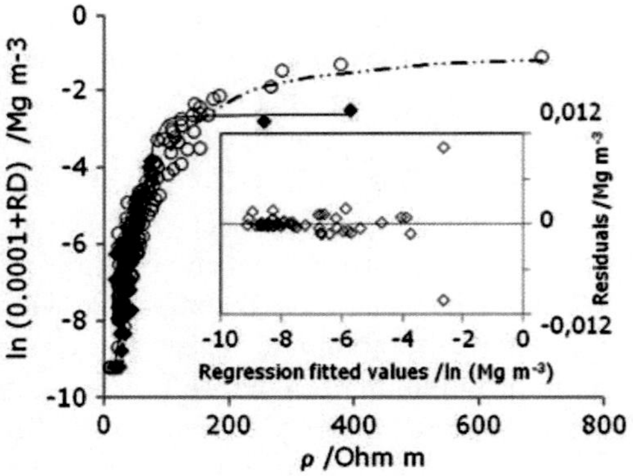

Fig. 11.8 (a) Root biomass per unit soil volume (RD) as a function of soil electrical resistivity (ρ) in two soil–plant systems. *Empty circles*: *Alnus glutinosa L*; *full diamonds*: olive and citrus in orchard; *grey dotted line*: logistic model from Amato et al. (2008): RD = ((0.32-0.0464850)/(1 + exp((ρ-241.01)/−109.48)))-0.046485; *black line*: three-parameter logistic model from Rossi et al. (2010): RD = (0.12409/(1+exp((81.33-ρ)/9.97743))) ($p < 2e^{-}16$); (b) residuals of the regression model as a function of fitted values for citrus. From Rossi et al. (2010)

tested on a separate data set (Fig. 11.4b) and predicted root biomass values (RDc) were found to be highly correlated with measured values (RDm) (Fig. 11.4b), and were not significantly different at the Wilcoxon Signed-Rank Test. Both logistic model and Wilcoxon Test were able to handle nonnormal heteroskedastic data.

An approach for data sets deviating from normality with the advantage of simplicity was introduced through the use of generalized linear models (glm) by Rossi et al. (2010). Generalized linear models include nonconstant variance and the model can be specified without prior knowledge of underlying distribution. They keep the simplicity of linear models (e.g., simple association between variables) but recur to the exponential family, allowing to choose between different variance–mean relationships. This greatly enhances model applicability since it includes different forms of heteroskedasticity and proves to be useful in all context in which the response distribution is difficult to be defined but mean–variance relations can be inspected graphically even with small data set. Just as in linear regression several approaches can be used to select the explanatory variables: individual variables may be dropped based on hypothesis testing or on model selection criteria like Akaike Information Criteria (AIC) that implies dropping individual variables and refit the model evaluating both goodness of fit and model complexity because it penalizes the increase of the number of predictor variables thus avoiding over-parameterization. The first insight on the relationship between response and predictor variables can be gathered through pairwise scatterplots and the computation of a correlation matrix that gives a measure of the co-variation between variables and hence gives a first indication of the presence of multicollinearity. To prevent multicollinearity a sub-selection of variables can be made using the variance inflation factors.

Rossi et al. (2010) analyzed ρ-RD data from a natural and an orchard system and in both they found that variance increased as a quadratic function of the mean, and the density plot showed right skewness. Such two characteristics together make the Gamma distribution an interesting alternative to data transformation. The Gamma distribution is used for continuous response variables that can only assume positive values, and in Gamma-glm the variance increases with the mean of the response variable (Faraway 2006).

Through their comparison of two different plant–soil systems Rossi et al. (2010) confirmed that ρ and RD are strongly related, but at large root biomass values the magnitude of the electrical response varied between systems and residuals were larger (Fig. 11.8); they therefore indicate that field procedures for estimating root biomass from ρ should include calibration at large biomass values.

11.3.3 Calibration

The goal behind the use of ancillary remote sensing data such as electrical resistivity is the possibility of using the sensor data (extensively acquired) as a proxy for the underlying soil attribute. This use requires that the relationship between the soil target attribute and the sensor data exists and is strong and consistent enough to be modeled through a regression approach. This may be true for root biomass as reviewed in Sects. 12.3 and 12.4. While model structure can be specified a priori, the estimate of model parameters always requires calibration through site specific sampling design. When a calibration procedure is adopted two issues must be addressed: how many samples are needed and where samples must be located in order to optimize the estimation of model parameters but at the same time keep survey costs low.

Given the skewed distribution of root and ρ data in the root-zone, random or regular schemes may be applied only with large numbers of sampling locations, otherwise the upper part of the range would have very low probability of being sampled because of its low frequency. In this case, important observations would be overlooked, and specifically in the range where calibration is most needed (see Sect. 12.4.2). For the calibration of multi-electrode data, the selection of sampling points should therefore be based on images with the surface response sampling (SRS) method. The goal of SRS is to select the smallest set of sampling locations for model parameters estimation and at the same time keep the model simplicity inherent to ordinary regression by eliminating the effects of spatially dependent error structure. The development of a SRS design follows a four step procedure described by Lesch (2005). In this approach the sampling location are selected based on the observed magnitude and spatial distribution of the covariate sensor data (i.e., resistivity image, conductivity map, yield map). Sampling locations are also chosen within distances that minimize the possibility of generating spatially dependent regression model residuals; this latter precaution allows the use of ordinary regression models instead of more complex spatial linear models or geostatistical models (Fitzgerald et al. 2006).

11.4 Special Configurations and Technical Adaptations for Root Studies

Technical bases of electrical resistivity tomography are the object of Chap. 11. Applying multi-dimensional geo-electrical techniques to root studies poses a series of methodological issues compared to traditional applications: geophysical techniques were developed for deep investigation and at a resolution of meters and over, while the exploration of root systems requires near-surface measurements and at a resolution of decimeters and less. Soil variability within the root volume includes a marked gradient of temperature and moisture conditions in a relatively short depth, associated with a high temporal variability of both properties. Moreover, other soil parameters that influence the electric signal (texture, salinity, bulk density) often exhibit large variability in the surface soil layers within short distances along both vertical and horizontal directions.

Also, root features relevant for ecology and management represent a continuum in soil therefore methods are not only required to identify single features by contrast with a background, but also to discern between levels. Therefore, in addition to qualitative evidence, quantitative information is needed. Applications for root methods therefore need adaptations and integration with auxiliary measurements.

Furthermore, root systems and processes are often studied in containers, therefore techniques need to be adapted to the needs of small surface but big depth and higher resolution.

11.4.1 Electrode Spacing, Resolution and Depth

Zhou and Dhalin (2003) described the effects of measurements error on the inversion process dividing the source of error that can possibly rise in field work in two types: errors related to inaccuracy in the electrode spacing and errors correlated with the magnitude of the observed potential. The latter arises for different reasons: a bad electrode contact, cable insulation damage, site background noise (telluric current and power line noise) and therefore may deviate at different sites, different times and with different data acquisition systems. It is also linked to electrode configuration. Some electrode arrays characterized by longer potential electrode distances (such as pole–dipole arrays) can be expected to be more prone to pick up noise. Geometry and spacing of the conductors used for current injection and detection determine spatial resolution, and imaging procedures may affect artifact projection as treated in Chap. 11.

One of the advantages of electrical techniques is flexibility in resolution, which can be easily modulated by varying the distance between electrodes from millimeters to meters.

Reducing the distance corresponds to increasing resolution but implies reducing the spatial coverage of the technique for a given number of electrodes, or increasing the time of image acquisition if more electrodes are used. Resistivity meters have a fixed number of channels therefore the maximum number of electrodes that can be practically managed in one acquisition is fixed. Tradeoffs between resolution and acquisition time for near-surface information are discussed by Petersen and Hagrey (2009).

The depth of measurement of a noninvasive imaging method with electrodes implanted at the soil surface is a fraction of the total length of the electrode array used in one acquisition or of the distance between the extreme electrodes (e.g., maximum depth = 1/5 of the length in some of the most commonly used configurations). Also, there is a loss of sensitivity and resolution with depth (Slater et al. 2000). For instance Srayeddin and Doussan (2009) using a surface array of 9.3 m in length could reliably track changes in water content that were quantitatively related to root uptake down to 60 cm only, while at greater depth the progressing water depletion could not be assessed. Longer arrays can be used in field settings; for instance Amato et al. (2008) reliably measure root biomass down to 1 m with an array of 11.75 m. In container studies, though, the available length for surface arrays (e.g., pot diameter or container length and width) is usually limited and does not allow to cover the desired depths. In a pot of 30 cm diameter Werban et al. (2008) were able to place a 30-cm surface array attaining only about 7 cm of depth coverage, and it was necessary to bury a second horizontal array in order to increase both depth and resolution within the root-zone. This setup, though, disturbs the soil region to be explored. Amato et al. (2009) approached the issue by using a borehole setup: four linear arrays of electrodes spaced at 1 cm were buried vertically in boreholes at the outer edges of the soil volume to be explored. The borehole technique allows to reach the whole desired depth even if the available pot surface is limited, and the same resolution is kept throughout the depth. Nevertheless, there is a loss of resolution with increasing distance from the arrays, therefore the targeted soil volume must not be too far from electrodes. One single borehole array yields 2-D images similar to surface configurations, whereas data from 2 or more boreholes may be combined (borehole–borehole) to increase coverage or to give 3-D images.

Limitations of this method are:

- Increased invasiveness compared to surface arrays, even though the soil volume object of study is not disturbed
- The need to insert appropriate materials in the hole to provide electrical contact between electrodes and the soil without excessively affecting measurements
- Time of data processing if multiple borehole arrays are used
- More time and skills required for installing electrodes

Based on modeling and pot measurements with micro-electrodes of 2–3 mm length, Petersen and Hagrey (2009) concluded that the highest overall resolution is attained through a combination of surface plus multiple boreholes since this set combines the advantages of surface surveys (high resolution of the upper boundary) and borehole–borehole surveys (mapping of the lower boundary of the root-zone).

11.4.2 Temperature Correction

Image acquisition is affected by temperature because ρ decreases about 2.02% per °C (Campbell et al. 1948). This effect may be overlooked in depth applications of geo-electrical techniques, but root-zone measurements are conducted in near-surface conditions where daily and seasonal temperature variations and gradients along the profile are large. Samouelian et al. (2005) recommend temperature correction through Campbell's equation (Campbell et al 1948) reviewing the literature on ρ variations with season, especially in coarse-grained soils. Amato et al. (2008) argue that temperature correction is needed for quantitative studies, reporting for a profile with thermal gradient as little as 2.6°C, difference in resistivity between top and bottom amounting to 80.9% of the bottom resistivity without temperature correction, and to 93.4% with temperature correction, with the consequent deformation of their RD–ρ relationship.

Minimizing the effects of temperature may be achieved by performing measurements at the same time of the day for daily time-sequences, or after equilibrating pot's temperature in controlled environment (Werban et al. 2008; Amato et al. 2009).

11.5 Open Issues

The interpretation of electrical resistivity images has advanced in discriminating between effects of roots and other factors, on the basis of statistical analysis and measurement of multiple parameters. Other approaches need to be adopted in order to exploit the full potential of the technique. Among them, pattern recognition and the study of mechanisms underlying the effect of plant roots on soil electrical resistivity, in order to provide the explanation of phenomena and to lay the basis for forward modeling of the root–soil interface electrical behavior. Plant tissues are structured systems made of anisotropic arrays of cells, intercellular spaces, and vascular elements. Cell walls have a large resistivity (Aubrecht et al. 2006): their multilayered structure contributes to their very large resistance on current flux unless the cell is seriously injured. Since water in the intercellular spaces has poor mobility, the only conducting structures in living plants are the pathways for ion movement in the vascular system, the water–ion uptake paths, and to a lesser extent the cell symplast. The proportion of water–ion conducting paths in a plant structure decreases with age, therefore in a large woody structure it can be quite small and resistive regions prevail (Hagrey 2006, 2007). Extremely high resistivities have been attributed to bark cells of the outer rings of root branches (Werban et al. 2008). Electrical resistivity is therefore highly variable within a plant tissue, and anisotropic: conduction along the plant axes of currents generated within the plant or at the plant surface is not comparable with conduction of electrical charges travelling in the soil and encountering external root layers and the soil–root interface.

Root effects have been proven to be strong enough to provide a quantitative response of ρ even in agricultural soils. The response is resistive and linked to root biomass and not length, and where root classes were discriminated the response was linked to coarse roots only. Results on fine roots and low biomass have shown no effect in some cases (Rossi et al. 2010), or a detectable and quantifiable resistive effect in cases where other soil condition was minimized (Amato et al. 2009). Even evidence of conductive volumes has been reported within root systems which were presumably largely composed of fine roots (Lazzari et al. 2008; Werban et al. 2008; Petersen and Hagrey 2009), although the spatial extent of conductive and rooted volumes were not completely coincident, or not measured, and the effect was not tested statistically nor was the influence of other factors discounted. Therefore we cannot draw definite conclusions as to whether or to what extent the conductive volume was due to fine roots or other soil properties independent on roots, or linked to root activity in such experiments.

Even for woody roots, though, effects need to be more thoroughly explained. Woody plant materials have large values of resistivity (Hagrey 2007), but Rossi et al. (2010) argue that:

- Only in some cases roots occupy a soil volume large enough that their effect may be viewed as simply due to the quantitative contribution of resistive material to soil
- The relationship of RD and ρ is not a straight line, hence there is no proportionality. Therefore the response of ρ to the presence of roots cannot be explained with their quantitative contribution only, but it must be due to other mechanisms that affect the flux of charges.

Rooted soil may be viewed as a complex of conductive soil volumes interacting with a resistive root matrix which is continuous and branched, and may be able to redirect and slow-down electrical charges because of its spatial arrangement more than to the actual occupied volume. Gaps at the soil–root interface are more likely to be found around large root structures (Carminati et al. 2009) and would also be efficient in interrupting the flux of electrical current. The lack of significant effect of fine roots in this experiment may be because of the small soil volume that they occupy or to their lower ability to represent efficient barriers to the flux of electrical charges compared with large structures. Also, fine root surfaces are the site of intense rhizosphere activities such as exudation, uptake, and microbial processes which may create local conditions of large conductivity linked to a greater concentration of electrolytes and electrically active surfaces (fine roots, bacteria, etc.).

11.6 How Can Multi-electrode Resistivity Imaging Contribute to Root Studies?

Resistivity tomography provides many opportunities for the advancement of root research, allowing in situ detection and quantification of single roots, and root/soil spatial variability at different scales from field to container settings.

11.6.1 Root Measurements

Multi-electrode images provide both qualitative and quantitative information, since a range of ρ has been proven to be related to the concentration of root biomass in soils, but further research is needed to explore the response of fine roots and elucidate mechanisms of the electrical behavior of rooted soils. The method can be applied in two and three dimensions, and being nondestructive the possibility of exploring a fourth dimension is granted through repeated measurements in time. This opens new alleys for studying root–soil interaction, since the technique is also sensitive to changes in other soil components like water, and Zenone et al. (2008) suggest that multi-temporal 3-D electrical imaging can be used in combination with changes of soil humidity to create "a sort of tracer experiment" to reconstruct root distribution and interaction with soil. Nevertheless, other soil features may interfere with measurements and especially low biomass values may be undetectable in variable soils. The method requires field calibration.

Acquisition of images is by far faster than any other field method for roots, of the order of minutes depending on the number and configuration of electrodes. As a consequence, time-scales of repeated measurements may range from minutes–hours to years or more. Roots are detected independent of their position, and the technique is flexible, but increasing spatial coverage decreases resolution so that single fine roots are not detectable, and sharp interfaces are smeared by data processing. Increasing resolution space coverage and depth are reduced, unless more invasive applications, like borehole, are used.

The strong response of ρ to roots also implies that below-ground plant parts can interfere with resistivity surveys conducted with other purposes.

Multi-electrode techniques allow to extract quantitative data and to visualize them with different restitution techniques, from tomograms in 2-D (e.g., Fig. 11.2) and 3-D (e.g., Fig. 11.4a, b, c) to renders of most densely rooted volumes through iso-surfaces (e.g., Fig. 11.4d, e, f).

11.6.2 Detection of Soil Features that Are Relevant for Roots

Electrical methods provide information about other soil components in addition to or instead of roots. This is considered one of the features that makes the method appealing for coupling with other techniques which do not provide soil information, like GPR (Zenone et al 2008; Vianden et al 2010). Also, in cases where 4-D measurements are made, the technique may provide information on both root structures and processes like water uptake and leaching.

Even when roots are not directly detectable because other soil features dominate electrical soil behavior, other properties related to the presence, spatial distribution, and functionality of roots may be mapped through electrical tomography, and indirectly provide information on root systems and below-ground interactions. This is mainly obtained through images of soil water uptake patterns providing a scale of view that is larger and multi-dimensional, and therefore more representative of field scale behavior than traditional techniques (Srayeddin and Doussan 2009). Also, information about deep patterns of root–water interactions may be obtained with this method in conditions where no other method would allow it, as concluded by Nijland et al. (2010) in a shallow rocky soil, where evidence of root water extraction down to 6 m and below allowed inference about deep root penetration in the bedrock.

11.6.3 Spatially Sound Sampling for Root Studies, Soil Studies or Calibration. Support for Differential Soil Management

Multi-electrode imaging provides the position of soil volumes with different root biomass and/or different soil properties, and therefore it represents a technique for exploring the spatial structure of soil or root data.

One of the advantages of resistivity imaging methods is speed. Amato et al. (2008) reported a total time of less than 60 min for data acquisition and inversion. Zenone et al. (2008) reported only 15–30 min for a 3-D block. An image available in the field within this short time is ideal as a guide for spatially sound sampling of root and root-related features or soil features that may be relevant for root studies and are useful for local calibration, or in general for spatial description.

This provides the basis for a revolution in approaches to root sampling and process description and fills a gap in root methods: providing a tool to unravel the spatial structure of data allows to design appropriate sampling strategies and to increase the efficiency and reduce the disturbance linked to sampling. The increasing use of remote sensing surrogate data for spatial prediction of soil properties has progressively introduced the use of prediction-based sampling strategies that are in fact purposely focused toward model estimation like SRS (Sect. 12.5) and make soil images the foundation of efficient sampling schemes. This is crucial for cases where there is a constraint in the number of samples (e.g., root biomass) that can be collected or locations where detailed tests (e.g., soil gas exchange) can be conducted because of disturbance, time, labor, cost, or other reasons. In such cases data sets are generally unsuitable for geo-statistical description and co-kriging techniques for spatial prediction due to the small number and spatial dispersion of samples or calibration points.

Quick spatial information on root and soil properties can also be used as inputs for differential soil management strategies.

References

Al Hagrey SA (2006) Electrical resistivity imaging of wooden tree trunks. Near Surf Geophys 4:177–185

Al Hagrey SA (2007) Geophysical imaging of root-zone, trunk, and moisture heterogeneity. J Exp Bot 58(4):839–854

Al Hagrey SA, Michaelsen J (2002) Hydrogeophysical soil study at a drip irrigated orchard, Portugal. Eur J Environ Eng Geophys Soc 7:75–93

Al Hagrey SA, Meissner R, Werban U, Ismaeil A, Rabbel W (2004) Hydro- bio-geophysics. The Leading Edge 23:670–674

Amato M, Ritchie JT (2002) Spatial distribution of roots and water uptake of Maize (Zea mays L.) as affected by soil structure. Crop Sci 42(3):773–780

Amato M, Basso B, Celano G, Bitella G, Morelli G, Rossi R (2008) In situ detection of tree root distribution and biomass by multielectrode resistivity imaging. Tree Physiol 28:1441–1448

Amato M, Bitella G, Rossi R, Gómez JA, Lovelli S, Ferreira Gomes JJ (2009) Multi-electrode 3-D resistivity imaging of alfalfa root zone. Eur J Agron 31:216–222

Amato M, Rossi R, Bitella G, Lovelli S (2010) Multielectrode geoelectrical tomography for the quantification of plant roots. Ital J Agron 3:257–263

Aubrecht L, Stanik Z, Koller J (2006) Electrical measurement of the absorption surfaces of tree roots by the earth impedance method: 1 Theory. Tree Physiol 26:1105–1112

Campbell RB, Bower CA, Richard LA (1948) Change in electrical conductivity with temperature and the relation with osmotic pressure to electrical conductivity and ion concentration for soil extracts. Soil Sci Soc Am Proc 13:33–69

Carminati A, Vetterlein D, Weller U, Vogel HJ, Oswald SE (2009) When roots lose contact. Vadose Zone J 8:805–809

Dahlin T, Leroux V, Nissen J (2002) Measuring techniques in induced polarisation imaging. J Appl Geophys 50:279–298

FAO-ISRIC-ISSS (2006) World reference base for soil resources. World soil resources reports. Food and Agriculture Organization of the United Nations, Rome

Faraway JJ (2006) Extending the linear model with R. Chapman Hall/CRC, Boca Raton, FL

Fitzgerald GJ, Lesch SM, Barnes EM, Luckett WE (2006) Directed sampling using remote sensing with a response surface sampling design for site-specific agriculture. Comput Electron Agric 53:98–112

Lazzari L, Celano G, Amato M, Al Hagrey SA, Loperte A, Satriani A, Lapenna V (2008) Spatial variability of soil root zone properties using electrical imaging techniques in a peach orchard system. Geophys Res Abstracts 10, SRef-ID: 1607-7962/gra/EGU2008-A-11689

Lesch SM (2005) Sensor-directed response surface sampling designs for characterizing spatial variation in soil properties. Comput Electron Agric 46:153–179

Loperte A, Satriani A, Lazzari L, Amato M, Celano G, Lapenna V, Morelli G (2006) 2D and 3D high resolution geoelectrical tomography for non-destructive determination of the spatial variability of plant root distribution: Laboratory experiments and field measurements. Geophys Res Abstr 8: 06749, Wien

Michot D, Benderitter Y, Dorigny A, Nicoullaud B, King D, Tabbagh A (2003) Spatial and temporal monitoring of soil water content with an irrigated corn crop cover using electrical resistivity tomography. Water Resour Res 39:1138

Nijland W, van der Meijde M, Addink EA, de Jong SM (2010) Detection of soil moisture and vegetation water abstraction in a Mediterranean natural area using electrical resistivity tomography. CATENA 81(3):209–216

Panissod C, Michot D, Benderitter Y, Tabbagh A (2001) On the effectiveness of 2D electrical inversion results: an agricultural case study. Geophys Prospect 49:570–576

Petersen T, Al Hagrey SA (2009) Mapping root zones of small plants using surface and borehole resistivity tomography. The leading edge 7:1220–1224

Pielou EC (1969) An introduction to mathematical ecology. Wiley, New York

Rossi R, Amato M, Bitella G, Bochicchio R, Ferreira Gomes JJ, Lovelli S, Martorella E, Favale P (2010) Electrical resistivity tomography as a non-destructive method for mapping root biomass in an orchard. Eur J Soil Sci 62:206–215

Samouelian A, Cousin I, Tabbagh A, Bruand A, Richard G (2005) Electrical resistivity survey in soil science: a review. Soil Till Res 83:173–193

Satriani A, Loperte A, Proto M, Bavusi M (2010) Building damage caused by tree roots: laboratory experiments of GPR and ERT surveys. Adv Geosci 24:133–137. doi:10.5194/adgeo-24-133-2010

Slater L, Binley AM, Daily W, Johnson R (2000) Cross-hole electrical imaging of a controlled saline tracer injection. J Appl Geophys 44:85–102

Srayeddin I, Doussan C (2009) Estimation of the spatial variability of root water uptake of maize and sorghum at the field scale by electrical resistivity tomography. Plant Soil 319:185–207. doi:10.1007/s11104-008-9860-5

Tabbagh A, Dabas M, Hesse A, Panissod C (2000) Soil resistivity: a non-invasive tool to map soil structure horizonation. Geoderma 97:393–404

Vianden MJ, Weihs U, Kuhnke F, Rust S (2010) Non-destructive tree root detection with geophysical methods in urban soils. Geophys Res Abstr 12:EGU2010-3963-3

Werban U, Al Hagrey SA, Rabbel W (2008) Monitoring of root-zone water content in the laboratory by 2D geoelectrical tomography. J Plant Nutr Soil Sci 71:927–935

Zenone T, Morelli G, Teobaldelli M, Fischanger F, Matteucci M, Sordini M, Armani A, Ferrè C, Chiti T, Seufert G (2008) Preliminary use of ground-penetrating radar and electrical resistivity tomography to study tree roots in pine forests and poplar plantations. Funct Plant Biol 35:1047–1058

Zhou B, Dahlin T (2003) Properties and effects of measurement errors on 2D resistivity imaging surveying. Near Surf Geophys 1:105–117

Chapter 12
Using Ground-Penetrating Radar to Detect Tree Roots and Estimate Biomass

John R. Butnor, Craig Barton, Frank P. Day, Kurt H. Johnsen, Anthony N. Mucciardi, Rachel Schroeder, and Daniel B. Stover

Abstract Ground-penetrating radar (GPR) is a nondestructive means of detecting buried objects with electromagnetic waves. It has been applied to detect coarse woody roots, estimate biomass, root diameter, and spatial distribution of roots. This chapter discusses the development of root assessment techniques, basic methodology, and examples of field applications where GPR was successful.

12.1 Introduction

Many root quantification methods are destructive (e.g., soil cores, pit, whole-plant excavation) and prohibit repeated measurements over time. Destructive sampling yields high quality data, though it is expensive and labor intensive. Frequently, this leads to small sample sizes that may not adequately capture spatial variability and lack statistical power to reveal subtle treatment differences. For those who study

J.R. Butnor (✉)
USDA Forest Service, Southern Institute of Forest Ecosystems Biology, Southern Research Station, 705 Spear Street, South Burlington, VT 05403, USA
e-mail: jbutnor@fs.fed.us

C. Barton
Forest Resources Research, NSW DPI, PO Box 100, Beecroft, NSW 2119, Australia

F.P. Day • R. Schroeder
Department of Biological Sciences, Old Dominion University, Norfolk, VA 23529, USA

K.H. Johnsen
USDA Forest Service, Southern Institute of Forest Ecosystems Biology, Southern Research Station, 3041 Cornwallis Rd, Research Triangle Park, NC 27713, USA

A.N. Mucciardi
Tree Radar, Inc., 512 Ashford Road, Silver Spring, MD 20910, USA

D.B. Stover
Earthwatch Institute, North America Regional Climate Center, Edgewater, MD 21037, USA

S. Mancuso (ed.), *Measuring Roots*,
DOI 10.1007/978-3-642-22067-8_12, © Springer-Verlag Berlin Heidelberg 2012

belowground productivity, root dynamics or root architecture, the possibility of measuring root parameters nondestructively is tantalizing, but may seem out of reach. Examples of nondestructive root quantification methods include: radioactive tracers, electrical resistance/conductance tomography, and ground-penetrating radar (GPR) (Kroon and Visser 2003; Johnston et al. 2004; Pierret et al. 2005; Butnor et al. 2008). GPR is particularly attractive since it utilizes off-the-shelf equipment that is highly portable, rapid, and can be used to scan multiple kilometers of transects per day. GPR has been shown to be a reliable means of detecting roots and estimating root biomass under suitable soil and surface conditions (Hruska et al. 1999; Cermak et al. 2000; Butnor et al. 2001; 2003; Barton and Montagu 2004; Dannoura et al. 2008; Hirano et al. 2009). To date, GPR has been employed to investigate belowground productivity in woody plant ecosystems and used to assess the influence of species, genetic crosses (Johnsen Kh Major and Maier 2003), elevated atmospheric CO_2 (Stover et al 2007), forest management practices (Butnor et al. 2003, 2005; Samuelson et al. 2008), and tree spacing (Samuelson et al. 2010) on belowground carbon allocation. Although relevant to the discussions in this chapter, fine details on the principles of GPR are beyond the scope of this chapter and described well elsewhere (e.g., Jol 2009; Daniels 2004). Therefore the purpose of this chapter is to describe how GPR has been used to detect roots, quantify root parameters including biomass and address methodological solutions for studies of root dynamics.

12.2 GPR Antenna Basics for Root Analysis

GPR antennas are ultra wide band (UWB) devices that propagate short pulses of electromagnetic energy into the ground and receive reflected signals on the soil surface. Each pulse consists of a spectrum of frequencies whose average value is the center frequency. As a rule of thumb low frequencies (<200 MHz) have deep penetration, but can only resolve large objects or features. While mid (200–1,000 MHz) and high (>1,000 MHz) frequency antennas cannot penetrate as deeply, they are able to resolve smaller objects. There is a tradeoff between resolution and penetration, but the nature of UWB antennas allows for a range of object sizes and depths to be detected. Whenever a pulse contacts an interface separating layers with different electrical conductance, a portion of the energy is reflected back to a receiver on the surface. The electromagnetic material property that creates the contrast and causes reflections is the dielectric (ε), which is a dimensionless quantity relating to a material's behavior when subjected to an electric field. The larger the difference between the dielectrics of two different materials, the stronger the radar wave reflection and vice versa. For example, the dielectric of water is 81 and that of an average soil is approximately 13, which produces a "dielectric contrast" of 6:1 which is large and will cause most of the radar wave energy to be reflected back to the surface antenna.

In the case of roots, the conductance of living, well-hydrated roots contrasts with drier soils. Each returning pulse is recorded as a waveform that oscillates between positive and negative phases, recording timing of phase changes and amplitude. Attributes of the buried roots are then modeled from waveform data. Roots that are dead or diseased will have lower moisture content and can be weak or nonreflective targets and usually undetectable. This trait is used in an inferential way to determine root health in urban root inspections. Even roots located in high water table soils are may be detectable (Gormally et al. 2010). Although living roots are detectable, it may only be possible to estimate a bulk property of the roots structure such as root density. It is important to understand that there are cases where living roots cannot be distinguished from dead, such as burnt logging debris and severed palmetto rhizomes which were still detectable 5 years after disturbance (Butnor et al. 2005).

For most root surveys, an antenna is connected to a survey wheel that meters out pulses according to distance traveled, usually 100–200 scans per meter. To collect data, the antenna is dragged along a transect maintaining close contact with the soil (Fig. 12.1). Air gaps of more than a few centimeter can impair signal propagation, especially with high-frequency antennas. Data are usually collected in linear transects or series of parallel transects and grid patterns (Fig. 12.1b, c), to give better spatial resolution.

Fig. 12.1 Photo of GPR unit and survey cart able to accommodate 400, 900 and 1.500 MHz antennas (**a**), close-up of a 1,500 MHz antenna (**b**) mounted on a sliding rail system (**c**) designed to collect closely spaced transects in a 2 m by 2 m grid. The arrow points to the location of the antenna

Creating representative root maps that show root density and individual roots that overlap horizontally, bifurcate, merge (root grafting), etc., requires considerably more effort and specialized offline signal analysis software. Properties of the GPR wave are used to "disentangle" the root mass and provide data to create detailed morphology maps. The key to this capability involves an understanding of the interaction between wavelength (λ) generated by the antenna, the concept of wavelength (λ), and the dielectric constant of the medium:

$$\lambda_c = V_{air}/f_c * (\varepsilon)^{1/2}, \tag{12.1}$$

where V_{air} is the velocity of the radar wave in air (30 cm/ns), f_c is the center frequency of the radar antenna (900 MHz penetrates up to 1 m, 1,500 MHz 0.3–0.6 m in most soils), ε (dimensionless), and λ_c is the wavelength at the antenna's center frequency. The wavelength depends upon both the antenna being used and the soil's electromagnetic capability and decreases when either the center frequency or the medium dielectric increase. GPR sensors have appreciable energy in a wide frequency range of approximately ±50% about the center frequency, f_c:

$$f_c - 0.5f_c \leq \text{Frequency Range of an UWB Radar Antenna} \leq f_c + 0.5f_c, \tag{12.2}$$

which means that the frequency range for a 900 MHz antenna extends from 450 to 1,350 MHz. Since there is a wavelength associated with each frequency in this range as per (12.1), the radar wave in the soil also contains a range of wavelengths with the minimum wavelength determined by the maximum frequency as per (12.2).

$$\lambda_{min} = V_{air}/1.5f_c * (\varepsilon)^{1/2} = 0.667\lambda_c. \tag{12.3}$$

This makes it possible to detect subsurface targets smaller than the diameter computed only from the antenna's center frequency (which is the usual practice). Root detection probabilities can be determined using the resolution (R) of the GPR antenna being used:

$$R \approx \lambda/3 = 0.333\lambda. \tag{12.4}$$

Resolution is determined by: (1) size – the smallest diameter target that can be detected and (2) space – the smallest distance between two targets that can be identified. The probability of detecting small targets and to separate individual targets within a cluster of targets increases as R decreases, which happens when λ decreases. So, from (12.1) and (12.3) it becomes apparent that R reaches a minimum, for a given f_c, as:

$$R_{min} \approx \lambda_{min}/3 = 0.222\lambda_c. \tag{12.5}$$

Consequently using an antenna with the highest possible center frequency, consistent with appropriate depth of penetration (since the depth of penetration varies inversely with increasing frequency), is a key component for roots detection and mapping. For a 900 MHz (0.9 GHz) antenna, the resolution is in the 1–1.5 cm range which means that roots of that size or larger, and root spacing of that size or larger, should be detectable if only the center frequency (and its associated wavelength) is considered. However, it is possible to detect smaller-sized roots (<1 cm) since the resolution is improved because of the UWB nature of the antenna sensor as shown in (12.2) and (12.5). The same values of R apply to root spacing (spatial discrimination).

12.3 History of Technical Developments for Root Analysis

Many of the early reports of root detection were unintentional (Bevan 1984; Truman et al. 1988; Farrish et al. 1990); roots were viewed as "noise" that complicated radar interpretation in otherwise useful scans in geophysical surveys (Doolittle and Miller 1991; Barker and Doolittle 1992). The degree to which tree roots introduce "clutter" or unwanted detections that interfere with target delineation has been quantified for military applications at Eglin Air Force Base in Florida (Niltawach et al. 2004). The first study dedicated to tree root detection focused on mapping the three-dimensional distribution of oak roots (*Quercus petraea*) in the Czech Republic (Hruska et al. 1999). A 450 MHz center frequency, antenna was able to detect roots larger than 3 cm to a depth of approximately 2 m. Transects were collected in a grid pattern. After considerable postcollection data processing, two-dimensional representations of reflected energy (radargrams) were manually plotted to create ground plans or root maps. A related study used the same methodology to study root survival in an urban environment (Cermak et al. 2000). These studies inspired and confounded researchers who were anxious to apply GPR survey methods to their own root dynamic studies; there were many more methodological questions raised than were answered relating to future application of this novel technique to belowground studies.

Root methodologies developed over the past 10 years have focused on improving signal processing and interpretation of reflection data acquired with commercially available bistatic (integrated transmitter and receiver) GPR antennas on the soil surface. The goal of signal processing is to reduce clutter and better represent the geometry of the desired target, in this case, roots (Daniels 2004). GPR data are commonly described and processed by the spatial coordinates they represent and referred to as A, B, or C scans (Daniels 2004). The A scan is a single time domain wavelet, characterized by phase changes (+/−) corresponding to electrical properties of the soil or reflective surfaces in the soil (Fig 12.2a). The B scan is a series of A scans collected along a transect. B scan processing integrates data in adjacent waveforms to create a representation of the XZ plane creating a slice or "virtual trench" (Fig. 12.2b). B scan processing focuses on accurately locating the

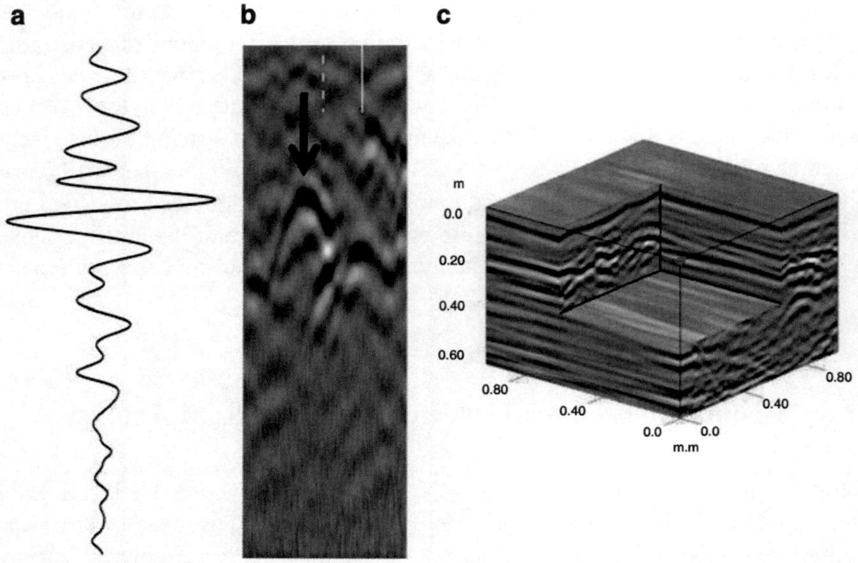

Fig. 12.2 Example of A scan, B scan, C scan. The *arrow* points to the location of the A scan on the B scan

position of an object and attempting to reconstruct the shape or relative size of the target. C scans (Fig. 12.2c) integrate B scans to make 3D or volumetric representations of buried objects (XYZ). An alternate class of processing is image analysis; B scan or C scan processing has taken place and software or algorithms are used to summarize high amplitude reflections, map edges, or calculate areas (Daniels 2004). All methods for analyzing root data use A scan and B scan processing, but they may be generalized as either image analysis-based (Butnor et al. 2001, 2003), predominately using relative high amplitude signal area on processed B scans, or waveform-based, focusing on geometric parameters of waveforms other than amplitude per se (Barton and Montagu 2004). Image analysis-based methods excel at assessing and quantifying root biomass distribution, while waveform-based methods detect roots using the B scan and estimate their diameter using zero crossing times of individual waveforms in the A scan. The choice of which method is most appropriate is dependent on the specific research question. Detailed methodologies and a discussion of relative strengths and weaknesses of both are covered in Sects. 12.4 and 12.5. Surface-based antennas are inadequate for assessing large vertical roots; this is especially true of roots comprising the taproot or root ball directly below a tree. Surface GPR antennas excel in locating reflective objects "suspended" in the soil matrix and not tap roots that transition from an aboveground cylinder (tree trunk) to belowground vertical roots. Methods to quantify vertical roots with borehole radar and travel-time tomography are described in Sect. 12.8.

12.3.1 *Image Analysis-Based Developments*

Using Southern pine forests as a model, Butnor et al. (2001) identified suitable soil characteristics and introduced the concept of using image analysis software to estimate root biomass from B scan data, instead of simply locating roots in otherwise "root free" soil. Roots with diameters as small as 0.5 cm were detected in the upper 30 cm of soil with a 1,500 MHz antenna. Using image analysis software to measure the area occupied by high amplitude reflections, root density could be predicted in field soils and root diameter could be predicted in a test bed when crossed at 90°. A key feature of the method is scaling relative root density with a small destructive sample collected with soil cores. Later work concentrated on improving methodology, especially postcollection data processing. Filtering algorithms were used to remove reflections and features in the B scans which were not caused by roots (Butnor et al. 2003). Consequently, the correlation between image analysis quantification and destructive samples improved markedly. Since changes in soil moisture and data collection parameters (computer settings) may affect calibration equations, it is necessary to have calibration cores linked with each field collection campaign for confident results. These concepts were employed to detect potentially pathogenic root fragments (2.5–8.2 cm diameter) after clearing a peach (*Prunus persica*) orchard for replanting (Cox et al. 2005). Using the methods described by Butnor et al. (2003), minor methodological improvements have been made and used to monitor belowground biomass in forestry experiments (Butnor et al. 2005; Samuelson et al. 2008; 2010) and climate change research (Stover et al. 2007). Studies by Hirano et al. (2009) confirmed that roots with high water content were readily detected, while those having <20% volumetric water content were not. Dead roots, woody debris, and very dry roots are not likely to be detected with GPR unless there is sufficient contrast with the surrounding soil matrix as observed by Cox et al. (2005) and Butnor et al. (2005).

12.3.2 *Geometry-Based Developments*

The concept of using B scans to identify the location of roots and A scan data to estimate their diameter was first demonstrated under controlled conditions in a sandy test bed (Barton and Montagu 2004). Unlike image analysis of B scans, the timing of phase changes in the A scan should be less dependent of temporal changes in soil moisture that alter the amplitude of reflected signals. Another waveform parameter that has been used to estimate diameter is the amplitude area (dB $\times$ ns) of the maximum reflected wave (Dannoura et al. 2008; Hirano et al. 2009). To date, all of the experiments using A scan data to estimate root diameters have been conducted in sandy test beds using buried excised roots or wooden dowels. The methods are most accurate when the root is crossed at a 90° angle. The methodology has not fully matured to allow confident assessment in the field at various angles. It is tempting to develop a biomass estimation based on the number of roots detected combined with their diameters along a transect, but the potential for underestimation is substantial. The detection rate of roots greater than 1 cm (with 1,500 MHz antenna) was reported to be 72%, while roots smaller than 1 cm dropped to less than 22% (Yamamoto et al. 2009), which would confound accurate estimates of biomass. Image analysis-based processing and calibration with soil cores mitigates this problem by not focusing on single roots, but the total root mass of a core sample.

12.4 Estimating Root Biomass

GPR is a useful tool for locating and delineating the extent of major discontinuities in electrical conductivity, but estimation of root mass is more complex. The approach to detecting tree roots and surveying root biomass has three steps: (1) optimizing dielectric, range and gain settings to collect quality data under field conditions, (2) postcollection data processing and calibration, and (3) scaling data calibrating radargrams (B scans) to root mass collected with soil cores.

12.4.1 Optimizing Antenna Settings to Collect Quality Data Under Field Conditions

To maximize data acquisition, it is important to clear areas to be scanned with GPR of any surface debris (i.e., logs, branches) so the antenna can be easily pulled along a transect. High-frequency antennas are susceptible to decoupling if there is an air gap of more than a few centimeter between the soil and antenna. This is important in forests that have rough surfaces and less so in agricultural or suburban sites. Deep or uneven litter cover may cause some loss of signal strength and should be raked aside (Butnor et al. 2005). A few short test scans are collected to determine the ε of the soil and determine gain settings. The dielectric value affects the speed of the electromagnetic energy through the soil and is determined by measuring the depth to a clear hyperbolic reflection and digging down to check the depth. This in turn calibrates the travel time of reflected energy into soil depth. Electromagnetic waves travel more quickly in air than water, so the soil moisture content, as well as mineralogy, has significant effects on travel time. Values for ε (unitless) range from 5 to 10 in sandy soil, 10 to 20 in clay and silt loams and >20 in wet soils. Seasonal changes in precipitation and consequently soil moisture may cause significant changes in ε for a given soil.

In order to view data in real time, signal amplification is needed and is referred to as adding "gain." The interaction between roots, root architecture, and soil properties can be quite dynamic, so the selection of gain settings is important. The amount of gain needed is dependent on soil conditions and is site-specific.

More amplification is needed for reflections from deeper depths, so gain is plotted along a time-varying curve. A number of "points" are used to define this curve; experience has shown that using as few gain points as necessary (i.e., 3) is helpful. Gain may be optimized in the field using auto-gain functions and then adjusted in high rooting areas near trees and low rooting areas away from trees to ensure detection while avoiding data loss due to over-amplification called "clipping." The antenna configuration is set, a survey wheel may be attached to the radar antenna and used to collect linear transect data, which may be used singly or compiled as parallel lines or grid patterns. Collecting data in concentric rings around trees is also useful in forest and arboricultural surveys (Zenone et al. 2008).

12.4.2 Postcollection Data Processing

12.4.2.1 B Scan Filtering

Data filtering is usually done with commercially available software [e.g., RADAN (GSSI Inc., Salem, NH; http://www.geophysical.com), EKKO (Sensors & software, Mississauga Ontario; http://www.sensoft.ca), GPR-SLICE (Geophysical Archaeometry Laboratory Woodland Hills, CA; http://www.gpr-survey.com), or TreeWin™ (TreeRadar, Inc. Silver Spring, MD; http://www.treeradar.com)]. The features of interest in the B scan are hyperbolic reflectors that indicate a root or other point anomaly. Filtration processing is used to remove nontarget clutter and focus on resolving the location and relative size of the roots (Fig. 12.3). Parallel bands in the radargrams are indicative of plane (horizontal) reflectors such as the ground surface, soil horizons, and bands of low frequency noise and are removed using a common filtration technique called background removal (Oppenheim and Schafer 1975). The location of roots is spatially corrected using a processing step called Kirchoff migration that identifies the core geometry of a hyperbolic reflector and reduces the impact of multiple reflections (Oppenheim and Schafer 1975; Berkhout 1981; Daniels 2004). This migration technique decomposes and compacts the geometry of the hyperbolas into a form that is closer to the actual feature. This may be substituted or combined with the Hilbert transformation (Butnor et al. 2003; Stover et al. 2007), which has similar goals as the Kirchoff migration, but instead of using geometry, the magnitude of the phase of the signal is transformed allowing subtle properties and objects to be revealed and reducing false "echoes" (Oppenheim and Schafer 1975). The filtered B scans are converted to 8-bit gray-scale image files for further analysis.

12.4.2.2 Image Analysis

Image analysis is used to summarize data from B scans that have already been filtered and optimized for root detection. This allows quantification of root density

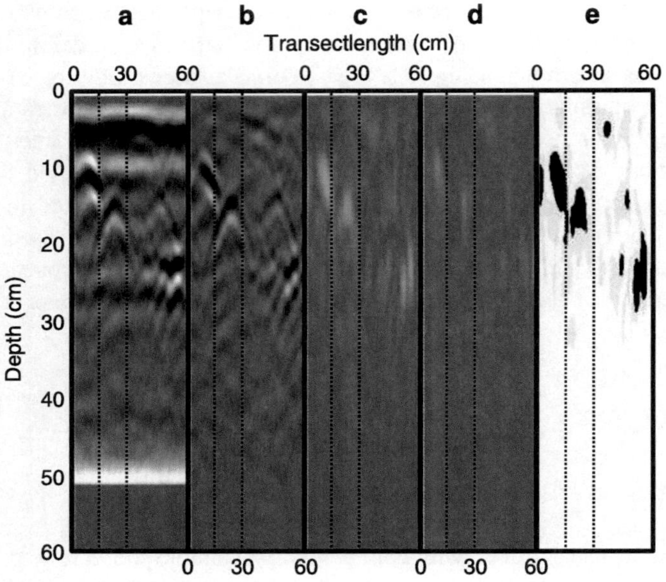

Fig. 12.3 B scan data filtering example showing: (**a**) raw scan, (**b**) background removal, (**c**) Hilbert transformation, (**d**) Kirchoff migration followed by Hilbert transformation, and (**e**) software-aided thresholding and selection of high amplitude pixels. Pairs of *dotted lines* show the relative volume of a 15-cm diameter soil core

along a transect (*X*-axis) or, considering depth (*Z*-axis), a virtual trench or slice into the soil profile. SigmaScan Pro Image Analysis Software (Systat Software, Point Richmond, CA) is particularly useful to automate image processing, but any premium software package would likely have similar features. The number or relative size of discreet detections could be counted with software, but the best information on root density is derived from the relative proportion of the images that represent tree roots. Since the orientation of roots is unknown, this is accomplished by counting the number of pixels that have crossed an intensity threshold. Intensity is a relative measure of how light or dark a pixel is, using 8-bit grayscale images, a value of 0 is black and 255 is white. The intensity threshold range will depend on soil and root properties and is fine tuned by focusing on small roots (~1 cm diameter) and through trial and error, arriving at the range (e.g., 60–200) that delineates detectable roots with minimal illumination of unwanted clutter. Then the software computes the number of pixels that fall within this range (Fig. 12.3e). The strength of the relationship between high amplitude pixel counts and actual root mass is assessed using validation root/soil cores. Areas of high root density are often missed by random coring, so it is recommended to use a combination of randomly selected soil core locations and GPR-guided selections of locations with high and low reflector density. For 1,500 MHz antennas, 15 cm diameter soil cores are well matched to the antenna footprint and are better suited than standard 7.5 or 10 cm cores. B scans are sectioned, leaving only the data

Table 12.1 Relationship between root biomass and "pixels within threshold" computed with image analysis techniques

Location	Correlation coefficient, *r*	Sample size, *n*	Species	References
Marston, NC	0.55	16	*Pinus taeda*	Butnor et al. (2001)
Marston, NC	0.81[a]	16	*Pinus taeda*	Butnor et al. (2001)
Bainbridge, GA	0.81	60	*Pinus taeda*	Butnor et al. (2003)
Sanderson, FL	0.80[b]	40	*Pinus taeda*	Butnor et al. (2005)
Kennedy Space Center, FL	0.71[b,c]	30	*Quercus* spp.	Stover et al. (2007)
Bainbridge, GA	0.80	84	*Pinus taeda*	Samuelson et al. (2008)
Hyogo Prefecture, Japan	0.83[a]	20	*Cryptomeria japonica*	Dannoura et al. (2008)
Maui, HI	0.80	48	*Pinus taeda*	Samuelson et al. (2010)
Saucier, MS	0.87	64	*Pinus elliotti*	Butnor (unpublished)
Saucier, MS	0.74	64	*Pinus palustris*	Butnor (unpublished)
Saucier, MS	0.74	64	*Pinus taeda*	Butnor (unpublished)

[a]Relationship between diameter and "pixels within threshold"
[b]Live root plus dead root biomass
[c]Original published equation was corrected

collected when the antenna was directly over the location of each soil core. Usually, linear regression is sufficient to get good correlation between GPR data and biomass harvested with cores (Table 12.1). Calibration equations are specific to the site and settings used during collection. Most surveys have been done as a single point-in-time collection to compare experimental treatments in forests (i.e., fertilizer applications, species, genotype, and spacing), so it is not clear whether these calibrations can be used over time for repeated measures. Temporal changes in soil moisture could interfere with the detection of small changes in root density. Studies are underway to better understand how soil properties, soil moisture, root density, and depth affect calibrations over time.

12.4.3 Scaling Data with Calibration Equations

Since root biomass calibrations are created using the soil core as the lowest common denominator, it is important to carry this through the scaling process. There is considerable point-to-point variation in root mass; simply summing all of the thresholded pixels in a complete transect will not yield useful information. The B scan must be partitioned into units that correspond to the width of a soil core and analyzed individually, effectively cropping out the area associated with the soil core from the transect data. With SigmaScan software, this process may be semi-automated using macro programs and summarized and assisted with known markers inserted into the radargrams from the survey wheel marker during data collection. Lateral root density may be estimated directly from transects, as long as

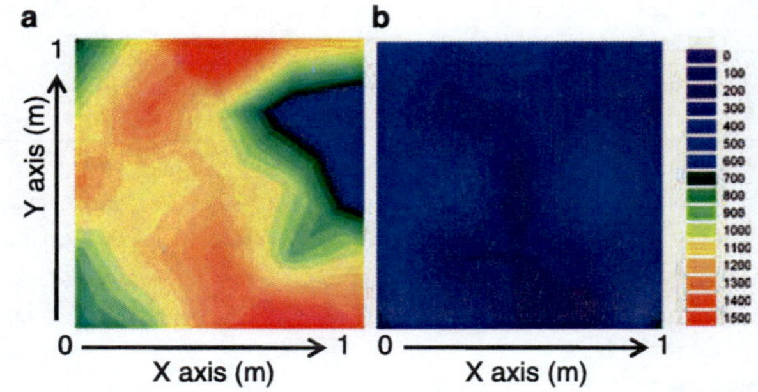

Fig. 12.4 C scans of root biomass (g/m^{-2}) in an undisturbed control plot (**a**), and root in-growth 30 months after all roots were removed from 1 m^3 pits in a loblolly pine plantation in Marston, NC

they are representative of the site. To map root biomass, a set of linear transects are arranged in parallel or a grid pattern that intersect at 90 angles. Roots can be complex targets; from one angle, the cross section may appear as a cylinder giving a clean hyperbola on the B scan, while the perpendicular approach may follow horizontally along the root or miss it altogether. Creating finely scaled grids under field conditions is difficult, since all intersecting points need to be properly interlaced. For that reason, they are usually limited to areas of 1 m^2 or less. An example of a 1 m by 1 m collection grid with 10 cm grid spacing (100 unique intersections) in a loblolly pine plantation (Carolina Sand Hills, NC) is shown in Fig. 12.4. Root density is displayed as a contour plot (C Scan); the control is an area that was undisturbed and has areas of high and low root mass (Fig. 12.4a). The other plot was excavated 30 months earlier and re-growth can be seen from the margins of the plot, while the center remains lightly colonized (Fig. 12.4b). In this example, the control plot has a mean root density of 982 g/m^{-2}, while the root in-growth plot has 331 g/m^{-2} 30 months after root removal. On sandy soils, roots as small as 0.5 cm may be detected with a 1,500 MHz antenna, so the 10 cm grid spacing is inadequate to map individual roots. In plantation forestry, analysis of root distribution and density shows how trees use resources and respond to management operations. It is necessary to cover larger areas than 1 m^2 grids, so parallel transects that are not interlaced are the best option for rapid sampling. In an example from a forestry experiment planted with loblolly pine, five 15 m scans spaced 0.25 m apart were collected on either side of a row of four trees, creating a 2.5 m by 15 m plot (equivalent to 1,000 soil cores) (Fig. 12.5). Data were collected in one polarity to save time (transects vs. grids). Figure 12.5a shows raw spatial data where each filled circle is a "virtual" soil core and Fig. 12.5b is a contour map created with krigged data. Both illustrations show the location of trees within a study area and the lateral root mass, concentrated near the tree and diminishing as the distance from the tree increases. This method only requires transect spacing of 25 cm, allowing a row of five or five trees to be sampled in approximately 20 min. The virtual core map seems somewhat easier for interpreting patterns than the contour. An artifact of this design may be that roots crossed at 90° resolve more distinctly than those following the path of the scan longitudinally. An alternative would be scanning concentric circles around trees. Roots radiating out from the tree would be contacted at similar angles, allowing the creation of polar plats with trees at the plot center.

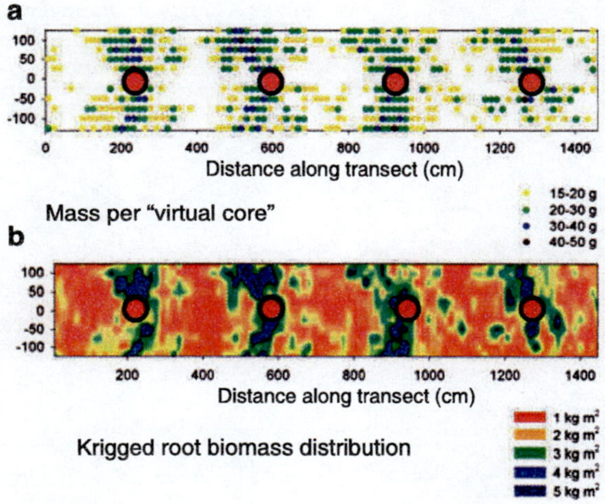

Fig. 12.5 A series of ten parallel transects were used to estimate root mass in a 10-year-old loblolly pine plantation in Bainbridge, GA. Modeled estimates are displayed as "virtual cores" (**a**) and processed with a spatial krigging algorithm to create a biomass contour map

12.5 Detecting Tree Roots and Estimating Their Diameter

At present, the use of GPR to detect roots and assess their size in the field is still in the developmental phase. However, a number of studies into the potential for GPR used in this way have been conducted (Hruska et al 1999; Cermak et al. 2000; Butnor et al. 2001; Barton and Montagu 2004). The aim of this section is to give a basic understanding of GPR in its application to detecting and determining the size of roots. The antenna propagates a wave into the ground and reflected signal is detected as an induced alternating voltage and this trace of alternating voltage with time is recorded. The antenna is then moved slightly and another pulse emitted (Fig. 12.6). For GPR systems used in this way, the antenna is designed so that the pulse propagates in the plane of the direction of travel along a transect with no directional information available, only the time taken for the pulse to return from all directions simultaneously. The radargrams can be viewed side by side and soil layers are seen as horizontal features across the traces while buried objects such as stones or roots show up as characteristic hyperbolas (Fig. 12.6). The apex of the hyperbola occurs when the antenna is directly above the buried object. The depth to the object can be determined from the time it takes the pulse to travel to the object and back and the speed of the pulse through the soil, which depends on a number of properties of the soil, including clay content, water content, and salinity. This approach has successfully been used to locate simple buried objects such as pipelines and landmines and in principle could be a way of determining the location of roots and mapping root systems; however, root systems tend to be complex with branching, interlocking, crossing over, and growing in all three dimensions. The resulting GPR radargrams tend to be complex and difficult to de-convolve into a reliable map of a root system.

Despite some current methodological limitations, progress has been made toward utilizing GPR to locate and determine the size of roots using simplified

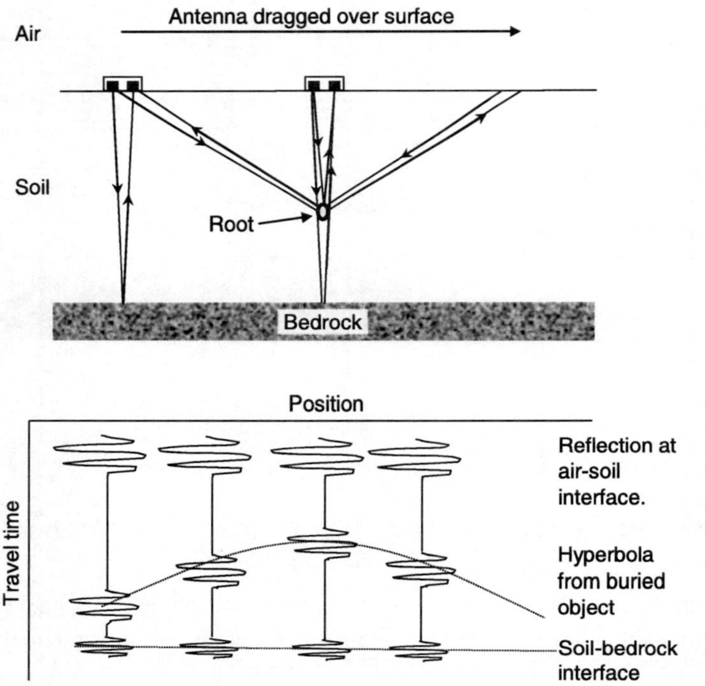

Fig. 12.6 Diagram showing how a GPR system is used to scan an area and the characteristic hyperbolic pattern that arises from reflections from a buried object

test beds. These controlled environment test beds allow users to explore the potential validation in site-specific conditions and gain better understanding of some of the parameters involved. The frequency of the emitted pulse has an inverse correlated influence on the size of the smallest object detectable: higher frequencies can resolve smaller objects, while low frequencies can only resolve larger objects. Wielopolski et al. (2000) demonstrated the use of a 1.5 GHz GPR to detect and visualize a small twig 2.5 mm diameter buried 25 cm deep in a sand box in the lab. One drawback is that high-frequency signals are attenuated rapidly and so can only be used to detect shallow objects; the maximum depth penetration of a 1 GHz antenna is about 1 m or less in typical soils (Conyers and Goodman 1997).

The transit time for the reflected signal provides information on the depth of the buried object, but there is additional information in the return pulse such as the signal strength and the shape of the waveform that may contain information about the size of the object. A study under controlled conditions where roots of known size were buried at known locations within a 4 m $\times$ 4 m $\times$ 2 m box of homogenous sand demonstrated a good correlation between parameters extracted from the waveform of the returned signal and the diameter of the root (Barton and Montagu 2004). Roots of the same size were buried at various depths and of different sizes (1–8 cm diameter) at one depth (50 cm), and three GPR antennas were tested (500,

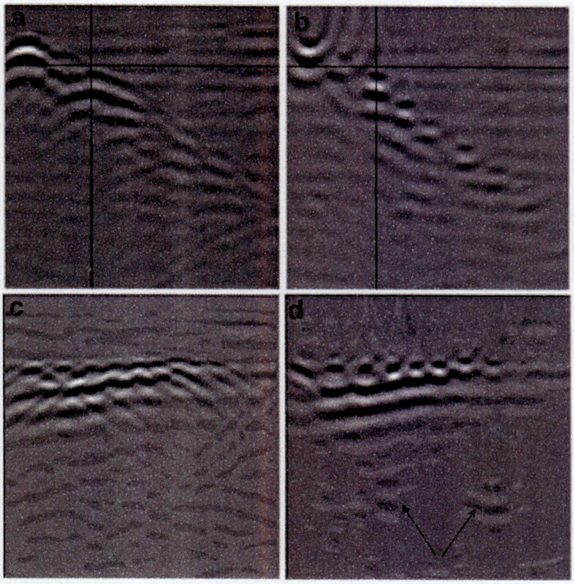

Fig. 12.7 (**a**) Shows the radar profile of a transect across the center of eight roots of the same diameter (5 cm) buried at various depths (15–155 cm). The characteristic hyperbolas are overlapping and interfering with each other making interpretation impossible. (**b**) Shows the same radar profile as (**a**) after processing with the migration algorithm. Now the eight roots are clearly visible. (**c**) Shows the radar profile of a transect across the center of nine roots of various size buried 50 cm deep. (**d**) Shows the same profile as (**b**) but after processing again the position of the roots is now much clearer. The *arrows* on D show the two roots buried at 155 cm which are now more visible. NB all profiles are from 500 MHz antenna (from Barton and Montagu 2004)

800, 1,000 MHz). An unsupervised maximum-convexity algorithm (Daniels 2004) was used to de-convolve the information contained in the hyperbolas, effectively focusing the signal back to the source of the reflector in order to determine the location of the roots (Fig 12.7). The GPR trace through the center of each root was then analyzed and the zero crossing times, the time taken for the signal to switch polarity as the waveform progressed, were extracted (Fig. 12.8). These timings correlated well with the size of the roots and a multivariate model was used to estimate root diameter from GPR data. Half of the dataset was used to calibrate the model, which was then tested on the other half of the dataset (Fig. 12.9). Surprisingly, the 500 MHz antenna gave the most accurate estimates of root diameter and under near optimal conditions it was possible to determine the diameter of the roots ranging in size from 1 to 8 cm with a root mean squared error of 0.56 cm. The calibration derived in this example is site-specific; it is not transferable to another site or applicable if the soil moisture content changes. Calibration may be required at the time of measurements until these effects are better understood. These early studies have shown some promising results but they have tended to be under optimal conditions with single roots placed in uniform soil. The goal of using GPR to map complex root systems in difficult real world conditions such as

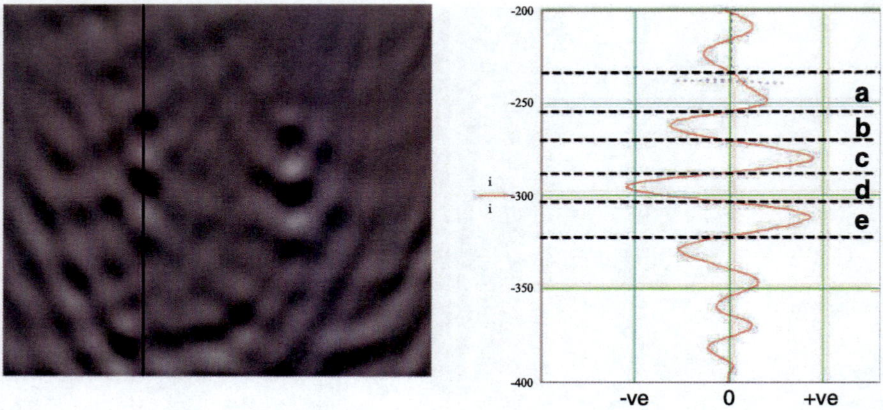

Fig. 12.8 (**a**) A close up of a processed 500 MHz profile showing two roots. (**b**) The trace through the center of the first root as indicated by the *vertical line* on panel (**a**). Parameters a–e are determined from the trace as the time (ns) between zero crossings of the waveform (from Barton and Montagu 2004)

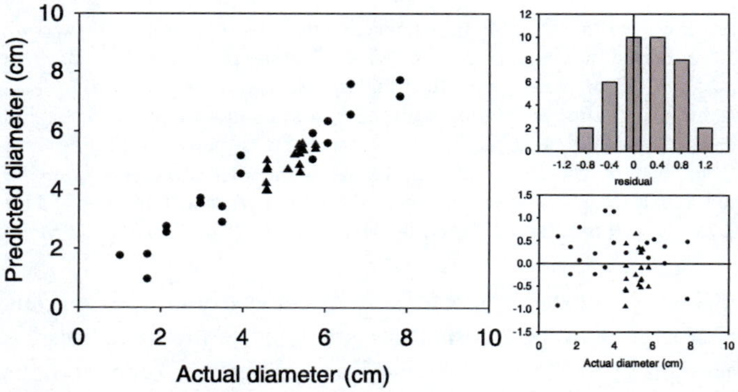

Fig. 12.9 Model testing: the predicted values of root diameter, using the half of the data not used to fit the model, plotted against the actual mean root diameters. The residual plots show that residuals are normally distributed with a slight positive bias (0.2 cm) and the RMSE = 0.56 cm (from Barton and Montagu 2004)

stony, multilayered soils will require advances in antenna technology and data processing algorithms such as those that are now commonplace in medical imaging.

12.6 Urban Roots Inspection

The benefits of conserving trees on urban and community development sites in both new neighborhood development and in the redevelopment of inner city areas are well known and increasingly quantified (Nowak and Dwyer 2007). Urban trees provide a number of economic (e.g., reduced heating and cooling costs, increased property value) and environmental benefits (e.g., reducing the urban heat island effect, reduced water runoff). Even with the best of intentions, existing trees are often inadvertently damaged or destroyed in the face of new development and land-use change. It is difficult to know where tree roots are located to avoid damage to them during construction and presents another useful opportunity to apply GPR in an urban setting.

12.6.1 Introduction

While GPR assessments in forests focus on determining biomass, the chief objective of urban roots inspection is to determine the 3D layout and density of structural roots (diameter $\geq$1 cm) as part of establishing the tree's structural integrity for assessing risk. A quantitatively determined root morphology map can be used to construct an optimal root protection zone instead of relying on rule of thumb to create a circular protection zone based on the tree's diameter. Urban roots inspection also presents the challenge of conducting data collection around obstacles, such as a tree located close to a structure (house, building, wall, and fence), landscaping occupying a considerable area under the canopy, and awkward tree locations such as adjacent to sidewalks and roads. Another critical challenge is that urban surfaces are often covered with concrete, asphalt, pavers, or bricks which GPR needs to penetrate to map the roots beneath. Additionally, unseen impediments including when the grade has been raised either by soil fill or by construction debris such as chunks of bricks and rocks, require special consideration. Scan transect plans need to be creatively devised that can cope with these obstacles resulting in an accurate root morphology map that can ultimately be used by certified arborists, developers, and governmental bodies to render risk assessment decisions to establish appropriate root protection zones. Using GPR to map urban tree roots has several advantages over existing methods; it is rapid and nondestructive causing no harm to the trees or environment, it allows repeated measurements over time and it allows observation of root distribution beneath hard surfaces.

12.6.2 Considerations Specific to Urban Root Inspections

GPR assists certified arborists in assessing the structural integrity of the structural roots and, hence, a tree's ultimate risk to human life and property in the urban forest. While GPR has been shown to locate tree roots in three dimensions in forest soils (Hruska et al. 1997, Butnor et al. 2001, 2003; Stokes et al. 2002), few studies have focused on root detection underneath pavement (Cermak et al. 2000). Although this technology has a long history in archeology and engineering to locate antiquities and utilities (Conyers and Goodman 1997), the practice of using it to map roots in urban soils, which can be compacted, layered and discontinuous, is comparatively new but shows significant promise (http://www.treeradar.com/CaseStudies.html). GPR-based root inspections can be used to make inferences about root health and the presence of diseased roots. The moisture within the woody roots causes the reflection due to the dielectric contrast with the soil matrix. Roots may lose moisture as they decay and become less reflective. Therefore, if root reflections drop significantly (including to zero) at some radius from the trunk in urban settings, it may be inferred that either root disease is likely present or that the roots have been severed by some past trenching operation.

As discussed earlier, the wavelength, and hence resolution, imposes a physical limit on the ability of the radar wave to discriminate tightly clustered roots into individual roots. But this concern is negated from both biomass and structural integrity points of view since the biomass of a tight cluster of small roots with an overall bundle diameter of "d" cm is identical as that of a single root of the same diameter. Yet another complicating factor for urban roots detection is the orientation of the roots relative to the direction of the transecting antenna. Ideally, GPR inspection is always attempted with the antenna crossing subsurface targets of interest in a perpendicular (90°) direction. Uniform, linear patterns are commonly seen in simple targets such as irrigation pipes that are laid out in a regular fashion. Roots, however, grow radially out from the trunk but they can change direction with increasing distance from the trunk. So it is not possible to guarantee a perpendicular transect which means the majority of the roots are usually transected at an oblique angle.

Additionally, "noise" generated by the nonuniform nature of urban soil matrix itself (e.g., voids, large rocks, construction debris, and multiple layered strata) create reflected signals that can mask reflections from roots, particularly small roots. This phenomenon is commonly referred to as a "signal-to-noise (S/N) problem." The application of high-frequency (small wavelength) UWB GPR antennas together with signal processing algorithms can significantly improve the S/N ratio of data recorded in both bare and covered ground, thereby greatly enhancing roots detection. Customized software packages, operating in a semi-automated way with the analyst, permit root clusters to be "teased apart" and individual roots determined, consistent with the physical limitations (e.g., TreeWin™ US Patent # 6,496,136).

12.6.3 Examples from Urban Tree Root Inspections

12.6.3.1 Equipment and Data Collection

A TerraSIRch™ GPR system equipped with a 900 MHz antenna on a modified cart with an integrated survey wheel from GSSI, Inc. was used to rapidly collect data in urban settings. The GPR system was set up to automatically record a radar wave for every 5 mm of movement along each scan line. This high data collection density,

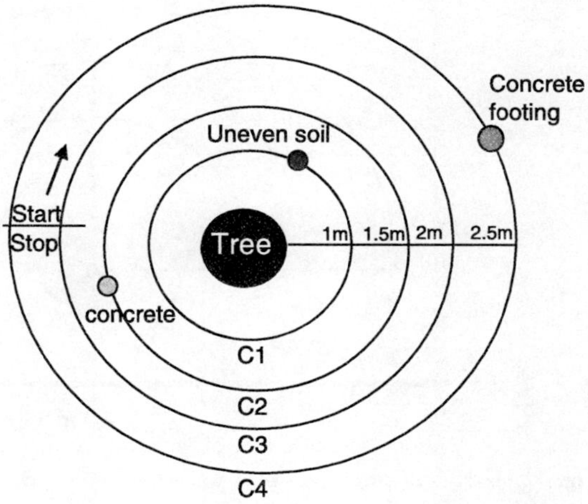

Fig. 12.10 Concentric 360° roots scan plan for trees located in relatively unobstructed areas. The radial spacing between scan lines is 0.5 m

along with the wave's beam width, permits a root, or root cluster, to be observed multiple times which enhances the detection probability during data analysis.

The scan pattern often needs to be creatively laid out to cope with existing obstacles, while inspecting as much of the surface as possible to accurately portray the underlying root structure. For trees in urban parks which are relatively isolated from obstacles, a series of concentric circular scans (Fig. 12.10) can be employed to collect data which is optimal to detect roots growing radially out from the target tree. The scan layout begins at the trunk with a 1 m radius and proceeds outward in 0.5-m radial steps. If the tree is close to a building, house or wall, a series of concentric semicircular scans can be conducted to collect necessary data (e.g., Fig. 12.11). If concentric scans are impractical, a series of parallel linear scans can be utilized. Ultimately, a combination of concentric circular, semicircular, and parallel line scans may need to be employed to maximize the data quality.

After data collection, the GPR radar transects were postprocessed with the TreeWin™ signal processing software package to produce the 2D "virtual trench" maps and the 3D top-down roots layout and density maps.

12.6.3.2 Data Processing

The high scanning density (every 5 mm of movement) ensures that a reflection from each subsurface root (or root cluster) target is recorded multiple times as the antenna passes over the subsurface target. The raw data (B-scan) is usually quite cluttered due to reflections from individual and clusters of roots, soil clutter, soil constituency, and soil stratification layers. Confounding factors, including high clay soils can be minimized by signal processing algorithms (filters) that are able to lower the amplitude of these extraneous factors while preserving the reflections due to roots. An example of the signal processing/filtering operation is presented in Fig. 12.12 where a typical B-scan is shown in the as-collected state and then after

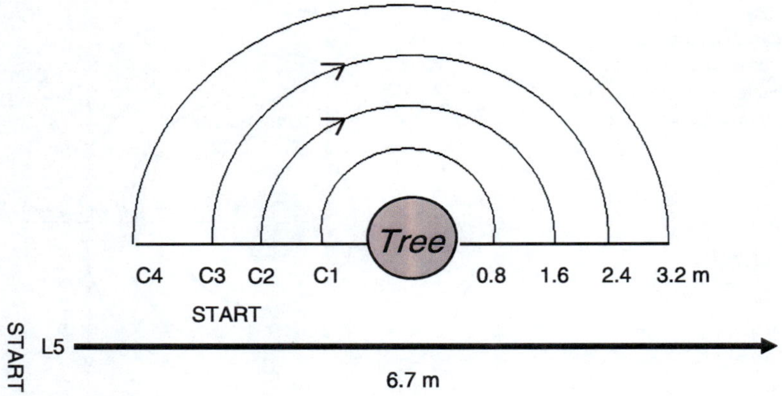

Fig. 12.11 Hybrid roots scan plan for a tree with a wall obstruction in an urban area. Four concentric semicircular scans begin and end on one side of the wall and are conducted on a grassy area. The fifth scan line is on the other side of the wall and is conducted on a concrete surface

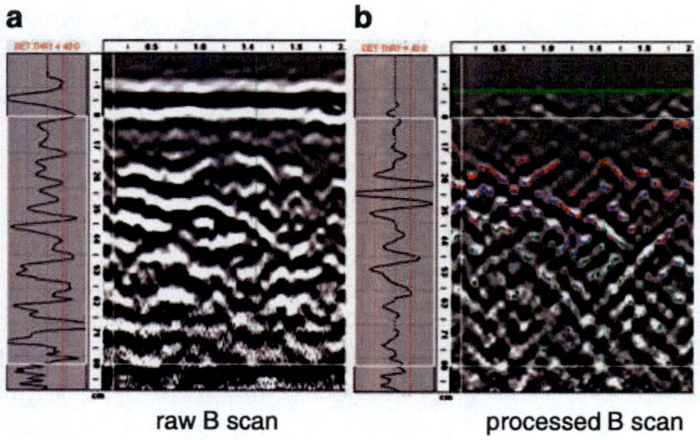

Fig. 12.12 Example of signal-to-noise enhancement of a single scan line using signal processing/filtering algorithms. The horizontal axis is the distance along the 2 m scan line. The vertical axis is depth (cm). The highlighted locations in Plot (**b**) denote candidate root detections based on stage 1 of a two-stage signal processing/filtering protocol

processing. The colors represent candidate root targets as automatically determined by the software.

12.6.4 Results and Root Mapping

Postprocessed radar data are visualized in a "virtual trench" format, as shown in Fig. 12.13, where each scan line produces a planar 2-D view. This generated view shows predicted subsurface structural root location, density, and depth similar to an

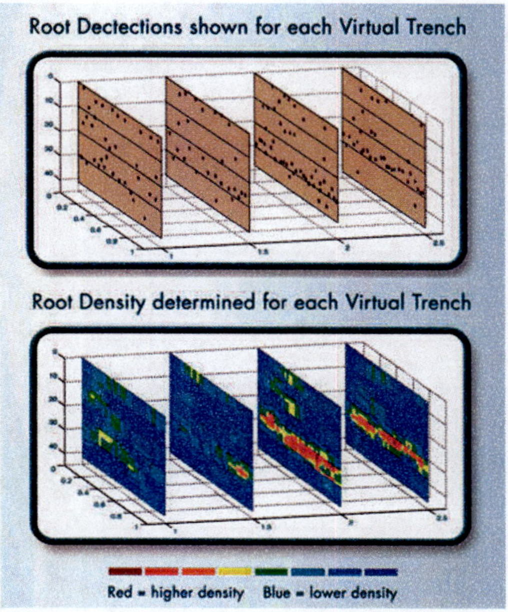

Fig. 12.13 2D "Virtual Trench" maps of a four-line roots scan. The dots in the top plot show the X, Y locations of the detected roots (horizontal axis is distance along the scan line; vertical axis is depth). The lower plot presents the same data as a 2D Root Density plot where the colors represent the number of dots per meter found for each of the 2D virtual trench maps

excavation trench. In this example, a high density of roots was detected at 30 cm depth between 2.0 and 2.5 m from the tree. Alternatively, root detections may be presented in polar plots with the tree at the center (Fig. 12.14). This projection provides layered information of the location of roots at specified depth categories (0–20, 20–40, and 40–60 cm). By identifying depth and density of roots near a tree, arborists can prevent tree damage from construction and public work upgrades. This information can also be utilized to pinpoint the location of root impacts to existing urban structures (i.e., damage to public utilities by root growth). Thus, such computer-generated maps are quite useful in guiding decisions with regard to urban forestry.

12.7 Complementing Minirhizotron Assessments with GPR in Long-Term Ecological Experiments

12.7.1 Need for Nondestructive Root Sampling Applications

Long-term experiments that involve belowground measurements pose a major challenge because these studies can tolerate only minimal disturbance to the soil and destructive sampling (i.e., pit excavations and soil core extractions) is severely limited. GPR and minirhizotrons offer viable options for quantifying root biomass throughout the course of a long-term experiment without the major disturbances that are associated with digging pits and cores. However, since GPR and

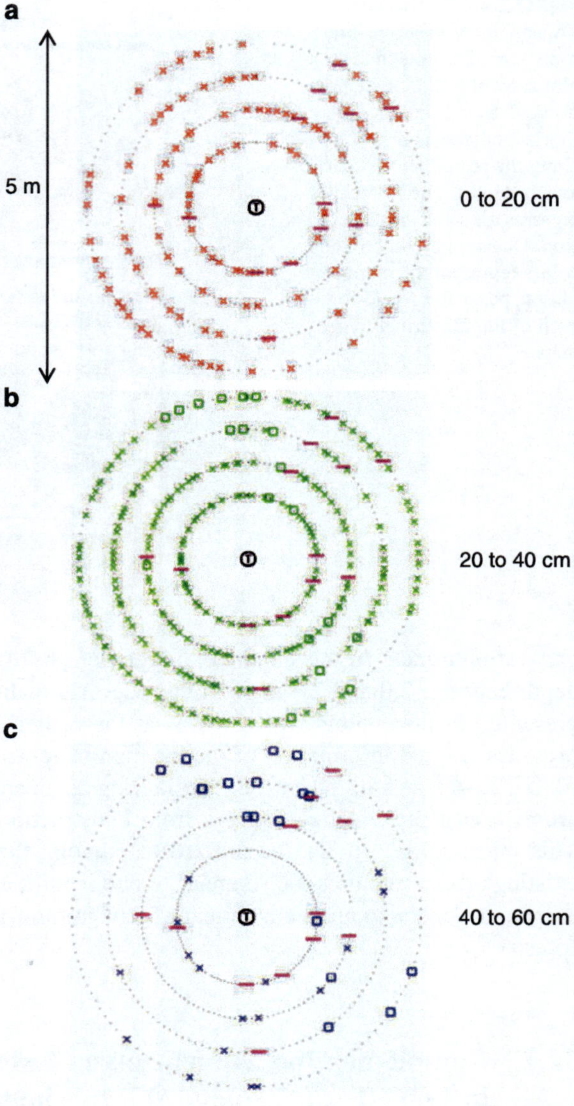

Fig. 12.14 3D plan view of predicted root locations of a *Fagaea fragrans* tree collected over turfgrass in Singapore City Park using the scan layout in Fig. 12.10. Roots are separated into three depth slices: 0–20, 20–40, and 40–60 cm

minirhizotrons estimate root biomass indirectly, they require independent validation using some destructive harvesting, and they also have shortcomings that must be recognized and accounted for. Depending on the goals of the research, simultaneous use of both methods may be recommended since GPR will not effectively capture fine roots and minirhizotrons are not capable of sampling coarse roots. However, there are issues involving overlap of coverage that need resolution; some of these were addressed in a long-term study of the effects of elevated atmospheric CO_2 in a shrub-dominated ecosystem (Schroeder et al. in preparation).

12.7.2 *Integrating GPR and Minirhizotrons to Study Impact of Elevated CO_2 in a Scrub-Oak Ecosystem*

This study was located at Kennedy Space Center on Merritt Island National Wildlife Refuge on the east coast of Florida, USA. The scrub-oak vegetation is maintained by fire and is dominated by woody evergreen species (*Quercus myrtifolia* Willd. (myrtle oak) and *Quercus geminata* Small (sand live oak)) that re-sprout after fire (Schmalzer and Hinkle 1992). Using open-top chambers (OTCs), portions of the ecosystem were exposed to 11 years of elevated atmospheric CO_2 following a controlled burn (Dijkstra et al. 2002). Starting 1996, sixteen (eight ambient and eight elevated [twice ambient] CO_2) OTC's, each enclosing an area of 9.4 m^2 (Drake et al. 1985) received fumigation with CO_2 which continued until 2007. Two nondestructive methods were employed to estimate root biomass in ambient and elevated CO_2 treatments: minirhizotrons to quantify fine roots (<2 mm diameter) and ground-penetrating radar to quantify coarse roots (>5 mm diameter). Roots >2 mm and <5 mm may be partially picked up by GPR but are still likely undersampled. The nondestructive estimates were compared with root biomass harvested with soil cores collected at the end of the study.

12.7.2.1 Data Collection

Two minirhizotrons were installed in each plot in 1996 prior to OTC construction (Day et al. 2006). Minirhizotrons were sampled every 3 months throughout the study to quantify fine root growth and dynamics. To compliment the fine root component, GPR was utilized in June 2007 to estimate coarse root biomass (Schroeder et al. in preparation) using a SIR-3000 radar unit equipped with a 1,500 MHz antenna (GSSI, Inc., North Salem, NH, USA). A 2 × 2 m fiberglass frame was positioned within the footprint of each plot and used to guide the scans every 16-cm in both the *x*- and *y*-direction. The 2-m long scans were processed using methods similar to Stover et al. (2007), described in Sect. 12.4.2. Twenty-five random intersections of *x*- and *y*-directional scans were selected per plot, and 15-cm wide subsamples (validation core size) were cropped from the corresponding 2-m long B scans. Pixels within an intensity range of 70–227 were counted for each cropped image (see Sect. 12.4.2.2). A regression equation (revised from Stover et al. 2007) relating pixel number and root biomass ($R^2 = 0.47$) was used to estimate biomass. Additionally, destructive coring was conducted at the end of the study to validate the minirhizotron and GPR estimates. Five soil cores were collected in each treatment plot with a 7-cm diameter corer to the depth of the water table (~200 cm). An additional five cores were taken for the 0–10 cm depth in each plot. Roots were extracted using a 2-mm mesh sieve followed by a 1-mm sieve, dried at 60°C for 24 h, sorted as fine (<2 mm diameter) or coarse (>2 mm diameter), and weighed to obtain biomass.

12.7.2.2 Results

Total root biomass for plots exposed to elevated CO_2 (final March 2007 minirhizotron biomass estimate plus the June 2007 GPR biomass estimate) was 7,772 g/m^2, compared to 6,749 g/m^2 for ambient plots ($P = 0.11$; Table 12.2). Coarse root biomass was not significantly different between treatments ($P = 0.26$; Table 12.1), nor was fine root biomass ($P = 0.31$; Table 12.1). Minirhizotron images likely undersampled the >2 mm diameter roots since only two roots in this size class appeared only in the elevated plot images and were excluded from the analysis (Brown et al. 2009). Fine root biomass estimates determined from minirhizotron data over the 11-year study period showed a repeating pattern of recovery and steady state following disturbance (Fig. 12.15), whereas GPR

Table 12.2 Root biomass (g/m^2; means ± standard errors) using three sampling methods: minirhizotron image analysis, ground-penetrating radar (GPR), and 7-cm diameter soil cores

	Fine roots (<2 mm diameter) Minirhizotron	Coarse roots (>5 mm diameter) GPR	Total	Fine roots (<2 mm diameter) Cores	Coarse roots (>2 mm diameter)	Total
Ambient CO_2	1,644 ± 173	5,105 ± 418	6,749 ± 422	1,230 ± 170	3,414 ± 168	4,643 ± 280
Elevated CO_2	1,942 ± 168	5,830 ± 487	7,772 ± 422	1,258 ± 359	4,255 ± 721	5,513 ± 697
	$P = 0.31^*$	$P = 0.26$	$P = 0.11$	$P = 0.94$	$P = 0.28$	$P = 0.27$

GPR estimates are to a depth of 60 cm; all others are to 100 cm (Schroeder et al. in preparation)
**P*-values represent results of nonparametric or ANOVA tests for difference in biomass between ambient and elevated CO_2 plots

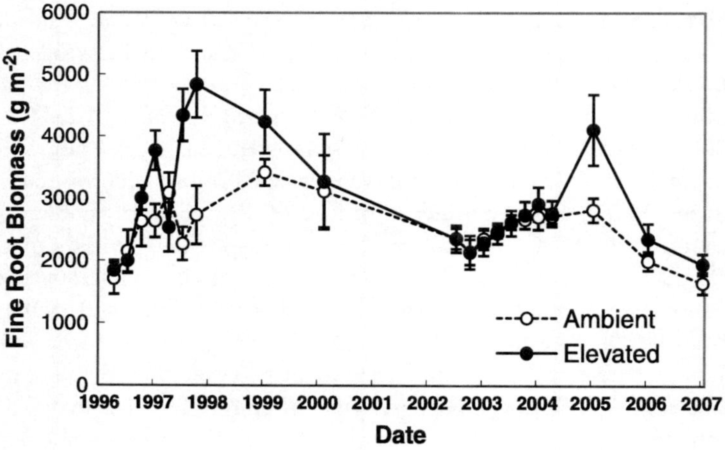

Fig. 12.15 Fine root (<2 mm diameter) biomass to 100 cm depth estimated using minirhizotrons over the 11-year study period (Schroeder et al. in preparation). Values plotted are means ± SE

estimates were only acquired in the final year of the study. Total biomass determined independently from cores was lower than the combined total from the minirhizotron and GPR sampling methods (Table 12.2). The CO_2 effect on total root biomass estimated from cores to a depth of 100 cm was not significant ($P = 0.27$), but the 870 g/m^2 mean difference between treatments was comparable to that observed for minirhizotrons and GPR (Table 12.2).

12.7.2.3 Interpretation

The combined total root biomass from minirhizotrons and GPR is likely an underestimate given the methodological gap (diameters <2 mm for minirhizotrons and >5 mm for GPR) in detecting roots in this study. Dense mats of near surface fine roots and clusters of fine roots may be detected as a larger object by GPR. Despite the potential for the GPR-minirhizotron method to underestimate biomass, it was still higher than independent sampling with soil cores (Table 12.2). Data presented here and previously collected from the site indicate that coring techniques, especially those using small diameter corers, undersample coarse roots. In a study comparing actual (whole-tree harvest extraction) and estimated (5-cm diameter soil cores) lateral-root density, soil cores underestimated root density by as much as half the actual value (Retzlaff et al. 2001). Our GPR-based estimates of coarse root biomass were greater than those obtained from coring but comparable to coarse root biomass directly sampled in a 1 × 2 m soil pit (Schroeder et al. in preparation). Thus, the GPR data accurately reflect destructive sampling on larger spatial scales (2 m^2) and suggest that the smaller area of cores (~0.02 m^2) underestimated coarse root biomass. Additionally, low frequency encounter of large roots during coring biases the data toward values of zero that may lead to underestimation of coarse root biomass (B. Hungate, unpublished data).

12.7.3 Conclusions

The hypothesis that there would be higher root biomass under elevated CO_2 after 11 years of enrichment was not supported; however, fine root biomass was at a low point at the end of the study and there was a substantial absolute difference in belowground biomass in elevated versus ambient CO_2 plots. Strong CO_2 effects on fine root biomass were seen after disturbance (i.e., fire in 1996 and droughts in 2000 and 2001) during periods of recovery, followed by steady state where CO_2 effects diminished. This suggests there are limits on the capacity of the soil to support fine roots resulting in a dynamic equilibrium ("root closure") (Day et al. 2006). Total root biomass was over three to five times greater than aboveground biomass in this system, reflecting the importance of belowground structures as a carbon reservoir allowing plants to re-sprout rapidly after fire. In summary, this study suggests that both minirhizotrons and GPR are effective and compatible in covering the range of root size classes but there are unknown areas of overlap that need resolution to better understand long-term ecosystem responses.

12.8 Imaging Root Systems with Borehole GPR

12.8.1 Description of Borehole Radar

Surface-based GPR can provide adequate resolution of lateral roots. However, most forest trees have a significant allocation to a root ball or large vertical taproots, which cannot be accurately assessed by surface measures. Borehole radar utilizes an antenna specially designed and dimensioned to be lowered into boreholes (~5 cm diameter) drilled into the ground, allowing investigation of vertically oriented targets. In reflection mode, the transmitter (Tx) and receiver (Rx) are lowered into a borehole and an electromagnetic pulse is propagated. A portion of the energy is reflected back to the receiver in a manner similar to conventional surface-based radar (Wänstedt et al. 2000), creating a vertical transect (Fig. 12.16). Data collected with a borehole transducer in reflection mode yield minimal information on vertical roots; the clearest delineation of roots occurs when they are crossed at an oblique angle and not followed parallel into the deeper horizons. Reflection GPR can give a relative assessment of rooting depth if enough holes are drilled. More detailed data may be collected in transmission mode by separating the Tx and Rx and locating them in opposite boreholes to create a crosshole configuration (Fig. 12.16) or placed on the soil surface (Fig. 12.16). By varying the depth or surface locations, a variety of ray paths can be created (Wänstedt et al. 2000). The simplest variable to measure and model is travel time (ns) between antennas. Existing borehole antennas were not designed for use in the near surface, so some experimentation is necessary to determine the minimum depth for effective wave propagation for a particular antenna. Near surface delineation may be improved by combining crosshole with borehole to surface configurations (e.g., Fig. 12.16). The use of borehole radar to study tree roots is still experimental (Butnor et al. 2006) and has not been widely adopted due to need for automated data collection equipment and processing software.

12.8.2 Borehole Data Collection and Processing

The methods for imaging roots directing beneath trees with high-frequency borehole radar are described below using an example from a preliminary study of Scot's pine (*Pinus sylvestris*) root systems (Butnor et al. 2006) in coarse sandy soil in northern Sweden.

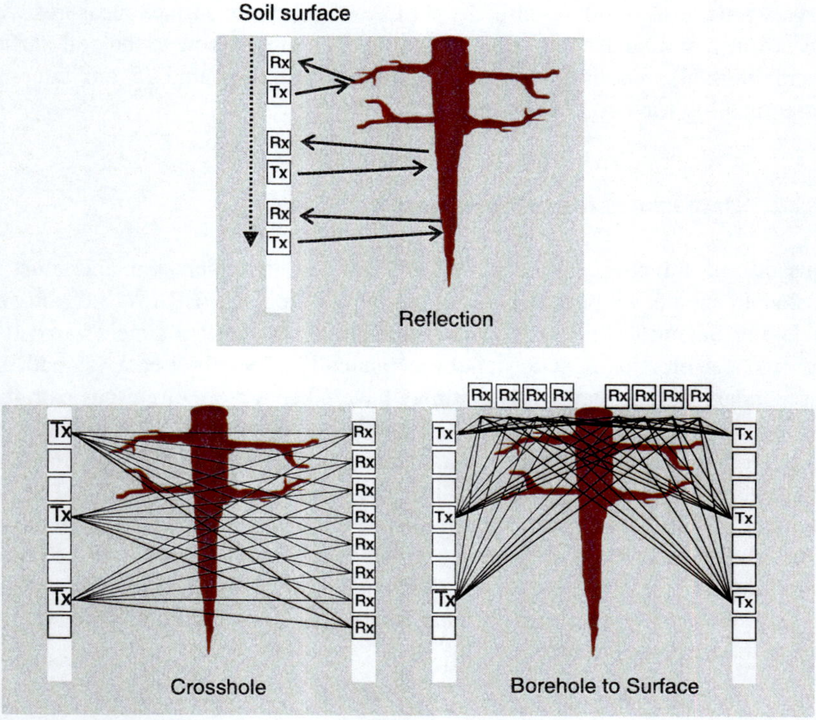

Fig. 12.16 In reflection mode, a bipolar antenna propagates an EM wave that reflects off of root surfaces and is detected by the integrated receiver. In transmission or single-shot mode, an EM wave is propagated in one direction and its arrival time is detected by a receiver in an adjacent borehole (crosshole) or on the soil surface (borehole to surface)

12.8.2.1 Equipment and Data Collection

The study site was in an uneven aged Scot's pine forest near Vindeln, Sweden typified by deep coarse sandy soil where trees ranged from 56 to 193 years old. At each of five test trees of different sizes, a 3 m transect was established on the surface. Near the beginning of each transect, a 5 cm soil auger was used to bore to a depth of 2 m; the subject tree was located at the midpoint (1.5 m) and another borehole was located at the end of the transect (3 m). In order to make travel-time measurements, a 1,000 MHz borehole transducer (Tubewave-1000, Radarteam AB, Boden, Sweden) was configured as a Tx and a second transducer was configured as an Rx. Although the radar unit (Sir-20, Geophysical Survey Systems Inc., Salem, NH, USA) was not designed to accept split signals on a single channel, a custom adaptor was constructed for the experiment. The Tx was operated in single shot mode, where an electromagnetic pulse was propagated and the time it took to cross the soil matrix and be detected by the Rx on the opposite side of the tree was measured. After each travel-time data point was collected, the antennas were manually repositioned (very labor intensive) to create unique ray paths on 5 cm intervals generating 1600 travel-paths per tree. Borehole to surface measures were collected in a similar fashion, though the Rx was moved across the soil surface (10 cm interval) and the Tx was manipulated below ground (5 cm interval), generating 2400 travel-paths per tree.

12.8.2.2 Data Processing and Tomographic Modeling

Thousands of travel-time data points are difficult to understand and must be modeled to create an interpretable tomogram. Using REFLEXW 3.0 software (Sandmeier Scientific Software, Karlsruhe, Germany), the first time of arrival of the propagated electromagnetic signal was manually selected for each ray path and any secondary or tertiary arrivals that may have taken a more circuitous path than those in the ray diagrams were ignored. A number of different analysis models were assessed to interpret the travel-time data, using a freshly cut log (27.5 cm diameter) inserted vertically in the soil (as a utility pole would be) for calibration. The best agreement was found with simple beam and weighted beam models. Use of curved ray models seemed to over-migrate, accurately showing the location of the object, but not correct size or proper shape. The end result of forward modeling was a tomogram that represents the physical properties of a two-dimensional plane between the boreholes (Sandmeier 2003).

12.8.2.3 Destructive Verification

After the five trees were sampled with GPR, the transect orientation was marked on the trunk for future reference. The trees were removed from the soil using a specially designed winch system that toppled the trees with minimal root breakage. Roots directly between the boreholes likely had the greatest influence on EM travel time, but there was no available guidance on the width of the detection zone. Based on preliminary findings from experimentation with the buried log, roots extending beyond 0.25 m on either side of the transect were discarded. The potential root volume beneath the measurement footprint was differentiated by depth and distance along the transect. As roots were harvested, they were separated into (15) 0.5 m^3 cells, so the distribution of root mass could be reconstructed and compared to the GPR-derived tomograms. It was necessary to develop a projection of the root mass data between the two boreholes to compare with the tomograms, by subdividing each cell into nine equal-sized units with the identical value and running a contour analysis in SigmaPlot 2001 (SPSS, Inc.).

12.8.2.4 Comparison of Tomograms to Excavated Trees

Using the buried log to calibrate the algorithms used in forward modeling was useful for guiding the root system analysis. Cross-hole tomography provided

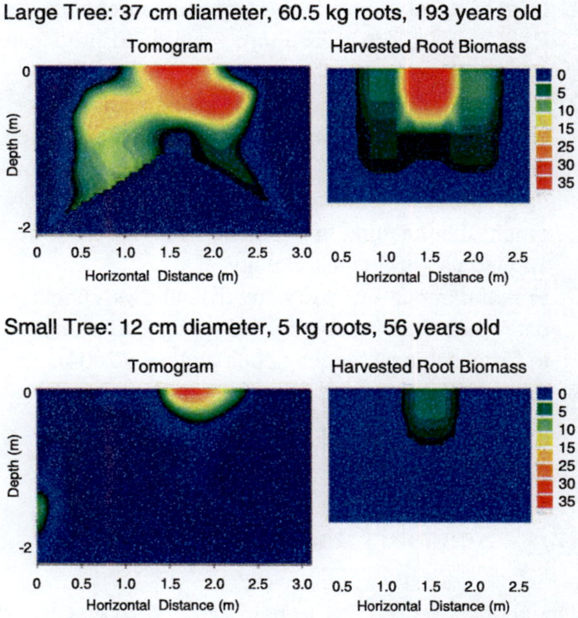

Fig. 12.17 Comparison of tree rooting area estimated from a borehole to surface tomogram compared to a projection of excavated root biomass for a large and small Scott's pine tree

excellent information on the depth of tree roots, but was less useful for imaging near surface features. For crosshole tomography to successfully image tree root structures, the boreholes need to be deep enough to achieve the pitch necessary to resolve the mass of roots near the surface. Borehole to surface measures provided the best information on the near surface, where the bulk of roots are found (0–0.3 m). This application of borehole GPR worked well on three of five trees. Figure 12.17 compares tomograms to excavated root biomass. Tree 1 shows a large mass beneath the bole, extending to either side. Directly underneath the tree is a region where root density is noticeably reduced. Tree 2 is the smallest tree, having only 5.1 kg of roots in the sampling zone. The tomogram clearly shows that this tree has the smallest root ball and limited rooting area within the measurement plane. While this was promising, two of the five trees were not accurately mapped in this example.

12.8.3 *Future of Borehole Radar*

Borehole radar has promise in woody root research. With some improvements, it could be useful to estimate belowground carbon storage in living roots and better understand where they acquire resources deep within the soil profile. However, several innovations are needed for wider adoption and utility of the methodology:

1. Development of new high-frequency borehole antennas that have more sensitive Rx's and Tx's that are more focused to propagate EM waves from a smaller point source.
2. Automation of data collection would also enhance the utility of borehole methodology. The time required to sample each tree manually was 5–6 h. The development of self-contained winch-based survey wheels, modular tracks upon which antennas could be moved on the surface and software to choreograph the antenna movements and measurements would revolutionize the applied use of this technology.
3. Forward modeling software that allows concurrent processing of travel-time data (including secondary and tertiary arrivals) and amplitude data is necessary to further this technique (Zhou and Sato 2004). This is not currently available in REFLEXW software.

12.9 Summary and Future Directions

The application of GPR to belowground studies has shown considerable promise. Groundbreaking studies have applied a technique commonly used in engineering, archeology, and geophysics to answer questions such as quantifying root mass, tree structure, and root diameter. However, there are a number of specific limitations to the technique including soil type, soil moisture, and salinity. In long-term studies where root excavations are not feasible due to study goals or time/labor limitations, GPR can provide an alternative to develop a better-informed picture of belowground structures and dynamics. There is a clear future for expanding the application of GPR to belowground studies as equipment and methodologies continue to improve. For example, development of automated scanning systems would make imaging root architecture more practical. For stand-level analysis in forests, a rapid method for creating root mass distribution maps that delineate the distribution and quantity of mass but do not attempt to detail the orientation of individual roots is the best option. In conclusion, GPR utilized concurrently with other belowground methods mentioned in this book will provide a powerful and robust insight on root systems that otherwise would be underestimated as historically observed.

Acknowledgements The research presented in Sect. 12.7 was funded by a grant from the U.S. Department of Energy (DE-FG-02-95ER61993) to the Smithsonian Institution with a subcontract (95-59-MPOOO02) to F. Day at Old Dominion University. The methods developed in Sect. 12.8 were made possible by in-kind support and expertise provided by Per Wikstrom, Radar Team, Boden, Sweden and Dr. Tomas Lundmark, SLU, Swedish University of Agricultural Sciences, Vindeln, Sweden.

References

Barker DB, Doolittle J (1992) Ground-penetrating radar – an archaeological tool. Cult Resour Manage 15:25–28

Barton CVM, Montagu KD (2004) Detection of tree roots and determination of root diameters by ground penetrating radar under optimal conditions. Tree Physiol 24:1323–1331

Berkhout AJ (1981) Wave field extrapolation techniques in seismic migration, a tutorial. Geophysics 46:1638–1656

Bevan BW (1984) Environmental effects on ground-penetrating radar. In: Expanded abstracts, 54th annual international SEG meeting. Society of Exploration Geophysicists, Atlanta, GA, pp 201–204

Brown ALP, Day FP, Stover DB (2009) Fine root biomass estimates from minirhizotron imagery in a shrub ecosystem exposed to elevated CO_2. Plant Soil 317:145–153

Butnor JR, Doolittle J, Kress L, Cohen S, Johnsen KH (2001) Use of ground penetrating radar to study tree roots in the southeastern United States. Tree Physiol 21:1269–1278

Butnor JR, Doolittle JA, Johnsen KH, Samuelson L, Stokes T, Kress L (2003) Utility of ground-penetrating radar as a root biomass survey tool in forest systems. Soil Sci Soc Am J 67:1607–1615

Butnor, J, Roth B, Johnsen K (2005) Feasibility of using ground-penetrating radar to quantify root mass in Florida's intensively managed pine plantations. Technical Report #38, Forest Biology Research Cooperative, University of Florida, Gainesville, 9 p

Butnor JR, Johnsen KH, Wikstrom P, Lundmark T, Linder S (2006) Imaging tree roots with borehole radar. Proceedings of the 11th international conference on ground penetrating radar, Columbus, OH, USA, June 19–22, 2006, 8 p

Butnor JR, Stover DB, Roth B, Johnsen KH, Day FP, McInnis D (2008) Using ground penetrating radar to estimate tree root mass: comparing results from two Florida surveys. In: Allred B, Daniels JJ, Ehsani M (eds) Handbook of agricultural geophysics. CRC, Boca Raton, pp 375–382

Cermak J, Hruska J, Martinkova M, Prax A (2000) Urban tree root systems and their survival near houses analyzed using ground penetrating radar and sap flow techniques. Plant Soil 219 (1–2):103–116

Conyers LB, Goodman D (1997) Ground-penetrating radar: an introduction for archaeologists. AltaMira Press, Walnut Creek London New Delhi

Cox KD, Scherm H, Serman N (2005) Ground-penetrating radar to detect and quantify residual root fragments following peach orchard clearing. Horttechnology 15(3):600–607

Daniels DJ (2004) Ground penetrating radar, 2nd edn. The Institute of Electrical Engineers, London

Dannoura M, Hirano Y, Igarashi T, Ishii M, Aono K, Yamase K, Kanazawa Y (2008) Detection of *Cryptomeria japonica* roots with ground penetrating radar. Plant Biosyst 142(2):1–6

Day FP, Stover DB, Pagel AL, Hungate BA, Dilustro JJ, Herbert BT, Drake BG, Hinkle CR (2006) Rapid root closure after fire limits fine root responses to elevated atmospheric CO_2 in a scrub oak ecosystem in central Florida, USA. Global Change Biol 12:1047–1053

Dijkstra P, Hymus G, Colavito D, Vieglais DA, Cundari CM, Johnson DP, Hungate BA, Hinkle CR, Drake BG (2002) Elevated atmospheric CO_2 stimulates aboveground biomass in a fire-regenerated scrub-oak ecosystem. Global Change Biol 8:90–103

Doolittle JA, Miller WF (1991) Use of ground-penetrating radar in archaeological investigations. In: Behrens CA, Sever TL (eds) Application of space-age technology in anthropology conference proceedings. Stennis Space Center Special Publication, NASA John C, pp 81–94

Drake B G, Rogers HH, Allen LH (1985) Methods of exposing plants to elevated carbon dioxide. Direct effects of increasing carbon dioxide on vegetation. U.S. Department of Energy Publication ER-0238, pp 11–31

Farrish KW, Doolittle JA, Gamble EE (1990) Loamy substrata and forest productivity of sandy glacial drift soils in Michigan. Can J Soil Sci 70:181–187

Gormally K, McIntosh M, Mucciardi A (2010) GPR detection and 3D mapping of lateral macropores: 1. Calibration. Soil Sci Soc Am J (in press)

Hirano Y, Dannoura M, Aono K, Igarashi T, Ishii M, Yamase K, Makita N, Kanazawa Y (2009) Limiting factors in the detection of tree roots using ground-penetrating radar. Plant Soil 319:15–24

Hruska J, Cermak J, Sustek S (1999) Mapping tree root systems with ground-penetrating radar. Tree Physiol 19:125–130

Johnsen Kh Major JE, Maier CA (2003) Selfing results in inbreeding depression of growth but not of gas exchange of surviving adult black spruce trees. Tree Physiol 23:1005–1008

Johnston CA, Groffman P, Breshears DD, Cardon ZG, Currie W, Emanuel W, Gaudinski J, Jackson RB, Lajtha K, Nadelhoffer K, Nelson D, Mac Post M, Retallack G, Wielopolski L (2004) Carbon cycling in soil. Front Ecol Environ 2(10):522–528

Jol HM (2009) Ground penetrating radar: theory and applications. Elsevier Science, Amsterdam

Kroon H, Visser EJW (2003) Root ecology, vol 168, Ecological Studies. Springer, Berlin, Heidelberg

Niltawach N, Chen CC, Johnson JT, Baertlein BA (2004) A numerical study of buried biomass effects on ground-penetrating radar performance. IEEE T Geosci Remote 42:1233–1240

Nowak DJ, Dwyer JF (2007) Understanding the benefits and costs of urban forest ecosystems. In: Kuser JE (ed) Urban and community forestry in the Northeast, 2nd edn. Springer, New York, NY, pp 25–46

Oppenheim AV, Schafer RW (1975) Digital signal processing. Prentice Hall, Englewood Cliffs, NJ

Pierret A, Moran CJ, Douddan C (2005) Conventional detection methodology is limiting our ability to understand the roles and functions of fine roots. New Phytol 166:967–980

Retzlaff WA, Handest JA, O'Malley DM, McKeand SE, Topa MA (2001) Whole-tree biomass and carbon allocation of juvenile trees of loblolly pine (*Pinus taeda*): influence of genetics and fertilization. Can J Forest Res 31:960–970

Samuelson LJ, Butnor J, Maier C, Stokes TA, Johnsen K, Kane M (2008) Growth and physiology of loblolly pine in response to long-term resource management: defining growth potential in the southern United States. Can J Forest Res 38:721–732

Samuelson LJ, Eberhardt TL, Butnor JR, Stokes TA, Johnsen KH (2010) Maximum growth potential in loblolly pine: results from a 47-year-old spacing study in Hawaii. Can J Forest Res 40:1914–1929

Sandmeier KJ (2003) REFLEXW Manual Version 3.0 for the processing of seismic, acoustic or electromagnetic reflection, refraction and transmission data, 346 p

Schmalzer PA, Hinkle CR (1992) Species composition and structure of oak-saw palmetto scrub vegetation. Castanea 57(4):220–251

Schroeder RE, Day FP, Stover DB, Pagel Brown AL, Butnor JR, Duval BD, Hungate BA, Dijkstra P, Siler TJ, Drake BG, Hinkle CR (in preparation) Root biomass in a Florida scrub-oak ecosystem exposed to 11 years of elevated atmospheric CO_2. Plant Soil (submitted)

Stokes A, Fourcaud T, Hruska J, Cermak J, Nadyezhdin N, Praus L (2002) An evaluation of different methods to investigate root system architecture of urban trees in situ: I. Ground penetrating radar. J Arboric 28:2–10

Stover DB, Day FP, Butnor J, Drake BG (2007) Application of ground penetrating radar to quantify the effects of long-term CO_2 enrichment on coarse root biomass in a scrub-oak ecosystem at Kennedy Space Center, Florida USA. Ecology 88(5):1328–1334

Truman CC, Perkins HF, Asmussen LE, Allison HD (1988) Some applications of ground-penetrating radar in southern coastal plains region of Georgia. The Georgia Agriculture Experiment Stations, College of Agriculture, University of Georgia, Athens, GA, Research Bulletin 362, 27 p

Wänstedt S, Carlsten S, Tirén S (2000) Borehole radar measurements aid structure geological interpretations. J Appl Geophys 43:227–237

Wielopolski L, Hendry G, Daniels J, McGuigan M (2000) Imaging tree root systems in situ. In: Noon DA, Stickley GF, Longstaff D (eds) Proceedings of the eighth international conference on ground penetrating radar. SPIE vol 4084, Washington, pp 642–646

Yamamoto R, Hirano Y, Dannoura M, Aono K, Igarashi T, Ishii M, Yamase K, Makita N, Ikeda T, Kanazawa Y (2009) Ground penetrating radar can detect roots of *Pinus thunbergii* in a coastal forest. International Symposium "Root research and applications" RootRAP, September 2–4, 2009, Boku – Vienna, Austria, 2 p

Zenone T, Morelli G, Teobaldelli M, Fischanger F, Matteucci M, Sordini M, Armani A, Ferrè C, Chiti T, Seufert G (2008) Preliminary use of ground-penetrating radar and electrical resistivity tomography to study tree roots in pine forests and poplar plantations. Funct Plant Biol 35:1047–1058

Zhou H, Sato M (2004) Subsurface cavity imaging by crosshole borehole radar measurements. IEEE Trans Geosci Remote 42:335–341

Chapter 13
Root Structure: In Situ Studies Through Sap Flow Research

Nadezhda Nadezhdina, Teresa S. David, Jorge S. David, Valeriy Nadezhdin, Jan Cermak, Roman Gebauer, and Alexia Stokes

Abstract Sap flow research highlights new perspectives to study in situ root structure of large trees. Several examples demonstrate the ability of the Heat Field Deformation method, HFD, to do this under natural and experimental conditions. Within the latter, localized irrigation, sink- or source-severing trigger sap flow responses that help us to understand the belowground parts of a tree, such as the presence of anastomoses between roots of different trees. The vertical profile of root density, as well as root size around a tree, can be derived from the stem sap flow radial profile. Increase of stem flow due to localized irrigation may be used to distinguish root locations near the corresponding stem sector. Responses of root or stem sap flow when exposing roots using an air-spade or following the severing of roots or branches help us to understand the relationships between different sapwood conducting layers and paths of water between sources and sinks.

N. Nadezhdina (✉) • V. Nadezhdin • J. Cermak • R. Gebauer
Institute of Forest Botany, Dendrology and Geobiocenology, Mendel University, Zemedelska 3, 613 00 Brno, Czech Republic
e-mail: nadezdan@mendelu.cz

T.S. David
Instituto Nacional de Recursos Biológicos, Av. da República, Quinta do Marquês, 2780-159 Oeiras, Portugal

J.S. David
Instituto Superior de Agronomia, Technical University of Lisbon, Tapada da Ajuda, 1349-017 Lisboa, Portugal

A. Stokes
Mixed Research unit in Peant Architecture, Botany and Bioinformatics, INRA, CIRAD, PS2 TA/A51 2 Bld de la Lironde, 34398 Montpellier cedex 5, France

S. Mancuso (ed.), *Measuring Roots*,
DOI 10.1007/978-3-642-22067-8_13,

13.1 The Dimorphic Root System of Large Trees: Measuring In Situ Methods

Tree roots are hidden under the ground and their structure and function is frequently an unknown in large trees. Terrain conditions, type and structure of soil, water and mineral contents and other factors, including leaf distribution, are responsible for the development of the tree root structure. Roots are tightly overlapping, occasionally making anastomoses and gradually adjust to changes in environmental conditions. The same species may have different root forms, depending on soil type. This is well documented on a classical study by Jan Jenik (1957) about the characterization of oak tree root systems. Different forms of root systems in oaks reflect the adaptation to local conditions and result in different functioning. The developed root structure always tries to optimize the functioning of the system under a certain soil. In the past, it could only be understood by exposing the whole root system.

In spite of the big differences between root systems of large trees, there is a common feature characterized by the presence of two main types of roots, the so called dimorphic root system (Kostler et al. 1968): superficial lateral roots occurring in shallow soil and sinker or tap roots penetrating down to deep soil layers. Such a root structure allows water uptake by roots from the soil horizons where it is easier to extract (Fig. 13.1). The dimorphic root structure is also important for the functional homogenization of the rooted bulk of soil: roots can help to return the system to equilibrium by recharging drier soil layers through hydraulic redistribution, HR, when any heterogeneity builds up (Fig. 13.1).

Root research is a rather demanding issue. Many studies dealing with large trees are still based on manual excavation, or on instrumentation monitoring only small parts of the root system. However, several instrumental methods are now available that measure, in a field, the root structure at the whole tree level. Structure-oriented methods include acoustic scanning (Milburn and Crombie 1983; Divos and Szalai 2003), ground-penetrating radar (Conyers and Goodman 1997; Hruška et al. 1999, Stokes et al. 2002, Hirano et al. 2009), isotope tracing (Burgess et al. 2000;

Fig. 13.1 Scheme of a dimorphic root system (superficial and sinker or tap roots) in large trees allowing two important functions: water supply (*left angle arrows*) and water redistribution (hydraulic lift, and hydraulic descent) within the whole rooted bulk of soil (*right circular arrows*)

Durand et al. 2007), electrical soil scanning (al Hagrey and Michaelsen 2002, Amato et al. 2009; Dalton 1995) and supersonic air stream (Smiley 2001). Sap flow, as the physiological process reflecting the activity of roots, has been used for structural studies during the last decades. Simultaneous measurements of soil water and the amount of water absorbed by trees (through sap flow) have been used to estimate rooting depth and the horizontal extent of the root system since the 1980s (Čermák et al. 1980, 1993a, b; Clothier and Green 1997). Sap flow has an additional significant advantage over the other root methods – the automatic monitoring with a high time resolution over long periods, which allows the collection of repeatable data on the dynamics and ecology of roots and on the mechanisms of the underlying processes for entire stands. A classical example is the long-term monitoring of tree transpiration performed in southern Moravia, aiming at a better understanding of the behavior of floodplain forests during regular floods and after flood mitigation through water management measures, which decreased the groundwater level (Čermák and Prax 2001) (Fig. 13.2). Roots were initially adjusted to the conditions of unlimited water supply from groundwater and, therefore, they were relatively small. Transpiration was almost topping potential evapotranspiration (Fig. 13.2 upper panel). Relative transpiration was more variable during this period of unlimited water supply due to more frequent rainy or foggy events. The following drop in groundwater level was of about 1–2 m and occurred in a layer of sandy gravel. Capillary rise in this type of soil is small and, therefore, trees became dependent on precipitation events and the water remaining in the heavy soil layer. Transpiration dropped down to a level of about 50% of potential evapotranspiration (Fig. 13.2 lower panel). Relative transpiration was less variable because weather conditions were then rather dry (few rain events) and long-term data were more dependent on groundwater table fluctuations. This study illustrates well the structural problems that the root systems of large trees may exibit in floodplains and the potentiality of the sap flow measuring techniques. However, to achieve this knowledge, periodical measurements of transpiration (in a daily basis are equal to sap flow) were needed for several decades.

Since new technologies were developed in sap flow research, new perspectives appeared to study, in situ, the root structure of large trees. Using the sap flow technique as a tool, we can distinguish how root functional parts are distributed around stems, in which direction (acropetal or basipetal, proximal or distal) water flows, how water is redistributed between different roots and soil layers or stems, etc. All this can be detected in an integrated way (single roots as well as the whole tree root system) and without soil excavations. Long-term behavior (years), as well as short-time responses (minutes) can be studied.

Experimental treatments can facilitate even more on the study of the structural functioning of roots. Severing a certain root should reflect in sap flow in stem and branches, and vice versa. Through this way, a link between different sources and sinks of water can be determined and how water pathways between them are distributed in the stem xylem. For example, in experiments reported by Nadezhdina and Cermák (2000a, b), the major lateral and sinker roots of adult lime (*Tilia cordata*) and spruce (*Picea abies*) trees were cut. Stem sap flow was

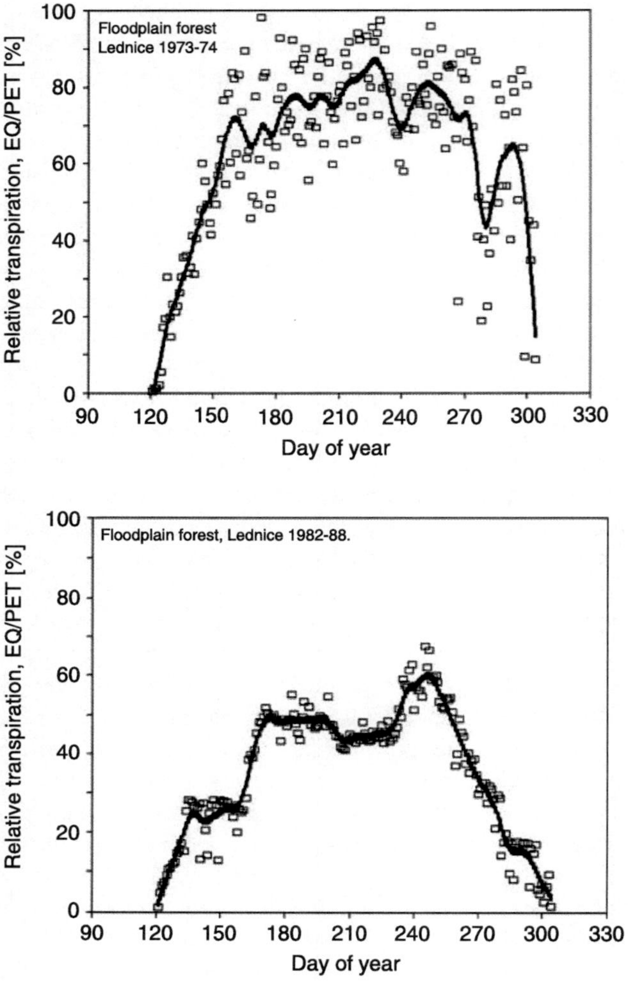

Fig. 13.2 Seasonal courses of relative stand transpiration (transpiration, EQ/potential evapotranspiration, PET) in a floodplain forest, Lednice (southern Moravia) during periods of initially unlimited water supply (*upper panel*), and after artificial lowering of groundwater table (by 1–2 m, outside the direct reach of root systems) (modified after Čermák et al. 2001)

immediately reduced in the outer sapwood just above the cut of the superficial roots. However, when deeper roots were cut, sap flow decreased in all layers of the stem sapwood. Therefore, Nadezhdina and Cermák (2000a, b) concluded that superficial lateral roots are probably connected to the outer stem xylem layers, whereas deep roots supply water more uniformly over the entire stem sapwood.

The ability of sap flow research to study root structure in depth will be highlighted and typified hereafter by several experiments carried out on different tree species and under different conditions using the Heat Field Deformation sap flow method.

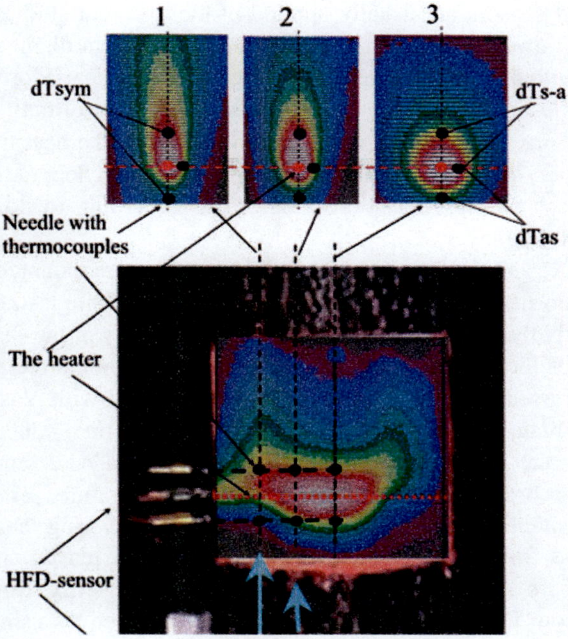

Fig. 13.3 Scheme and photo of a tree stem with a "radial" infrared image (IR) of the heat field generated around the linear heater. IR image is shown for a stem smoothed surface in the same scale as it was visible by the infrared camera. Radial sap flow sensor was installed in parallel to the smoothed stem surface. Six thermocouples were inserted in each needle, but only three are shown for simplicity. Thermocouples in a certain xylem depth "sense" the deformation of heat field in the corresponding tangential section of the stem. Frontal infrared images are shown in the upper part of the figure as well as the position of both differential thermocouples that measured temperature differences dT_{sym} and dT_{as}. $dT_{s\text{-}a}$ is calculated as the difference between dT_{sym} and dT_{as}. *Light arrows* symbolically mark flow rates in the corresponding xylem depths

13.2 The Heat Field Deformation Sap Flow Measuring Technique

The heat field deformation sap flow measuring technique, HFD, (Nadezhdina et al. 1998, 2006; Steppe et al. 2010) can be used in structural studies by installing sensors either in lateral superficial or sinker roots of different size, or in tree stems (at base or at breast height). The method is based on the measurement of the deformation of the heat field around a linear heater, radially inserted in a tree stem or root. Deformation of the heat field is evaluated as a function of the ratio between two temperature differences, recorded in the axial and tangential directions from the heating needle by two pairs of differential thermocouples: symmetrical and asymmetrical. Further details on the method are given elsewhere (Nadezhdina 1999; Nadezhdina et al. 2006, 2008).

Two types of sensors are usually applied in the different studies: single-point sensors for sap flow measurement in small roots (2–6 cm in diameter) and multi-point sensors (with 6 thermocouples per needle, 6–12 mm apart) for measuring sap flow radial profiles in stems or big coarse roots. Application of multi-point sensors allows synchronous measurements of the deformation of the heat field, generated by the same heater, in different "tangential sections" along xylem radius (Fig. 13.3). Heat field in each measured section is deformed according to flow rates in the corresponding xylem layer.

Both the heater and the thermocouples of each sensor are mounted into stainless steel hypodermic needles of 1–1.5 mm in diameter (for single- and multi-point sensors, respectively). The first measuring point is usually at the depth of 20 mm from the head of the needle and its depth below cambium can be regulated by a plastic pipe "dressed" on the needle. Needles are coated with Vaseline or other products to facilitate insertion and to provide a better thermal contact with wood. Sensors in stem are covered with polyurethane foam padded around the head of each needle, in order to homogenize ambient temperature. The sensors are protected from overheating, due to solar radiation, by covering them with aluminum sheets, and from rain or stemflow with watertight plastic sheets wrapped around the sensors and attached to the stem with sealing wax. Root sensors are covered by silicon. The sensors inserted in roots are covered by a small plastic box in order to avoid mechanical damage. All root sensors are covered by soil after 1–2 days of records after verifying the good functionality of the system. Temperature data can be recorded by different data loggers (Unilog Midi-12, EMS Company, Brno, Czech Republic; CR10X, Campbell Scientific, Shepshed, UK; DL2-e, Delta-T Ltd. England), usually every 10 s with averages at different intervals (30 s to 15 min) according to the objective of the trials. The heating system is powered by car batteries, solar panels or mains.

Sap flow per a certain stem or root section "i" ($q_{i,}$ g cm^{-1} h^{-1}) is calculated as:

$$q_i = 3,600D\frac{(K + \mathrm{d}T_{\mathrm{sym}} - \mathrm{d}T_{\mathrm{as}})}{\mathrm{d}T_{\mathrm{as}}} \cdot \frac{Z_{\mathrm{ax}}}{Z_{\mathrm{tg}}} \tag{13.1}$$

where $\mathrm{d}T_{\mathrm{sym}}$ and $\mathrm{d}T_{\mathrm{as}}$ are the temperature differences (°C) recorded by the symmetrical and asymmetrical pair of the thermocouples, respectively, in the measured tangential section of a stem or root; Z_{ax} is the axial distance between any end of the symmetrical thermocouple and the heater, equal to half of Z_{sym} (1.5 cm); Z_{tg} is the tangential distance between the upper end of the asymmetrical thermocouple and the heater (0.5 cm); D is thermal diffusivity of fresh wood (cm^2 s^{-1}) taken as 2.5×10^{-3} cm^2 s^{-1} (Marshall 1958); and K is the difference between $\mathrm{d}T_{\mathrm{sym}}$ and $\mathrm{d}T_{\mathrm{as}}$ under conditions of zero-flow [$K = (\mathrm{d}T_{\mathrm{sym}} - \mathrm{d}T_{\mathrm{as}})_0$]. K depends on sensor geometry (distances between thermocouples and the heater), basic wood properties, surrounding conditions and power supply voltage. K values are determined at each measuring location, which guarantees the accuracy of zero and very low flow measurements.

In structural studies often the qualitative variation and the direction of sap flow are important. Therefore, primary data of sap flow per section q_i (a surrogate of sap flow density) are used for the analysis without further conversion.

13.3 Examples of Studies on Root Structure Based on the HFD Method

All experiments and results reported in this chapter were obtained by the HFD sap flow measuring technique applied to roots or stems of trees in different sites of the Czech Republic and Portugal. In these sites, climate is temperate varying from continental to mediterranean and the studied species include coniferous (*Pinus nigra*, *Picea abies* and *Pseudotsuga menziesii*) and hardwoods, either evergreen (*Quercus suber*) or deciduous (*Tilia cordata*, *Sorbus intermedia*). Site names and characteristics, such as the studied periods, soil and climate types, species (and organs studied), and types of HFD sensors used are described in Table 13.1.

13.3.1 Location and Orientation of Roots Near a Stem Sector

The location, orientation of roots and its correspondence with specific stem sectors can be studied in experiments with localized soil irrigation at different azimuths and distances around stem. These treatments are only efficient when the soil is dry. One example of this is a study carried out in a lime tree (*Tilia cordata* Mill., DBH = 15 cm), where one multi-point HFD sensor was inserted, at breast height, in the east sector of the tree stem, and irrigation was applied sequentially to two soil sectors at different azimuths (Fig. 13.4). The first localized irrigation in soil sector 1 (south-east) did not induce any change in stem sap flow, whereas the second localized irrigation (east) caused an immediate increase in sap flow, even in late day hours. Results showed that roots of sector 1 were not connected to the east stem sector, but many roots of sector 2 seemed closely related to the measured stem sector. By this simple, non-destructive and rather precise way it would be possible to find out to which extent roots can spread from the measured stem position. Localized irrigation should then be applied starting from a wider sector. This approach could be useful in urban environments.

13.3.2 Root Size/Density Around Stems (Maximum Flow Magnitude)

The radial profile of stem sap flow in different azimuths seems to depend on root distribution around tree stems. This was analyzed on a large mountain ash

Table 13.1 Characterization of the study sites and sap flow measuring conditions

Figures	Site	Study period	Soil type	Climate type[a]	Species, organs	Type of sensor[b]
Figure 13.4	Sobesice (Czech Rep.), 49°15′N, 16°37′E, 368 m asl	August 1997	Illimerized soil on weathered granodiorite	Continental, annual temperature 7.8°C, annual precipitation 580 mm	*Tilia cordata*, stem	Multi-point, (10 mm)
Figures 13.5, 13.6, and 13.9	Brno (Czech Rep.), 49°11′N, 16°38′E, 210 m asl	August 2002	Deep loess-clay soil of quarternary origin	Continental, annual temperature 7.8°C, annual precipitation 580 mm	*Sorbus intermedia*, stem	Multi-point, (10 mm)
Figures 13.8 and 13.10					*Pinus nigra*, stem	Multi-point, (10 mm)
Figure 13.7	Olomucany (Czech Rep.), 49°19′N, 16°40′E, 485 m asl	September 2009	Modal cambisol	Continental, annual temperature 7.5°C, annual precipitation 625 mm	*Pseudotsuga menziesii*, stem	Multi-point, (10 mm)
Figure 13.11	Lezirias (Portugal), 38°50′N, 8°49′W, 15 m asl	September 2008	Deep sandy-loam	Mediterranean, anual temperature 16°C, annual precipitation 750 mm	*Quercus suber*, big roots[d]	Multi-point, (6 and 12 mm in different roots)
Figure 13.12	Jedovnice (Czech Rep.), 49°20′N, 16°45′E, 450 m asl	June 2002	Gley-soil	Continental, annual temperature 7.5°C, annual precipitation 620 mm	*Picea abies*, Small root,[c] stem	Single-point, Multi-point (10 mm)

Sites are presented according to the order of fgures in the text. Nomenclature of vascular plants by Kubát et al. (2002)
[a]Annual temperature and annual precipitation are long-term means
[b]Numbers in brackets indicate distance between thermocouples along a measuring needle (sensor)
[c]Small roots: diameter between 2 and 7 cm
[d]Big roots; diameter greater than 10 cm

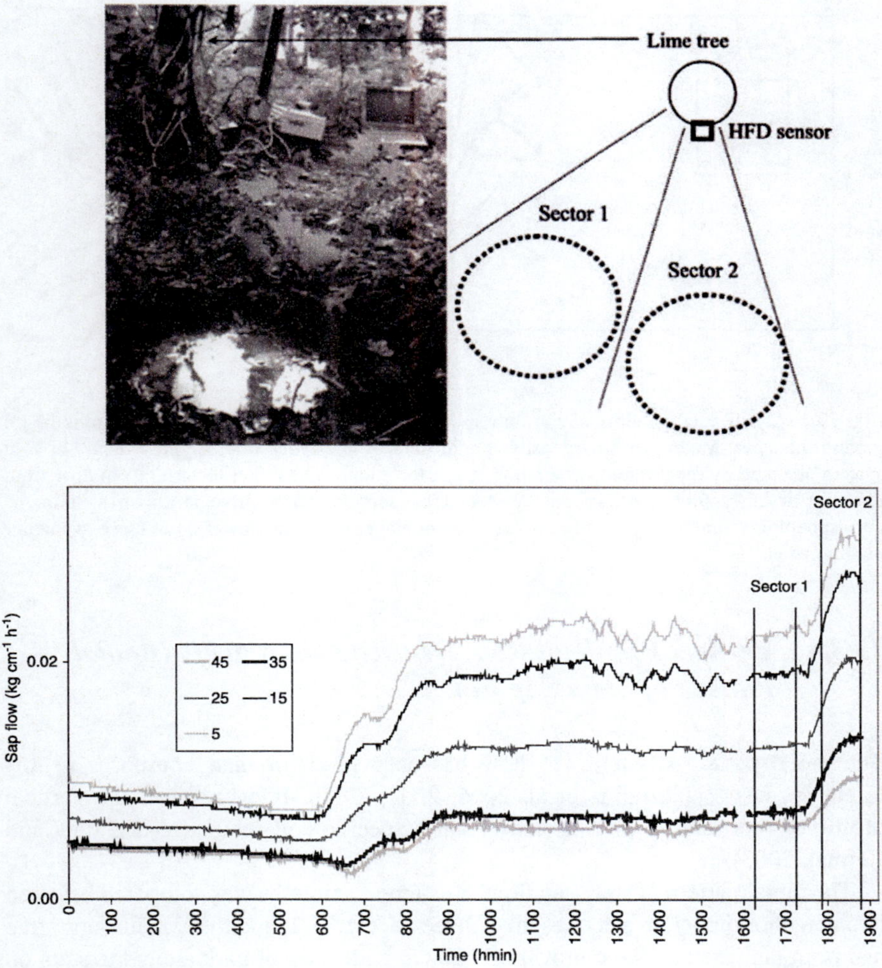

Fig. 13.4 *Upper*: Scheme and photo of one lime tree where a HFD sensor was installed in tree stem at DBH (east side) and localized irrigation was applied sequentially at two different soil sectors, 1. (SE) and 2. (E) (105 l each), in late afternoon. *Low*: Sap flow dynamics recorded by the HFD sensor during the experiment. Numbers in legend refer to depth below cambium (increasing with distance from cambium)

tree (*Sorbus intermedia* (Ehrh.) Pers., DBH = 40 cm, mountain ash tree 1) in an urban situation, where root distribution was previously assessed by the Ground Penetrating Radar technique (GPR) (Stokes et al. 2002) (Fig. 13.5) and subsequently examined by root excavation using an air-spade. Data showed that the magnitude of sap flow rate in stem sectors was clearly dependent on the distribution and density of superficial roots around tree stem: stem sap flow was higher close to bigger roots (SE), and lower close to smaller/shorter roots (Fig. 13.5).

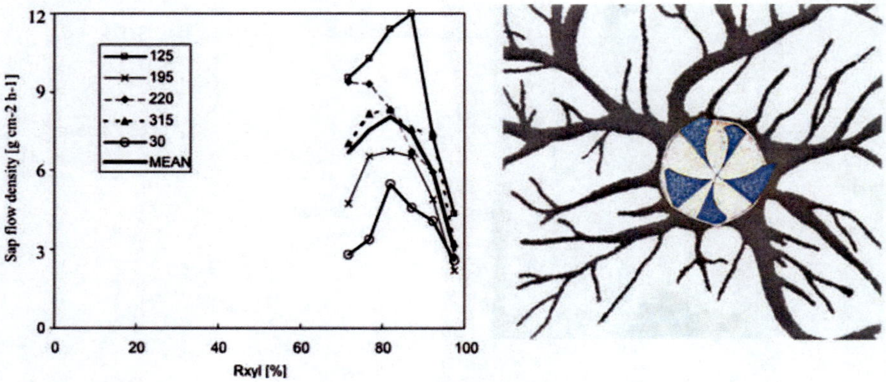

Fig. 13.5 *Left*: Radial patterns of sap flow in a mountain ash tree at five different azimuths (°) around the stem. Mean sap flow density is indicated by the *black line*. *Right*: Schema of root system assessed by the Ground Penetrating Radar technique. The radial pattern of sap flow from left part of this figure is drawn on the stem cross-section. Each curve is drawn next to the corresponding azimuth. Higher flows occur close to bigger roots, and lower flows close to smaller or short roots

13.3.3 Vertical Profile of Root Density/Root Activity (Radial Pattern of Stem Sap Flow)

The radial pattern of stem sap flow has been used for many purposes in different species (Nadezhdina et al. 2006, 2009, 2010). It may reflect the vertical profile of root density or root activity in a specific soil sector (Nadezhdina and Cermak 2003).

The radial pattern of stem sap flow in different azimuths was compared between two neighbor mountain ash trees, tree 1 (see Sect. 15.3.2) and tree 2. This latter tree had been damaged by the removal of a 20 cm wide strip of bark along the stem on the east side, resulting in a wound infected by a saprophytic fungus (*Exidia glandulosa*, Phragmobasidiomycetes). In the infected zone sap flow was substantially reduced, increasing only in the deeper layers (data not shown). In all other azimuths sap flow peaked close to cambium. Variations in the radial pattern of mean sap flow across the stems of both trees were large. Subsequent excavation using an air-spade showed that, in the most superficial soil layers, tree 2 had a dense network of fine roots interlaced, whereas tree 1 had both coarse and fine roots. The fungus attack in tree 2 damaged one coarse root, but several small roots developed afterwards. The differences in root structure (density) between trees were reflected on a narrower peak of mean stem sap flow in tree 2 closer to cambium (Fig. 13.6).

The radial pattern of sap flow in stems may also be used to analyze root activity at different soil depths (Nadezhdina et al. 2007; Cermák et al. 2008), as illustrated in Fig. 13.7 for a Douglas fir tree. Sap flow radial profiles across the

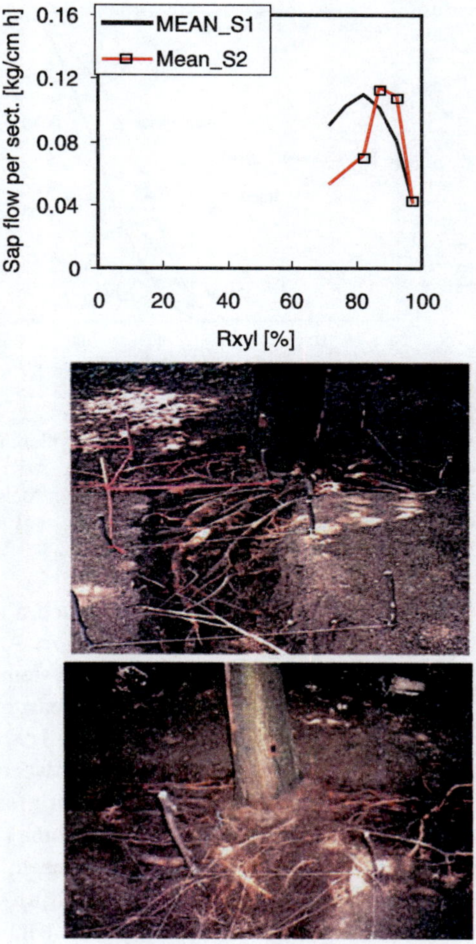

Fig. 13.6 *Upper*: mean radial profiles of sap flow in the stem of two mountain ash trees: tree 1 (see Fig. 13.5) and tree 2. Tree 2 was damaged from one side by fungus attack. *Middle and low*: Photo of transects excavated by air-spade around mountain ash trees (tree 1, middle, and tree 2, lower)

entire sapwood depth are usually asymmetrical, but it is possible to decompose them into two curves with different peak location (see Cermák et al. 2008, developed originally for gas-chromatography, see Novak 1975). These peaks are usually observed close to cambium for superficial roots and deeper in xylem for sinker roots, providing information on root water uptake patterns according to root type.

13.3.4 Partial Surface Root Opening to Study Surface Water Absorption

The excavation of superficial soil, uncovering roots, associated to the monitoring of stem sap flow may provide insight into the effect of soil mobilization techniques on

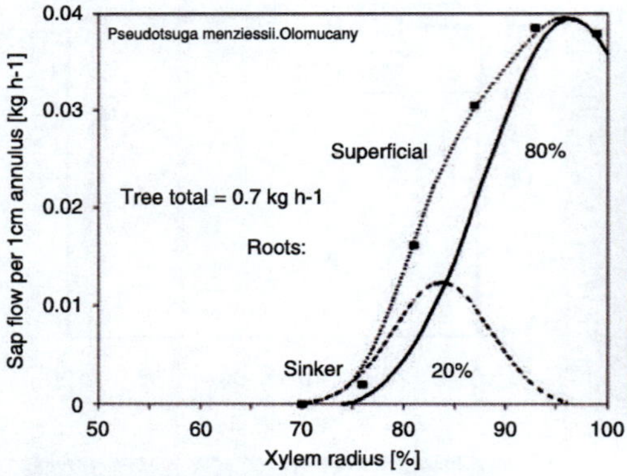

Fig. 13.7 *Left*: Radial pattern of sap flow rate per sapwood annuli in a mature Douglas fir tree represented by a *dotted asymmetrical curve* with *full squares*. This curve was decomposed in two different ones with peaks close to cambium (*solid line*) and in the inner xylem (*dashed line*), indicating sap flows from superficial and sinker roots, respectively. Sapwood depth reached about 30% of stem xylem radius. *Right*: Photo of the tree root system opened by an air-spade at the end of the study

tree water uptake. One example of such a study was provided by an experiment conducted in a Black pine (*Pinus nigra* Arn.) tree (DBH of about 40 cm) with several multi-point sensors installed in stem. In order to determine the part of the stem connected to a particular root, measurements were continuously taken up to 40 cm depth, before, during and after soil excavation, with an air-spade. Excavation uncovered roots colonizing the soil between the azimuths 330° and 150° (mainly east side), at a distance from tree stem greater than 2.5 m. Sap flow was almost invariable on the southern side of the trunk (195°), due to the lack of excavation at this position (Fig. 13.8). However, due to excavation at the 15° azimuth, flow decreased as roots were thin and dried up, influencing water uptake (Fig. 13.8). At the 330º azimuth (boundary between excavated and non-excavated sides) sap flow increased considerably, probably in compensation for the decreased absorption at the open side. This example shows that trenching and digging may influence tree water absorption, even if no mechanical damage is inflicted to the coarse roots.

13.3.5 Root Severing to Study Root Connection to Stem Xylem

The HFD method relies on the continuous application of heat, making it suitable to measure sap flow with a high resolution time (seconds). It can thus be applied to study the short-term responses of stem sap flow to different experimental treatments (with 30 s or 1 min interval), namely root severing. For example, monitoring stem sap flow in different xylem layers during root removal, allows determining how each root is geometrically linked with the conducting pathway in the tree stem. One

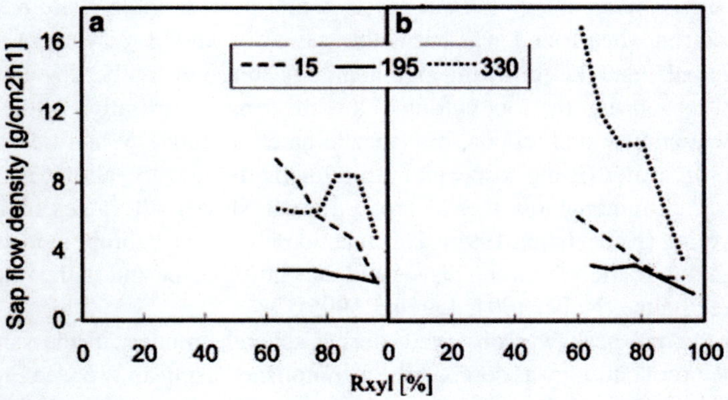

Fig. 13.8 Responses of the radial pattern of sap flow around the stem of a black pine tree (**a**) before root system excavation and (**b**) during excavating of roots with an air-spade. Sap flow increased significantly in inner sapwood layers at 330°, at the boundary between the excavated and non-excavated sides of the tree, indicating possible compensation for the decreased absorption in the open side

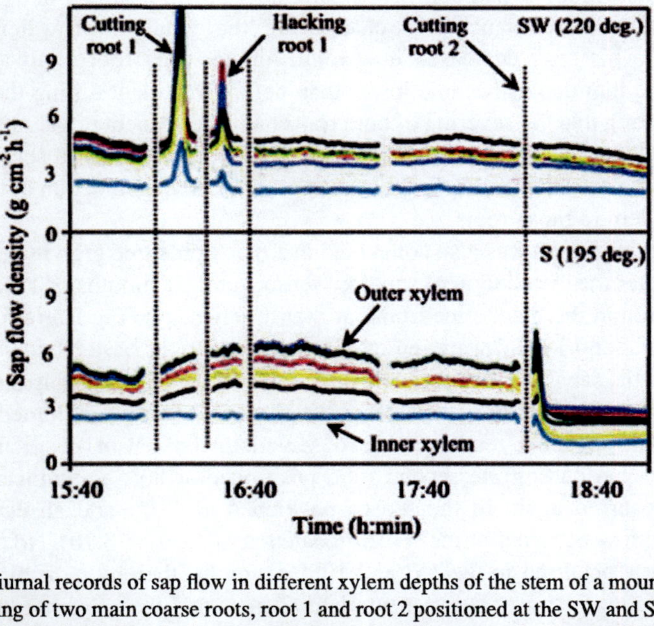

Fig. 13.9 Diurnal records of sap flow in different xylem depths of the stem of a mountain ash tree during severing of two main coarse roots, root 1 and root 2 positioned at the SW and S sides of tree stem, respectively, just below both sensors. The cutting effect was only observed in the stem sap flow sensor positioned above the corresponding root

example of such a study was conducted in a mountain ash tree, where two big lateral roots (both around 13 cm in diameter), located at S (195°, root 1) and SW (220°, root 2) azimuths, were cut and the stem sap flow profiles in the corresponding azimuths were measured (Fig. 13.9). The cutting effect was only observed in the stem sap flow sensors above the cut roots. A first peak in the SW stem sap flow sensor occurred when root 1 was cut with a saw. The abrupt increase of sap flow (within seconds) and the following exponential decrease to another level or zero is the typical response to the root cutting below the sensor (separation of the source) (Nadezhdina and Cermák 2000a, b; Nadezhdina et al. 2004). When tree root was severed under water (using a special cutting methodology) the abrupt increase of flow was even higher. Flow then decreased much slower till values higher than before severing (Nadezhdina 1999). This reaction is due to an abrupt increase in the water potential gradient between leaves and the cutting plane and to the removal of root-soil resistance (Nobel 1991; Larcher 1995).

Then a second peak was observed, during several minutes, in the same stem sensor when root cutting was completely accomplished using an axe (see Fig. 13.9). This additional peak of flow after hacking the partially cut root 1 may have been caused by the release of some stored water into the axial flow. Sap flow returned afterwards to a steady level but, at a depth of around 30–40 mm below the cambium, to values lower than those recorded prior to root removal. This illustrates that the water pathways in the stem were only partially connected to the cut root at the position of this sensor. A similar peak of sap flow was recorded by the S sensor when the lateral root 2, just below it, was removed with a saw. This is due to the releasing of root resistance and changes in the gradient of water potential (Fig. 13.9). This peak decreased in several minutes after root removal, and sap flow in all xylem depths became lower than before treatment during the next few days, indicating that the severing of both roots had a permanent effect on the supply of water to the tree stem. In both cases, reactions occurred only at sensors placed above the cut roots, thus indicating a very strong sectorial connection of that part of the stem xylem to those roots.

A similar treatment was also conducted in a black pine tree growing at the same location. After the installation of sap flow sensors at the azimuths of 15°, 80°, 110° and 195° around the black pine trunk, at breast height, two big lateral roots (with diameters of 8 and 10 cm) were removed with a saw on the eastern side of the pine stem, below the sap flow sensors at 80° (EN) and 110° (ES). On cutting the first root, positioned at 110°, sap flow in the stem above this root decreased immediately, but no significant sap flow response was observed in the sensor positioned at 80° (Fig. 13.10). On cutting the second root, positioned at 80°, a significant drop in sap flow occurred again in the sensor positioned at 110°, and an even greater decrease in flow occurred at the sensor located at 80° (Fig. 13.10). No changes in stem sap flow occurred in the 15° and 195° azimuths, thereby indicating that the axial conducting pathway between the lateral roots and the proximal parts of the stem were closely connected. As could be seen from the staining experiment carried out at this tree, the conducting pathway in Black pine spiraled by about 15° to the south, which is quite common in pine sp. (Vite and Rudinsky 1959; Rudinsky and Svihra 1970). Therefore, changes in flow at 110° should be expected when a root at 80° is cut, but not vice-versa.

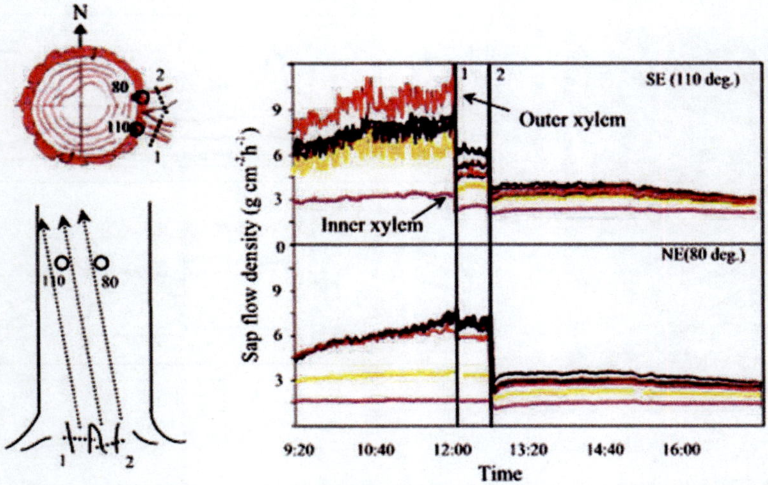

Fig. 13.10 *Left*: Scheme of the positioning of sap flow sensors inserted, at breast height, in the stem of a black pine tree, at the 80° and 110° azimuths, and of the sequence of severing treatments in two coarse roots (marked by numbers 1 and 2 below sensors). The hypothetically reconstructed axial pattern of stem sap flow is indicated according to root-severing experiments. *Right*: Diurnal records of sap flow in the stem of a black pine tree measured in different sapwood depths during the root severing experiment. The stem sap flow at the 110° azimuth decreased in response to cutting of both roots, whereas sap flow at the 80° azimuth decreased markedly only after the cutting of root 2

13.3.6 Responses of Root Sap Flow to Sink-Severing

Sap flow naturally ceases in any conductive plant organ when it is mechanically damaged, e.g. by the cutting of vessels or tracheids. This process is fast and usually follows a non-linear pattern (Rychnovska et al. 1980). Sap flow reaction to cut depends primarily on the measuring location relative to cut placement. When the cut organ (relative to sensor position) represents a sink (foliage), the usual negative water potential of foliage changes to values close to the atmospheric pressure at the cut surface. Due to this change, roots (holding their original water potential) start to drive water down. In the case of big roots, where xylem is connecting different parts of foliage to different absorbing roots, responses to gradual branch severing can be more complex, as observed in a *Quercus suber* tree (Fig. 13.11). The parts of the root xylem partially connected to the severed branch responded in a classical way through an abrupt, short (seconds to minutes) decrease in flow. This was followed by a stabilization that may even be at the pre-cut level, if other connected and not yet severed branches compensate for the loss in the driving sink (for example due to a better illumination after the severing of the neighboring branch). If the monitored root xylem is not connected to the severed branch, no changes occur in flow rate (see cuts 2, 3, and 6 for the upper graph in Fig. 13.11). An opposite response-type occurred in some cases when the sink was severed: initially flow

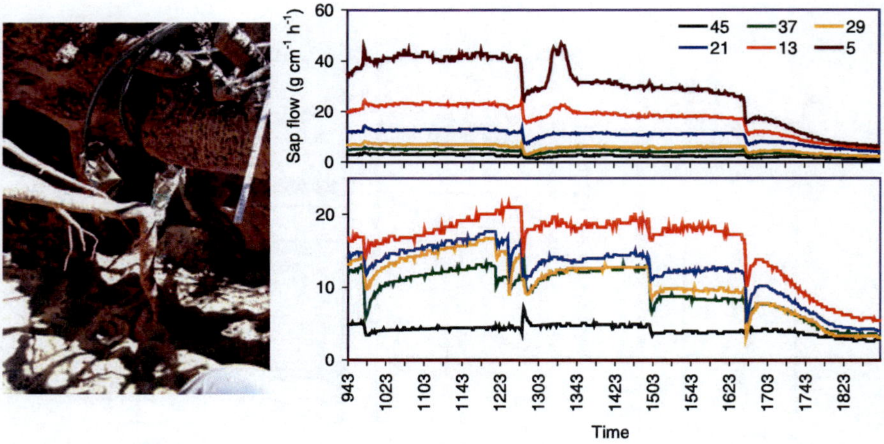

Fig. 13.11 *Left*: Photo of tap and big lateral roots of a *Quercus suber* tree with multi-point sensors. *Right*: Fragment of sap flow dynamics of two of the roots during gradual tree branch cutting. *Vertical dashed lines* indicate timing of branch severing. Numbers in legend refer to depth below cambium (increasing with distance from cambium)

abruptly increased (see cuts 1 and 5 for M1 in Fig. 13.11). This can probably happen when the measured root xylem layer is not connected directly to the severed branch, but indirectly (maybe tangentially or radially) through a neighbor xylem layer. This latter xylem layer could initially suffer a decrease in flow and an increase in water potential in direct response to sink-severing, creating a radial/tangential water potential gradient that caused the short-lived increase in flow observed in the measured xylem layer (not directly connected to the harvested branch). This can only happen when severing occurs under high evaporative demand providing a competing sink (compared to root) for the remaining water from the severed branch.

Type of reaction and response timing will depend on distance between measured and severed places, as well as on environmental conditions and experiment type. In a case of whole sink sudden severing through stem winching 2 m above the ground (Nadezhdina and Cermák 2000b), an abrupt failure of sap flow was observed in coarse roots, up to negative values, but only in the compressed side of the stem and in sapwood layers with highest flows before treatment.

13.3.7 Studying Anastomoses Using Root Severing

The presence of anastomoses (grafts) between roots of the same tree or of different trees may also be detected through the analysis of sap flow profiles in tree stems when roots are subject to severing. Multi-point sap flow sensors were inserted at the south azimuth of the stem bases of two neighbor Norway spruce trees (433 and 475) and at a south orientated root (475). Root severing of the surface of the root

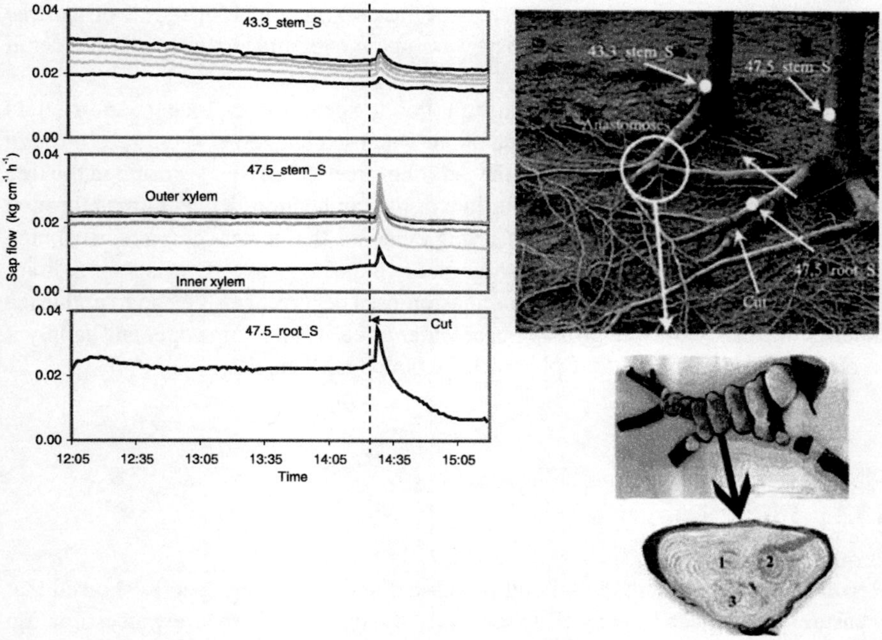

Fig. 13.12 *Left*: Sap flow rate in different sapwood depths of the south azimuth (S) of the stems of Norway spruce trees 433 and 475 and of one root of tree 475. The flow in the root clearly responded to root severing performed 12 cm upstream the root sensor. Responses of different magnitudes were also recorded in the stem of both trees. Tree numbers corresponds to its circumferences at breast height. *Right upper*: View of the root system of spruce trees 433 and 475 (exposed by air-spade after completion of sap flow measurements) with positioning of sap flow sensors at stem bases and on the severed root of tree 475. Clear anastomoses of roots in the measured sectors of both trees are well visible. *Right left*: Intraspecific root graft. On detail you can see the number of root which form root anastomosis (number 1 and 2 are roots from tree 433 and number 3 is root from tree 475). Joint tree rings are well visible

performed 12 cm upstream of the sap flow sensor, resulted in a pronounced asymmetric peak in root flow (Fig. 13.12), similarly to the usual reaction to root cutting, as described above. The root flow then decreased gradually within 30 min to a level of about 30% of the value prior to cut. Severing of this root also resulted in a flow peak in the stem sensor positioned above the severed root on the same tree (475). The flow was highest in outer xylem layers and lowest in the inner layers. After the peak, sap flow in the stem decreased only slightly in the outer xylem. Sap flow in the southern (S) stem section of the tree 433 also responded to the severing of the root of tree 475, showing a similar but smaller peak in all sapwood layers. However, the magnitude of this peak resumed pre-treatment levels after stabilization (around 15 min). The occurrence of the flow peak was simultaneous in both stems, although 2–3 min lagged behind the peak in the cut root. The occurrence of simultaneous stem sap flow peaks in both trees was explained by anastomosis between roots of both trees in the proximal zone of root S of the tree 475 (see Fig. 13.12). Root anastomosis was cut out later and analyzed in more detail (see Fig. 13.12 lower Right).

Root anastomoses occur frequently in woody species (DesRochers and Liefers 2001). However we did not find data about water exchange between trees through such roots. Response of sap flow in the neighbor tree 433 to root severing in the tree 475 provides a clear evidence of the functional connection between trees through anastomosis. This experiment gives strong evidence that water can be redistributed from one tree to another through a network of connected roots. This is an important feature of the evolutionary reason for anastomoses occurrence. We can hypothesize that in some cases one tree growing near water (creek) can pump water and supply it to other trees situated in drier places (hilly bench near creek).

13.4 Final Remarks

Results from experiments in several broad-leaf and coniferous species showed that measurements of sap flow by HFD sensors at the tree stem base allow an estimation of the root spatial distribution both horizontally and in depth in the soil. Considering the depth in soil, results showed that surface roots supply water mostly to the outer layers of stem sapwood, whereas deep roots supply water more evenly over the entire sapwood area.

The possibility of studying the behavior of intact roots in situ by such a functional-based, non-destructive technique (as sap flow) may constitute a significant upgrade in field studies on the structure of tree root systems.

When a destructive approach is possible, the immediate responses of sap flow to source-severing may significantly improve knowledge on tree hydraulic architecture and on the characterization of some root properties (such as the magnitude of the root/soil resistance to water flow, their water capacity, etc.). Responses of sap flow in roots to sink-severing also seem useful in studies on the impact of mechanical stress on trees. Overall, cutting experiments can provide very useful information to studies on root/soil, root/crown and resistances/conductances relationships, as well as on the modeling of tree hydraulic architecture.

The experimental studies reported in this chapter were partially supported by the projects:

- Czech national projects: MSM 6215648902 IGA LDF 12/2010.
- Portuguese Foundation for Science and Technology project: POCI-PPCDT/AGR/59152/2004.
- Part of this study was funded by the John Z. Duling Grant Program (International Society of Arboriculture).

References

Al Hagrey SAA, Michaelsen J (2002) Hydrogeophysical soil study at a drip-irrigated orchard. Portugal, Europ J Environ Engineering Geophysics 7:75–93

Amato M, Bitella G, Rossi R, Gomez JA, Lovelli S, Gomes JF (2009) Multi-electrode 3D resistivity imaging of alfalfa root zone. Eur J Agron. doi:10.1016/jeja.2009.08.005

Burgess SSO, Adams MA, Turner NC, Ward B (2000) Characterization of hydrogen isotope profiles in an agroforestry system: implication for tracing water sources of trees. Agric Wat Manage 45:229–241

Čermák J, Prax A (2001) Water balance of the floodplain forests in southern Moravia considering rooted and root-free compartments under contrasting water supply and its ecological consequences. Ann For Sci 58:15–29

Čermák J, Huzulak J, Penka M (1980) Water potential and sap flow rate in adult trees with moist and dry soil as used for assessment of the root system depth. Biol Plant 22:34–41

Čermák J, Matyssek R, Kučera J (1993a) Rapid response of large, drought stressed beech trees to irrigation. Tree Phys 12:281–290

Čermák J, Matyssek R, Kučera J (1993b) The causes of beech decline on heavy soils after sudden reduction of stand density (in Czech). Lesnictví-Forestry 39:175–183

Cermák J, Nadezhdina N, Meiresonne L, Ceulemans R (2008) Scots pine root distribution derived from radial sap flow patterns in stems of large leaning trees. Plant Soil 305:61–75

Clothier BE, Green SR (1997) Roots: the big movers or water and chemical in soil. Soil Sci 162:534–543

Conyers LB, Goodman D (1997) Ground penetrating radar. An introduction for archeologists. Altamira Press, a Division of Sage Publications, Inc. Walnut Creek, London, New Delhi

Dalton FN (1995) In-situ root extent measurements by electrical capacitance methods. Plant Soil 173:157–165

DesRochers A, Liefers VJ (2001) The coarse-root system of mature *Populus tremuloides* in declining sites in Alberta. Can J Veg Sci 12:355–360

Divos F, Szalai L (2003) Tree evaluation by acoustic tomography. In: Beall FC (ed) Proceedings of the 13th international symposium on nondestructive testing of wood. University of California, Berkeley, CA, USA, pp 251–256

Durand JL, Bariac T, Ghesquiere M, Biron P, Richard P, Humphreys M, Zwierzykovski Z (2007) Ranking of the depth of water extraction by individual grass plants, using natural 18O isotope abundance. Environ Exp Bot 60:137–144

Hirano Y, Dannoura M, Aono K, Igarashi T, Ishii M, Yamse K, Makita N, Kanazawa Y (2009) Limiting factors in the diction of tree roots using ground penetrating radar. Plant Soil 319:15–24

Hruška J, Čermák J, Šustek S (1999) Mapping of tree root systems by means of the ground penetrating radar. Tree Phys 19:125–130

Jenik J (1957) Root system of pedunculate and sessile oaks (in Czech). Rozpravy Československé akademie věd, řada matematických a přírodních věd. Academia, Praha, roč.67

Kostler JN, Bruckner E, Bibelricther H (1968) Die Wurzeln der Waldbaume. Paul Parey, Hamburg. pp 284

Kubát K, Hrouda L, Chrtek J jun., Kaplan Z, Kirschner J, Štěpánek J (2002) Klíč ke květeně České republiky. (Kye to the Flora of the Czech Republic). Academia, Praha

Larcher W (1995) Physiological plant ecology. Springer, Berlin, Heidelberg, New York, Barcelona, Budapest, Hong Kong, London, Milan, Paris, Tokyo

Marshall DC (1958) Measurements of sap flow in conifers by heat transport. Plant Physiol 33:385–396

Milburn JA, Crombie DS (1983) Sounds made by plants a novel applications of acoustic emission analysis. Bull Aust Acoust Soc 12:15–20

Nadezhdina N (1999) Woody plant behavior and stress assessment based on sap flow measurement. Application in forestry and horticulture. Associate Professor. Thesis at the Mendel University of Agricuture and Forestry in Brno, Czech Rep

Nadezhdina N, Cermák J (2000a) Responses of sap flow in spruce roots to mechanical injury. In: Klimo E, Hager H, Kulhavy J (eds) Spruce monocultures in Central Europe: problems and prospects, EFI Proceedings No 33, pp 167–175

Nadezhdina N, Cermák J (2000b) Responses of sap flow rate along tree stem and coarse root radii to changes of water supply. In: Stokes A (ed) Proceedings of the supporting roots of trees and woody plants: form, function and physiology. Kluwer, Dordrecht, pp 227–238

Nadezhdina N, Cermak J (2003) Instrumental methods for studies of structure and function of root systems in large trees. J Exp Bot 54:1511–1521

Nadezhdina N, Cermak J, Nadezhdin V (1998) Heat field deformation method for sap flow measurements. In: Čermak J, Nadezhdina N (eds) Measuring sap flow in intact plants. Proceedings of the fourth International Workshop, Zidlochovice, Czech Republic, IUFRO Publications, Brno, pp 72–92

Nadezhdina N, Tributsch H, Čermák J (2004) Infra-red images of heat field around a linear heater and sap flow in stems of lime trees under natural and experimental conditions. Ann For Sci 61:203–213

Nadezhdina N, Čermak J, Gasparek J, Nadezhdin V, Prax A (2006) Vertical and horizontal water redistribution within Norway spruce (*Picea abies*) roots in the Moravian Upland. Tree Physiol 26:1277–1288

Nadezhdina N, Cermak J, Meiresonne L, Ceulemans R (2007) Transpiration of Scots pine in Flanders growing on soil with irregular substratum. For Ecol Manag 243:1–9

Nadezhdina N, Ferreira MI, Silva R, Pacheco CA (2008) Seasonal variation of water uptake of a Quercus suber tree in Central Portugal. Plant Soil 305:105–119

Nadezhdina N, Steppe K, De Pauw DJW, Bequet R, Cermak J, Ceulemans R (2009) Stem-mediated hydraulic redistribution in large roots on opposing sides of a Douglas-fir tree following localized irrigation. New Phytol 184:932–943

Nadezhdina N, David T, David J, Ferreira I, Dohnal M, Tesar M, Gartner K, Leitgeb E, Nadyezhdin V, Čermak J, Jimenez MS, Morales D (2010) Trees nerver rest: the multiple facts of hydraulic redistribution. Ecohydrology 3(4):431–444

Nobel PS (1991) Physiochemical and environmental plant physiology. Academic, Harcourt Brace Jovanovich Publishers, San Diego, New York, Boston, London, Sydney, Toronto

Novak J (1975) Quantitative analyses by gas-chromatography. Marcel Dekker, New Jork, p 176

Rychnovska M, Cermak J, Smid P (1980) Water output in a stand of *Phragmites communis* Trin. A comparison of three methods. Acta Sci Nat (Brno) 14:1–27

Rudinsky JA, Svihra P (1970) Structures of sap flow in coniferous species and their ecological and phylogenetic relationships (in Czech). Struktury transpiracneho prudu vody ihlicnatych stromov a ich ekologicke a fylogeneticke vztahy. Lesnicky Casopis 16:143–156

Smiley ET (2001) Air Excavation to Improve Tree Health. Tree Care Industry, May 44–48

Steppe K, De Pauw DJW, Doody TM, Teskey RO (2010) A comparison of sap flux density using thermal dissipation, heat pulse velocity and heat field deformation methods. Agric For Meteorol 150:1046–1056

Stokes A, Fourcaud T, Bruska J, Cermak J, Nadezhdina N, Nadezhdin V, Praus I (2002) An evaluation of different methods to investigate root system architecture of urban trees in situ: ground penetrating radar. J Arboricult 28:1–9

Vite JP, Rudinsky JA (1959) The water conducting systems in conifers and their importance to the distribution of trunk-injected chemicals. Contrib Boyce Thompson Inst 201:27–38

Chapter 14
Root Function: In Situ Studies Through Sap Flow Research

Nadezhda Nadezhdina, Teresa S. David, Jorge S. David, Valeriy Nadezhdin, Jan Cermak, Roman Gebauer, Maria Isabel Ferreira, Nuno Conceicao, Michal Dohnal, Miroslav Tesař, Karl Gartner, and Reinhart Ceulemans

Abstract Sap flow measured by the Heat Field Deformation technique, HFD, is sensitive to flow responses to small changes in water potential gradients within the tree hydraulic systems. When these changes occur abruptly, under experimental treatments (severing, localized irrigation, heavy loading), sap flow movement can be used as a marker to study root functionality, for example root ability to redistribute water and withstand heavy machinery pressure. Experiments also showed that a compensation mechanism may operate in trees, with a temporary increase in the absorbed water due to a preferential use of one part of the root

N. Nadezhdina (✉) • V. Nadezhdin • J. Cermak • R. Gebauer
Institute of Forest Botany, Dendrology and Geobiocenology, Mendel University, Zemedelska 3, 613 00 Brno, Czech Republic
e-mail: nadezdan@mendelu.cz

T.S. David
Instituto Nacional de Recursos Biológicos, Av. da República, Quinta do Marquês, 2780-159 Oeiras, Portugal

J.S. David • M.I. Ferreira • N. Conceicao
Instituto Superior de Agronomia, Technical University of Lisbon, Tapada da Ajuda, 1349-017 Lisboa, Portugal

M. Dohnal
Faculty of Civil Engineering, Czech Technical University in Prague, Thákurova 7, 166 29 Prague, Czech Republic

M. Tesař
Institute of Hydrodynamics of the ASCR, v.v.i., Pod Paťankou 30/5, 166 12 Prague, Czech Republic

K. Gartner
Department of Forest Ecology and Soil, Research and Training Centre for Forests, Natural Hazards and Landscape, Seckendorff-Gudentweg 8, Vienna 1131, Austria

R. Ceulemans
Department of Biology, University of Antwerpen, Campus Drie Eiken, Universiteitsplein 1, 2610 Wilrijk, Belgium

S. Mancuso (ed.), *Measuring Roots*,
DOI 10.1007/978-3-642-22067-8_14,

system when another part is damaged or when a water source is lost. Long-term measurements of root sap flow allow distinguishing between water uptake from shallow and deep rooted trees, at different exposures at a forest edge and from healthy and infected trees. Root sap flow can be used as an indicator of tree stress or of the prevailing mechanisms used by trees to survive drought, under irrigation or rain-fed conditions.

14.1 Sap Flow Measurements as a Tool to Study Root Function In Situ

Dimorphic root habit in large trees maximizes their two most important functions in tree survival and productivity: water uptake and hydraulic redistribution (HR). Both can be characterized by the direction and magnitude of flow in the sapwood and can be detected by the sap flow technique. Traditionally, water uptake was evaluated by the measurement of transpiration in stems. However, the development of new tools for sap flow measurement made it possible to monitor sap flow also directly in roots (Burgess et al. 2001; Nadezhdina et al. 2006a). Simultaneous measurements of sap flow in roots and stem xylem provide a better insight into the seasonal water uptake patterns and the absorption from different soil horizons (Nadezhdina et al. 2008). They also allow the monitoring of the origin, magnitude and duration of hydraulic redistribution (HR) in roots (Nadezhdina et al. 2008) and stems (Burgess and Bleby 2006; Nadezhdina et al. 2009). HR was shown to be of special importance for tree survival during drought. Different types of HR contribute to maintain transpiration during depletion/recharge cycles and different phases of tree life (Burgess et al. 1998; Brooks et al. 2002; Moreira et al. 2003; Hultine et al. 2004; Ryel. 2004; Meinzer et al. 2004; Nadezhdina et al. 2006a, 2008, 2010; Warren et al. 2007).

The HFD method (Nadezhdina et al. 1998) responds fast and is highly sensitive to any change in the driving force within the xylem water pathway (Nadezhdina and Cermak 2000a, b, 2003; Nadezhdina et al. 2004, 2006a, b; Stokes et al. 2000), being especially adequate for studies on root functioning. Detailed measurements of sap flow variation, for example radial patterns in the sapwood, enable a faster evaluation of the extent and availability of different water sources.

Continuous sap flow measurements in roots allow the direct detection of the water uptake pathways and the response rates to the changing driving forces. These responses are more pronounced in smaller roots, whereas in large roots they are dampered by flow integration from different root branches linked to different soil horizons. The measurement of radial sap flow profiles in big roots and stem bases contributes to the functional understanding of the root water uptake patterns, both in time and in space.

In functional studies (especially of HR), the positioning of the sensors on the lateral roots is important (Fig. 14.1). The small roots far from the stem (s2, s3) are more suitable to study HR. It is less likely to detect HR in taproots or when the

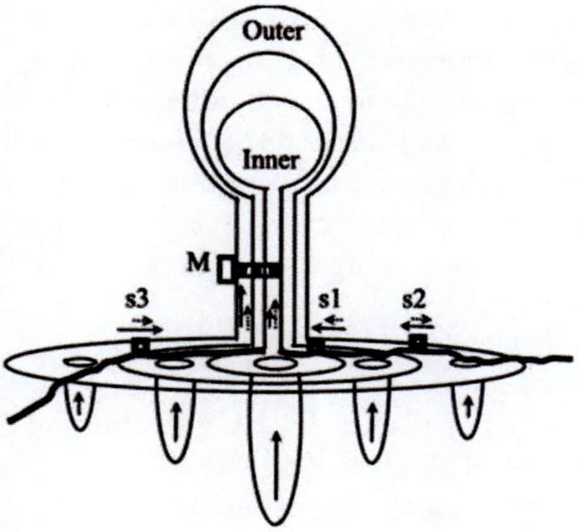

Fig. 14.1 Scheme of a model tree, showing the hydraulic connections between superficial and deep roots connected to different parts of the canopy through stem xylem layers. M is a multi-point sap flow sensor allowing detection of flow in different layers of stem or root xylem; s1, s2 and s3 are single-point sensors installed either in deep (s3) or in superficial (s2 and s1) roots

sensor is installed in a small root near stem (s1), because a sinker root may be connected to it after sensor positioning. Anyway, roots are good places for sensors, but the collected information at each point is only a part of the whole root system story. Stem is a more integrative place of measurement and allows the study of the functioning of the entire root system. Our results showed that the outer sapwood seems more important for water transport under favourable tree growth conditions, that is when water mainly comes from the surface soil layers where fine roots assure water and nutrient absorption. The outer xylem links these superficial roots to the well-illuminated (more active) foliage. On the other hand, the deep roots are presumably connected to the inner (shaded) foliage through the inner stem xylem layers (Nadezhdina 2010).

In this chapter, we will show different examples on the use of the HFD method to study the physiology of tree roots, with emphasis on water uptake and hydraulic redistribution.

14.2 Examples of Studies on Root Function Based on the HFD Method

Trees are "closed" hydraulic systems and, therefore, changes in one sector will reflect at the whole system level. These changes can be studied by monitoring the variations in the rate and direction of sap flow. All experiments and results reported in this chapter were obtained by the Heat Field Deformation (HFD) sap flow measuring technique applied to roots or stems of trees in different sites of the

Czech Republic and of other countries across Europe (Austria, Belgium, Germany and Portugal). Climate in all experimental sites is temperate, varying from continental to maritime and mediterranean, and the studied species included coniferous (*Picea abies* and *Pseudotsuga menziesii*) and hardwoods, either evergreen (*Quercus suber* and *Olea europaea*) or deciduous (*Acer platanoides, Quercus robur, Quercus petrea, Tilia cordata*). Site names and characteristics, such as the studied periods, soil and climate types, species (and organs studied) and used HFD sensor types are described in Table 14.1.

Details on the heat field deformation sap flow measuring technique and on type of sensors were described in the previous chapter (see Chap. 13). For functional studies in roots it is very important to measure bi-directional flows, a condition fulfilled by the application of HFD sensors (with a symmetrical pair of differential thermocouples). The same sensor configuration is used to record reverse flows, but flow direction (and thus, changing temperature gradients) should be taken into account. When values of dT_{sym} become negative, equation (13.1) (see Chap. 13) is transformed to:

$$q_i = -3600\mathrm{D}\frac{(-K + dT_{\mathrm{as}})}{(dT_{\mathrm{sym}} - dT_{\mathrm{as}})} \cdot \frac{Z_{\mathrm{ax}}}{Z_{\mathrm{tg}}} \tag{14.1}$$

As in structural studies, in root functioning research, often the qualitative variation and the direction of sap flow are important. Therefore, primary data of sap flow per section q_i (a surrogate of sap flow density) are used for the analysis without further conversion.

14.2.1 Sensitivity of Root Sap Flow to Abrupt Changes in Water Potential Gradients

The HFD method is highly sensitive to small changes in the hydraulic system of large trees. It can be used to analyse short-term responses of sap flow to small changes in water potential gradients in the soil–leaf continuum. One example of such a situation was observed when localized irrigation was applied on the eastern side of a maple tree, where stem sap flow was being recorded every minute by a multi-point sap flow sensor inserted at breast height in the same azimuth. An abrupt short-term response of sap flow was observed in response to the localized irrigation (Fig. 14.2), similarly to that found during root severing experiments (see Figs. 13.9 and 13.12). This peak in stem sap flow may be due to a sudden increase in the soil–leaf water potential gradient and to a decrease in root–soil resistance, under high diurnal evaporative demand and when a fast wetting of the dry soil occurs.

Table 14.1 Characterization of the study sites and sap flow measuring conditions

Figure	Site	Study period	Soil type	Climate type[a]	Species, organs	Type of sensor[b]
Fig. 14.2	Sobesice (Czech Rep.), 49°15′N, 16°37′ E, 368 m asl	August 2001	Illimerized soil on weathered granodiorite	Continental, annual temperature 7.8°C, annual precipitation 580 mm	*Acer platanoides*, stem	Multi-point, (10 mm)
Fig. 14.6		May–July 2001			*Tilia cordata*, small roots[c]	Single-point
Fig. 14.12		July–August 2005			*Pseudotsuga menziesii*, big roots[d]	Multi-point, (10 mm)
Fig. 14.13		August 2001			*Tilia cordata*, big root	single-point
Fig. 14.3	Vranov (Czech Rep.), 49°18′N, 16°37′E, 503 m a s l	May 2003	Gley-soil	Continental, annual temperature 7.5°C, annual precipitation 620 mm	*Picea abies*, big root	Multi-point (10 mm)
Fig. 14.4	Rajec (Czech Rep.), 49°26′N, 16°42′E, 632 m a s l	June 2005	modal cambisol	Continental, annual temperature 6.5°C, annual precipitation 717 mm	*Picea abies*, small roots	Single-point
Fig. 14.5	Liz, Sumava Mts (Czech Rep.), 49°04′N, 13°41′E, 850 m a s l	June 2007–Sept. 2009	Oligotrophic forest Eutric Cambisol on weathered Proterozoic biotite paragneisses	Temperate, annual temperature 6.3°C, annual precipitation 861 mm	*Picea abies*, small roots	Single-point
Fig. 14.7	Ferreira do Alentejo, (Portugal) Irrigated: 38°1.34′N, 8°10.84′W, 97 m asl; Non-irrigated: 38°0.95′N, 8°6.58′W, 150 m asl	June–November 2010	Irrigated: Luvisol; Non-irrigated: Vertisol	Mediterranean, annual temperature 16.0°C, annual precipitation 525 mm	*Olea europaea*, small root	Single-point
Fig. 14.8	Pouzdrany (Czech Rep.), 48°57′N, 16°36′E, 182 m a s l	April–May 2010	Fluvic Cambisol	Continental, annual temperature 9.5°C, annual precipitation 475 mm	*Quercus robur*, small roots	Single-point

(continued)

Table 14.1 (continued)

Figure	Site	Study period	Soil type	Climate type[a]	Species, organs	Type of sensor[b]
Fig. 14.9	Rio Frio, Palmela (Portugal), 38°38′N, 8°51′W, 30 m asl	March–September (2003)	Cambisol	Mediterranean, annual temperature 16°C, annual precipitation 750 mm	*Quercus suber*, small roots, stem	Single-point, Multi-point (10 mm)
Fig. 14.10	Nothern Upper Swabia (Germany), 48°01′N, 10°27′E, 600 m a s l	July–August 2004	Luvisol	Continental sub-montane, annual temperature 7.7°C, annual precipitation 775 mm	*Picea abies*, stem	Multi-point (10 mm)
Fig. 14.11	Furstenfeld (Austria), 47°05′N, 16°05′E, 320 m asl	August–September 2003	Stagnic Luvisoil	Continental, annual temperature 8.8°C, annual precipitation 765 mm	*Picea abies*, stem	Multi-point, (10 mm)
Fig. 14.14	Bilovice (Czech Rep.), 49°14′N, 16°40′E, 314 m a s l	September 1997	Clay soils with Devonian limestone as a parent rock	Continental, annual temperature 7.2°C, annual precipitation 561 mm	*Picea abies*, big root	Multi-point (10 mm)
Fig. 14.15	Wilrijk (Belgium) 51°10′N, 4°22′E, 16 m a s l	August 2004	Moderately wet sandy loam soil with particulated soil texture B horizon	Temperate maritime, annual temperature 9.6°C, annual precipitation 776 mm	*Quercus petrea*, big root	Multi-point (10 mm)

Sites are presented according to the order of Figures in the text. Nomenclature of vascular plants by Kubát et al. (2002)
[a]Annual temperature and annual precipitation are long-term means
[b]Numbers in brackets indicate distance between thermocouples along a measuring needle (sensor)
[c]Small roots: diameter between 2 and 7 cm.
[d]Big roots: diameter greater than 10 cm

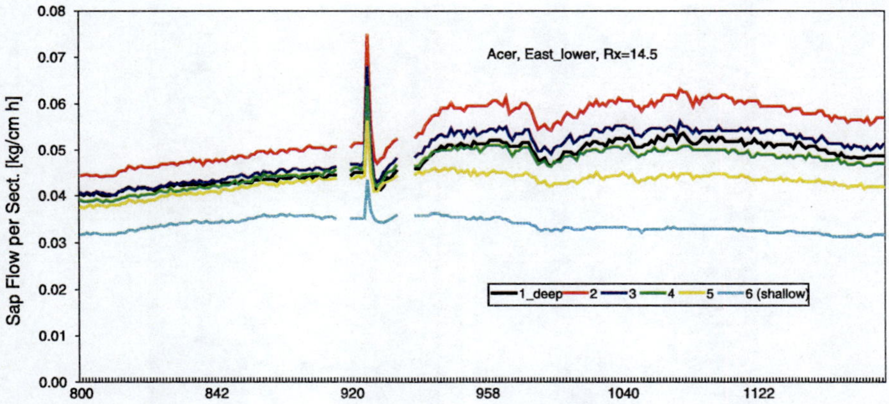

Fig. 14.2 Sap flow dynamics in different xylem layers of a maple stem sector, measured at breast height in the east azimuth, subject to localized soil irrigation in the same azimuth and close to the stem. Abrupt changes of water potential gradients in the soil–plant–atmosphere continuum resulted in short-term flow peaks, before a new equilibrium in the system was established. Data were recorded at 1 min time-step

14.2.2 Sensitivity of Root Sap Flow to Changes on Pressure Over the Soil

Sap flow measurements can also be used to evaluate the influence of soil compaction, due to heavy machinery pressure, on root function. Several experiments were carried out with spruce trees under different soil types and soil water conditions. Sap flow was monitored in big and small roots and also at the base of tree stems, before, during and after (up to 2 days) machinery movement. The area where tree roots had been under pressure was later exposed by an air spade and the uncovered roots analysed. All roots beneath the trucks were marked in different colours (Fig. 14.3), and the distances from the stems and depths belowground were measured. Sap flow changed during and after treatment only in the most superficial roots, up to 10–12 cm below soil surface (Nadezhdina et al. 2006b). Deeper roots were well protected by soil. In big roots, flow increased suddenly during truck movement (as observed during sawing experiments) due to the high mechanical pressure, but decreased afterwards to a level lower than before treatment (see Fig. 14.3). This effect of sap flow lowering continued through the following day after treatment. Sap flow changes were recorded in the outer root xylem layers (till 30 mm). Studies in different soil types showed that the influence of heavy machinery is more pronounced under soft and wet soil conditions.

14.2.3 Localized Irrigation to Study Root Ability to Redistribute Water

The application of localized irrigation to a dry soil close to a tree stem, under high evaporative conditions, is a very efficient and simple way to study root ability to

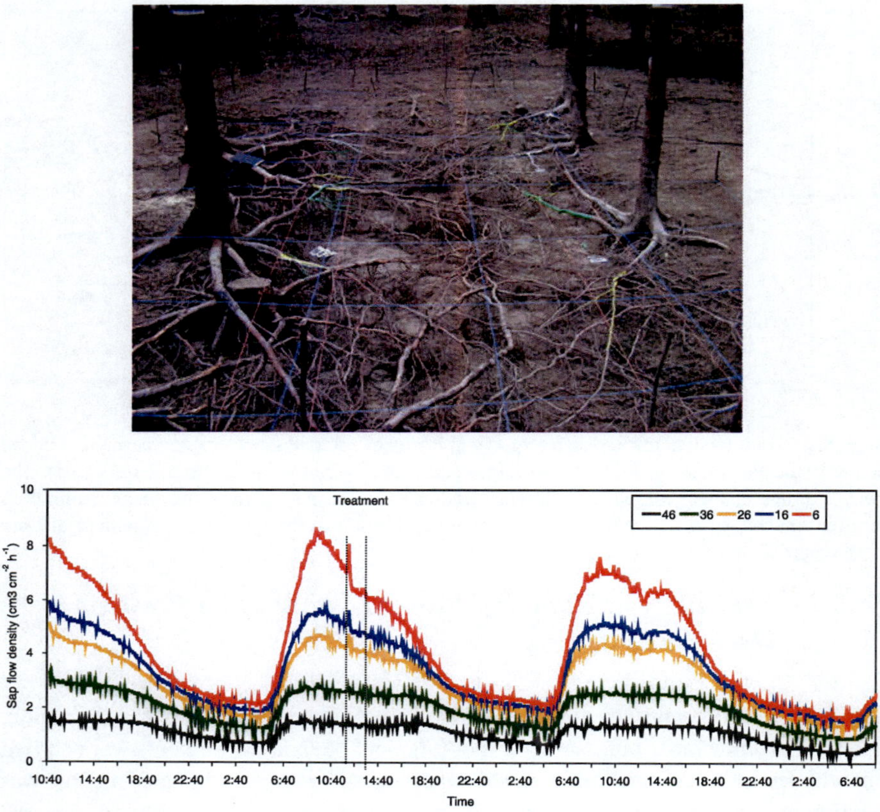

Fig. 14.3 *Upper*: View of the experimental plot Vranov after root exposure by an air-spade. Roots that were under mechanical pressure are marked by different colors. *Lower*: Sap flow dynamics in one of the measured roots (marked in *white* in the *left side* of the treated roots). Numbers in legend refer to depth below cambium (increasing with distance from cambium)

redistribute water. One example for a small spruce tree root is shown in Fig 14.4. After localized irrigation, sap flow increased in root 6 which was the closest to the irrigated soil. This extremely fast response of the watered root to irrigation was accompanied by a decrease in flow in the nearby non-watered root 5 and also to a lesser extent in root 4. Reverse flow was recorded in root 5 during night and also during the whole of the next day with drizzly weather, when leaf water potential was very high. By comparing flows in days with dry and wet surface soil, it is possible to recognize root position in soil even without soil excavation: root 6 was superficial (flow increased after irrigation and prolonged rain) and the other roots were deeper (flow decreased in those roots after rain because water was easily available from superficial roots).

This experiment shows the ability of spruce roots to quickly re-switch their absorption activity, in response to water availability. Therefore, the whole tree

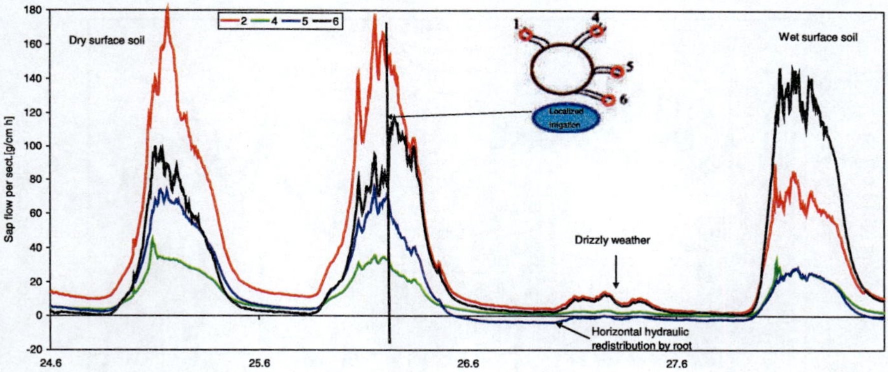

Fig. 14.4 Sap flow dynamics in four small roots of a spruce tree, where localized irrigation was applied to the dry soil in a hot day (timing of irrigation is marked by a *vertical line*)

may survive under dry conditions if some of the roots can uptake water from a permanent groundwater reservoir.

14.2.4 Monitoring of Root Water Uptake to Study Long-Term Root Functioning

Due to its simple construction, the HFD sensor can be easily installed in roots providing direct measurements of water uptake. Sap flow records are especially valuable when carried out for long periods. Sensors can robustly work at the same place during several months or even years. One example of a long-term study (3 years) of water uptake by two spruce roots in a temperate forest at the Sumava Mountains (Czech Republic) is presented in Fig. 14.5. Sap flow sensors were inserted at a 50 and 80 cm distance from the tree stem on roots r1 and r2, respectively. Root diameter was similar: 6.3 and 6 cm in r1 and r2, respectively. The long-term records of root sap flow showed that root water uptake pattern changed according to soil moisture and weather conditions: during winter root water uptake was greatly reduced and increased under warmer weather. When the surface soil was wet, during summer 2007, sap flow was higher in the more superficial root (r1). However, after a very wet winter and till mid-summer 2008, sap flow was higher in the deeper root r2 because soil water content was higher in the 60 cm soil layer than in top soil. The dry winter and spring periods of 2009 restricted root water uptake in both roots. The relationship between maximum sap flow in both roots showed that differences in water uptake between roots occurred during high evaporative demand conditions, when sap flow was highest. Under such conditions, water uptake was re-switched to the root that was in the wetter soil layer. Furthermore, water uptake from these two roots also differed in the timing of hydraulic redistribution occurrence (Nadezhdina et al. 2010).

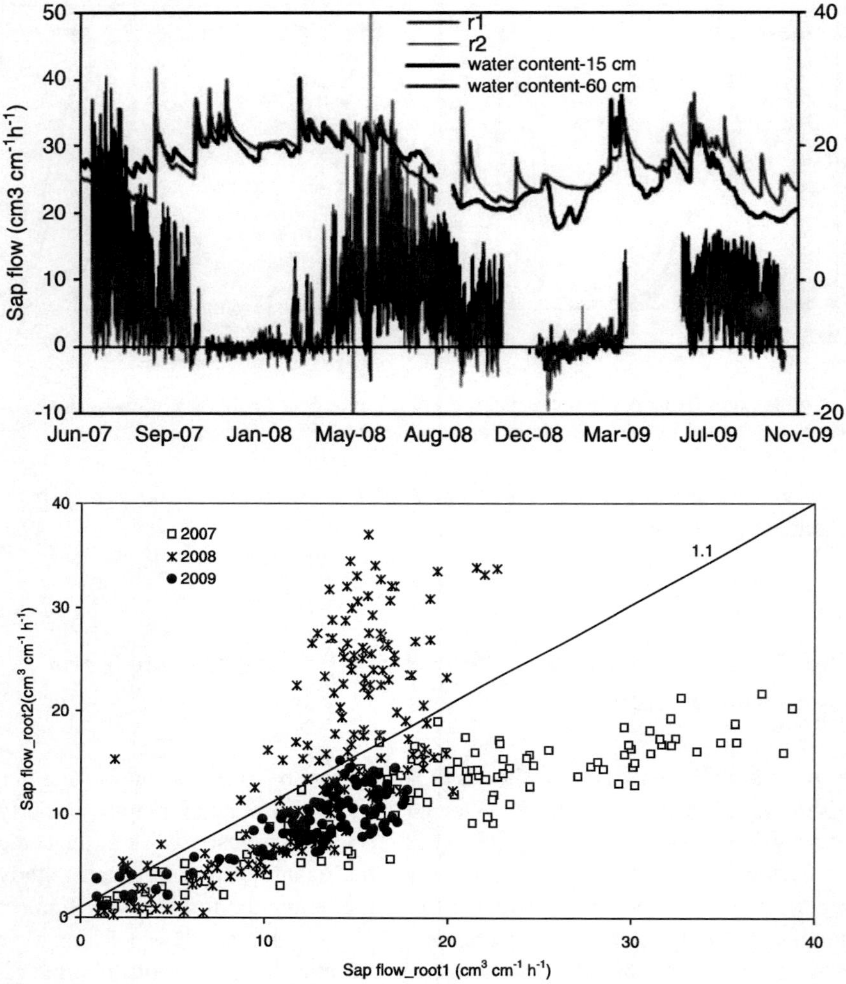

Fig. 14.5 Long-term (3 years) sap flow dynamics in two small roots of a spruce tree. Soil water content in two soil layers (15 cm – *black* and 60 cm – *gray*) are also shown. *Lower panel*: Relationship between maximum sap flow in both roots for three growing seasons

14.2.5 Monitoring of Root Water Uptake to Study Root Functioning at the Edge of a Forest

Long-term records of sap flow in roots (Fig. 14.6) can also provide insight into root functioning of trees in different social positions in a forest. An experiment was conducted in a lime tree growing at the edge of a forest in Sobesice. Trees were absent from the northern and eastern sides of the studied tree, but close on the southern and western sides. Three single-point HFD sensors were installed in three

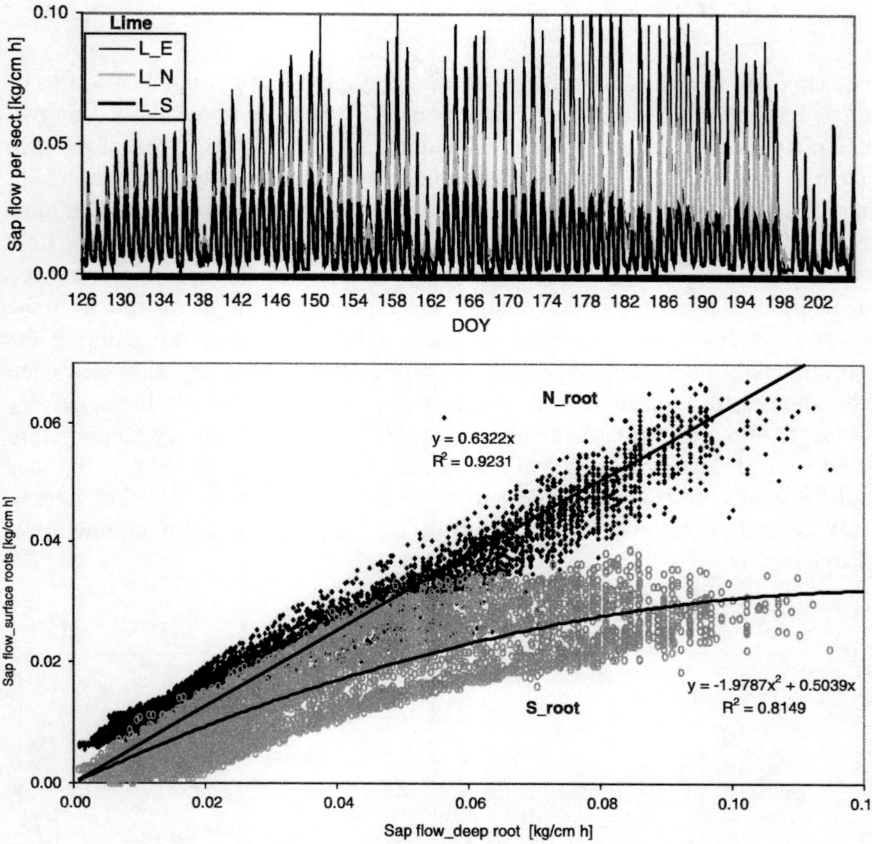

Fig. 14.6 *Upper panel*: Sap flow dynamics in roots of the lime tree recorded during May to July of 2001 at the Sobesice forest. The sampled tree was growing at the northern edge of the forest. East-oriented big root was going deep to the soil, whereas two other smaller roots, south and north oriented, were close to the surface. *Lower panel*: The relationship between sap flows recorded in the deep east-oriented root and in two surface roots

roots close to the tree stem base: two roots were superficial and south and north oriented, and the third one, bigger and oriented to east, reached deeper soil horizons. At the beginning of the season, sap flow in the supeficial roots was of the same magnitude. However, from the middle of June 2001 onwards, flow in the southern root gradually decreased, whereas flow in the roots from the open northern and eastern sides, where no other trees competed for water, increased in response to higher evaporative demands. Upon the onset of the heavy rain period (end of August), root behaviour resumed to that of the beginning of the season.

14.2.6 Study of Differences in Water Uptake by Non-irrigated and Irrigated Olive Trees

Root sap flow dynamics may also provide insight into the water uptake patterns of irrigated and non-irrigated trees. Single-point HFD sensors were installed on two small olive roots on each of two experimental plots, one irrigated by localized drip system and the other non-irrigated. Edapho-climatic conditions were similar in both plots. The seasonal dynamics of root sap flow showed that olive roots in both plots have different water uptake strategies (Fig. 14.7). In the beginning of June, sap flow in the dry plot peaked in both roots and gradually decreased in a more pronounced way in the shallower root. Negative night flows started to occur in this root from the beginning of June and the magnitude and duration of reverse flows during the night gradually increased. This shows that olive trees in the dry plot used deep water sources for transpiration and watered the superficial roots in the upper dry soil layers through vertical hydraulic redistribution, VHR. High night flows were recorded in the deep root, which was gradually decreasing as deep soil water availability also decreased. Changes in water uptake patterns occurred in autumn (DOY 281) after rain: sap flow increased in the shallow root and nighttime flows became positive.

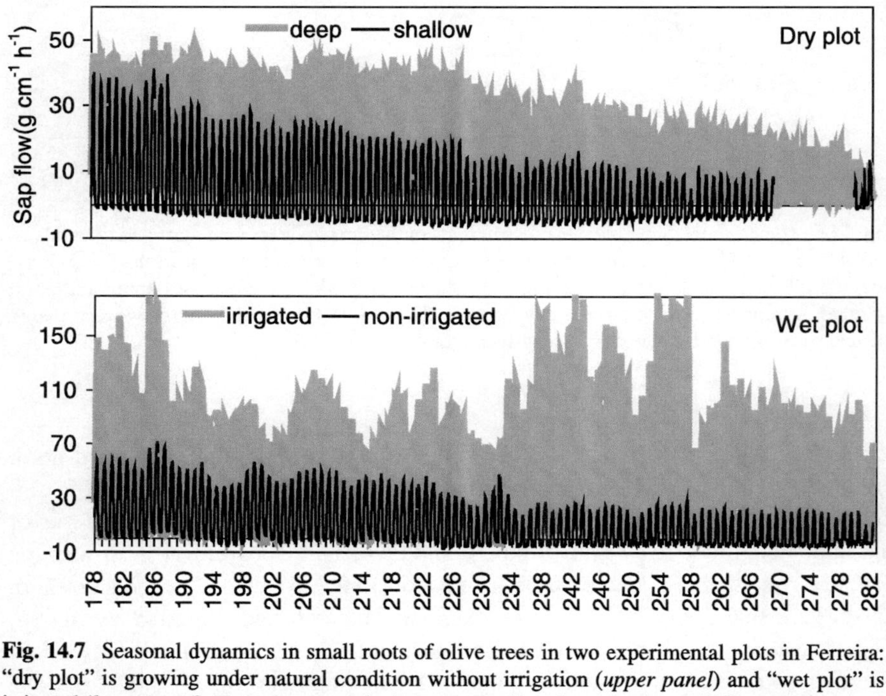

Fig. 14.7 Seasonal dynamics in small roots of olive trees in two experimental plots in Ferreira: "dry plot" is growing under natural condition without irrigation (*upper panel*) and "wet plot" is irrigated (*lower panel*). Two roots at each plot were monitored to demonstrate differences in water uptake: deep and shallow roots in the dry plot and irrigated and non-irrigated roots in the wet plot

At the wet plot, sap flow was much higher in the irrigated root due to the surplus of water supply. Variations in time of water uptake in the irrigated root were solely due to changes in the atmospheric evaporative demand in summer (June, August and September). Night flows only occurred in this root under high evaporative demand conditions, being otherwise equal or close to zero. Negative night flows started to occur in the non-irrigated root from July onwards and their magnitude and duration were highest from August up to the end of studied period. We think that the reverse night flows in the non-irrigated root are in this case due to the horizontal hydraulic redistribution, HHR, from the wet/irrigated side where water was easily available. From the middle of August onwards the non-irrigated root and the roots at the dry plot did not follow the atmospheric evaporative demand contrary to what happened in the irrigated root (see Fig. 14.7). Restrictions in soil water were then dominant in the water flux pathway.

We hypothesize that olive trees growing under different conditions of water availability (irrigated or non-irrigated) will develop differently structured root systems. Irrigated trees should have a large number of superficial roots that will be able to absorb water from nearby drips and feed not only transpiration but also other roots in non-irrigated locations via HHR. On the other hand, non-irrigated trees should develop a large number of deep roots to tap water during the summer drought from the deep soil. They will also develop a lignotuber in order to be able to store water for severe drought periods.

14.2.7 Study of the Differences Between Water Uptake in Healthy and in Infected Trees by Hemiparasite

Oak trees are rather often infected by the hemiparasite *Loranthus europaeus*. In order to understand the differences in the water regime of healthy and infected trees, we analysed root water uptake in three small roots of neighbor *Quercus robur* trees. Single-point sensors were inserted in the roots at the end of April and sap flow was monitored during the 2010 growing season. We found that root water uptake by the infected tree (193) was higher than that of the healthy tree (205) (Fig. 14.8). This could be explained by a lower water potential in leaves of the hemiparasite hosted in oak tree compared to oak leaves. The higher water uptake of the infected tree may lead to host drying during limited water supply. We also found that root water uptake of both trees was related to leaf development: sap flow started at least 8 days earlier in roots of the infected tree than in roots of the healthy one (see Fig. 14.8).

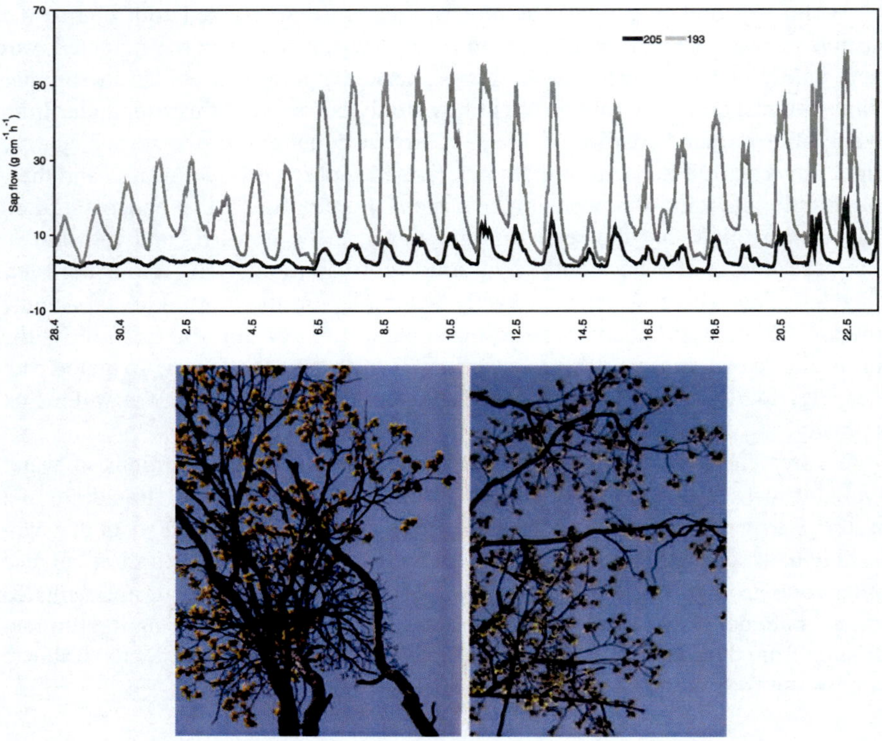

Fig. 14.8 Sap flow dynamics in roots of two neighbour *Quercus robur* trees. Each line represents mean sap flow of three roots of infected (193) and healthy (205) trees. Numbers in the legend refer to tree circumference at breast height. *Lower panels*: The leaf development is delayed almost one week on healthy oak (*right*) compare to infected oak tree (*left*)

14.2.8 Monitoring of Sap Flow in Roots and Different Stem Xylem Depths to Study the Ability of Root Systems to Maintain Transpiration During Drought

During drought periods, when soil water content in some horizons drops below a certain limit, the functioning of the root system drastically changes, adjusting to the other possible sources of deeper water and through mechanisms of water redistribution in soil via roots. Trees homogenize moisture in the whole bulk of soil occupied by roots much faster than what would occur in non-colonized soil.

As was mentioned in Chap. 13, roots from different soil horizons are presumably connected to different layers of stem xylem. Therefore, not only sap flow in roots but also in stem along its xylem radius may give integrated information on the water supply by the whole root network. Several examples for trees with different types of root systems (deep and shallow rooted) are presented below.

14.2.8.1 Deep Rooted Tree

Quercus suber roots may reach several meters in depth (Nadezhdina et al. 2008). The ability of these species to support high transpiration during the summer drought was studied in Portugal in 2003. Sap flow sensors were installed in several small roots and also at tree stem.

When the whole bulk of soil occupied by roots was wet, flows in both root types (shallow and deep) as well as in different stem xylem layers were proportionally related (upper panels in Fig. 14.9). With gradual topsoil drying due to high evaporative demand, flow in the deep root continued to increase whereas a progressive daytime flow decrease and a nighttime reverse flow increase was evident in the shallow root, indicating hydraulic redistribution via roots. Sap flow also gradually increased in the inner stem xylem and decreased in the outer one (middle panels in Fig. 14.9). After the heavy rain at the end of August, reverse flow stopped in the

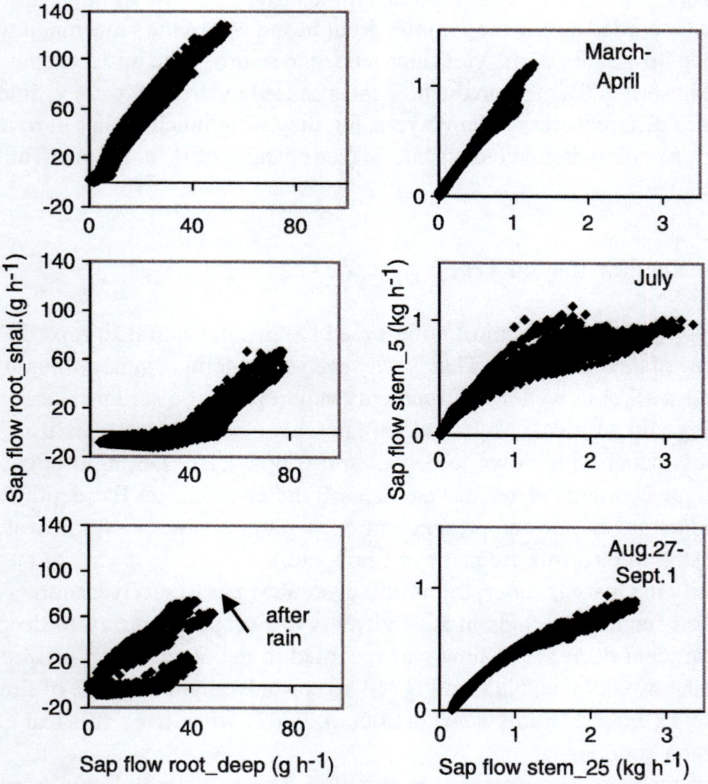

Fig. 14.9 Relationship between sap flows in two small roots (shallow and deep - *left*) and two xylem layers of a *Quercus suber* stem (*right*), during different periods of the 2003 growing season: wet soil (*upper panels*); drying topsoil (*middle panels*); end of drought – following rain at the end of August (*lower panels*)

Table 14.2 Equation of regression between the small roots (deep root is x, shallow root is Y) and between the outer (5 mm below the cambium) and the inner (25 mm below the cambium) stem xylem (the inner xylem is x, the outer xylem is Y) calculated for each month of the 2003 growing season

Month	Roots equation r^2	Stem equation r^2
March	$Y = 2.163 \times 0.97$	$Y = 0.862 \times 0.99$
April		$Y = 0.779 \times 0.98$
May	$Y = 2.008 \times 0.96$	$Y = 0.586 \times 0.95$
June		$Y = 0.526 \times 0.93$
July	$Y = 1.838 \times 0.91$	$Y = 0.431 \times 0.89$
August		$Y = 0.350 \times 0.90$
September	$Y = 0.963 \times 0.79$	$Y = 0.411 \times 0.91$
	$Y = 0.470 \times 0.64$	
	$Y = 0.241 \times 0.47$	
	$Y = 1.594 \times 0.97$	

shallow root and daily flow increased significantly. Sap flow in the deep root was half of that recorded during the summer drought and was of the same magnitude as in March. Sap flow in the stem xylem also tended to return to the initial proportionality observed in spring. Changes in the flow rates caused by drought were visible both in roots and in different stem xylem layers, but they were much higher in roots due to the occurrence of hydraulic redistribution (see changes of r^2 for roots in Table 14.2).

14.2.8.2 Shallow Rooted Tree

Norway spruce, one of the most widespread species in Central Europe, belongs to the shallow plate root system. That is why even short-term summer drought (lasting for several weeks) in wetter environments may represent a serious hazard to plant functioning and survival. Nevertheless, lateral roots of spruce trees usually have sinker roots penetrating down to deeper soil (Fig. 14.10). During drought, spruce tree survival can depend on the number of sinker roots, on the depth to which sinkers penetrate and on the presence of deep water sources (level of water table, type and structure of soil, frequency of rain, etc.).

Similarly to *Quercus suber*, the ability of spruce trees to survive summer drought was studied for long periods in several plots in Europe. Contrary to deep rooted trees, significant decrease of flow was recorded in the outer xylem of spruce trees during drought with a stable (see Fig. 14.10) or only slight increase of flow in the inner xylem (see Fig. 6 in Nadezhdina et al. 2007) when trees still had access to deeper water source.

If deep water is also scarce, transpiration does not depend any more on the atmospheric driving forces and can drop up to zero, as observed in spruce trees during the severe drought of 2003 (Furstenfeld, Austria – in Nadezhdina et al. 2010). Tree survival then depends on physiological adjustment to extreme water stress. Sap flow in stem xylem may well typify this situation (Fig. 14.11): at the end

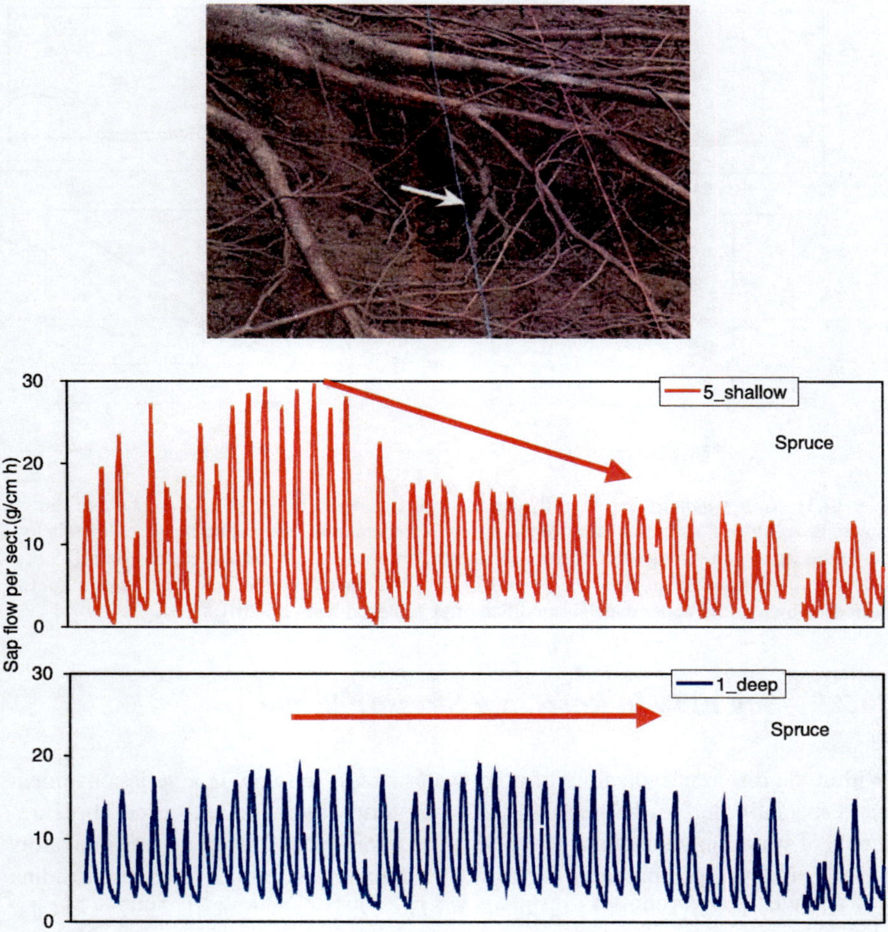

Fig. 14.10 *Left*: Roots of a spruce tree opened by the air-spade technique. *White arrow* indicates sinker roots. *Right*: Sap flow dynamics in the stem of a spruce tree measured in the outer (*upper panel*) and inner (*lower panel*) stem xylem

of drought from day 239 till 242 flow in stem was extremely low and similar in the inner (45–55 mm) and outer (15–25 mm) xylem, day and night, whereas after rain flows were much higher and clearly different between layers during the day. During the drought, night flows occurred only in the inner xylem and reverse flow was recorded in the outer xylem, indicating the occurrence of hydraulic redistribution mediated by the crown and soil (Nadezhdina et al. 2010). Relationships between flows in the outer and inner xylem against flows in the middle xylem layer during severe drought and for wet soil (see Fig. 14.11) clearly show the ability of sap flow measurements in stems to characterize different water uptake patterns within the root system.

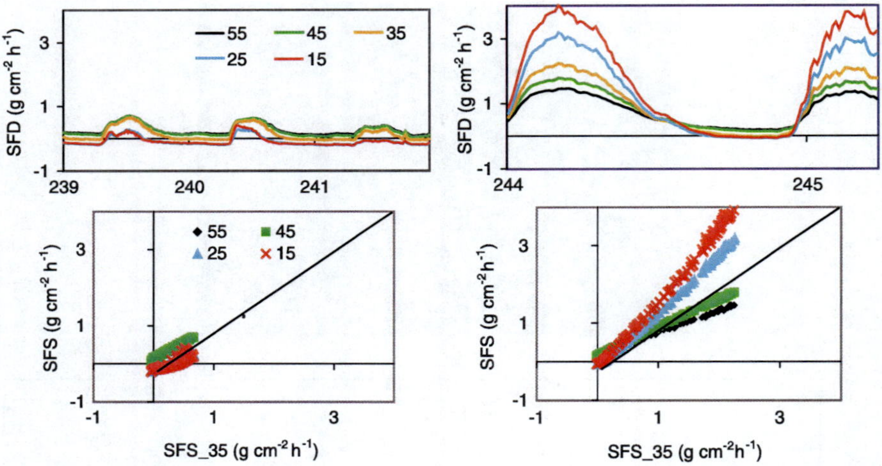

Fig. 14.11 *Upper panels*: Sap flow dynamics in different xylem layers of a spruce tree during severe drought (*left*) and after rain events (*right*). Numbers in legend indicate xylem depth below cambium (mm). *Lower panels*: Relationships between flows in the middle stem xylem (35 mm below the cambium) and in the inner (45 and 55 mm) and outer (25 and 15 mm) stem xylem for the periods shown in the upper panels. (modified after Nadezhdina et al. 2010).

14.2.9 Sap Flow in Roots as a Stress Indicator

Within the different indicators of plant water stress, sap flow is a very convenient one especially due to the possibility of long-term automatic and direct measurements of water movement through plants. Different types of water stress indicators have been developed based on sap flow measurements (Cermák 1986; Nadezhdina 1988, 1989, 1992, 1999a, b; Cabibel and Do 1991; Cohen et al. 1997; Ginestar et al 1998; Massai et al. 2000; Fernandez et al 2001; Remorini and Massai 2003). Sap flow in roots is particularly informative to identify drought periods, since soil drought can be very easily characterized by the occurrence of reverse flow in superficial roots (Burgess et al. 1998). Beyond the application of localized irrigation during drought can increase the informative nature of sap flow in roots as a stress indicator.

We can see this on an example of sap flow measurements in two opposite roots of Douglas-fir recorded during drought, followed by localized irrigation of one root and its drying during continued hot weather, and finally after rain events. Sap flow in both roots was in equilibrium under homogeneous soil conditions (see dry and wet periods in Fig. 14.12). When heterogeneities appeared due to the application of localized irrigation, the system quickly changed to another state (within hours) and then gradually returned again to the previous state (within 9 days in our case) (see both middle panels in Fig. 14.12). This can be roughly represented as a hysterezis loop (Fig. 14.12), where deviation from the baseline, length of period, needed to return to equilibrium, and existence and duration of HR can be used as an indicator

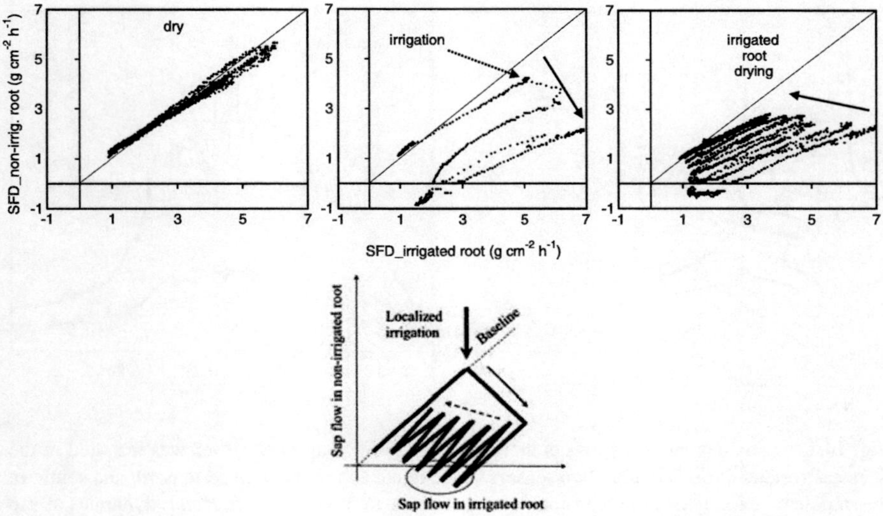

Fig. 14.12 *Upper panels*: Relationship between sap flow density, SFD, in big opposite irrigated (south oriented) and non-irrigated (north oriented) roots of a Douglas-fir tree for different periods: dry (*left*), during and 1 day after localized irrigation (*middle*); following gradual soil drying (*right*). Data refer to the 25 mm layer below cambium.
Lower panel: Scheme demonstrating the use of localized irrigation as stress indicator based on analyses of the relationship between sap flow in irrigated (*X*-axes) and non-irrigated (*Y*-axes) roots (roots from opposite side have to be analysed to have maximum response). *Full thin arrow* indicates direction and amplitude of changes of fluxes after localized application of water. Amplitude and inclination from baseline (characterizing relationship between sap flow in both roots under homogeneous conditions) would depend on stress level caused by dry soil and also would depend on quantity of applied water and ability of studied species to redistribute water horizontaly. *Dashed arrow* indicates gradual return to the new balance in soil from both tree sides. Each zigzag indicates daily circle of fluxes with higher value for midday transpiration and negative value for redistributed water. The more zigzags and numbers (nights) with reverse fluxes the severer stress conditions occurred

of water stress level. These features should be species-specific. Thus, similar tests on different species growing in the same conditions can be used to characterize their plasticity or physiological response to drought.

14.2.10 Compensation Mechanisms in Roots

The unique ability of the HFD method to record very fast responses in the tree hydraulic system can be used for the understanding of mechanisms of tree survival under sudden destructions. We studied for example which compensation mechanisms occurred in roots when some roots or their certain parts are destroyed.

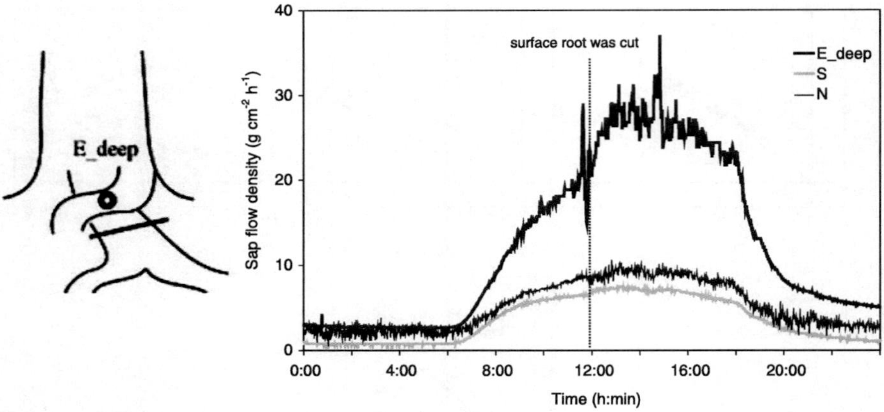

Fig. 14.13 *Left*: scheme of two roots of *Tilia cordata* tree. Sap flow sensor was installed in the deep east-oriented root. Another two sensors were situated on roots oriented to north and south. In the middle of a sunny day the big superficial east-orientated root was cut. *Right*: dynamics of sap flow in different roots shows immediate increase of flow in the eastern deep root after root cutting with a small flow increase in other measured roots. Note high peak flow occurrence during severing of neighbouring root

14.2.10.1 Compensation Mechanism Between Different Roots

Two big coarse roots of a lime tree (*Tilia cordata*, DBH = 27 cm) were situated on its eastern side and had visibly different penetration patterns: one was superficial and the second one was going deep in soil (Fig. 14.13). Single-point sensor was installed on the deep root and also in two other roots oriented to North and South. When the big superficial root was cut from the eastern side of the lime tree, an immediate increase in flow was recorded in the deep eastern root. Flow also increased in the two other measured roots, but substantially less than in the deep root from eastern side. This may indicate that both roots from the eastern side were connected to a common part of the crown whereas roots from south and north were connected with other crown parts. High peak of flow during severing of neighbouring root also confirms a common sink for both roots (see also Sect. 13.3.5). Flow in the east deep root was also higher in the following day compared to pre-cut values. Therefore, the abrupt loss of a significant part of the water source could be compensated for by an increase in flow in the other roots connected to the same sink. The amplitude and duration of such compensation could be limited by soil water availability.

14.2.10.2 Compensation Mechanism Between Different Xylem Layers of the Same Root

The radial profile sensor was installed on big coarse root of a spruce tree and its higher order roots were gradually cut (Fig. 14.14). Each cut caused flow redistribution between xylem layers of the big root. Therefore, a compensation mechanism

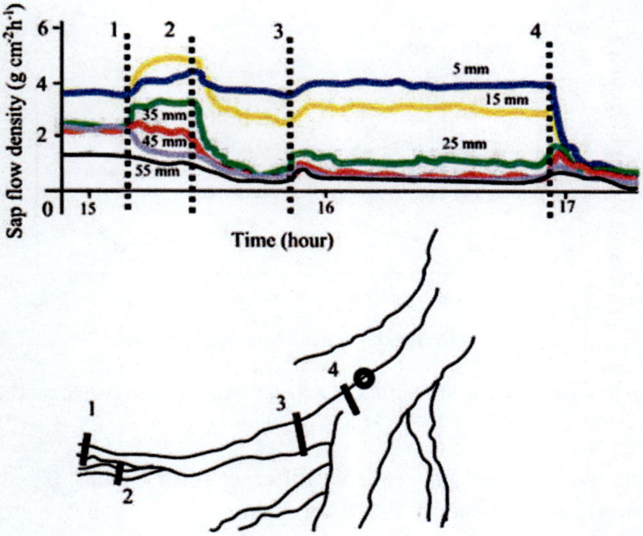

Fig. 14.14 *Upper panel*: sap flow dynamics in different xylem layers of a large spruce tree root. Numbers near colored lines indicate xylem depths below cambium; numbers near vertical dashed lines indicate cuttings of different branches of monitored root. *Lower panel*: scheme of a big spruce root with three complete cuttings different branches (1, 2 and 3) and one partial cut to the depth of sensor insertion (6 cm)

may operate in the same large root allowing an increase in water absorption by one part of the root system when another part of the root system is damaged or loses its water source. This mechanism represents an important safety feature at root level for tree survival.

Ability of root xylem to re-switch absorption between xylem layers was also checked in a ring-porous tree – *Quercus robur* L. Sap flow was measured by a multi-point sensor in a big root of an oak tree. After several days of measurements, part of the root just below the sensor was cut by saw till a depth of 23 mm below the bark. An immediate decrease in flow was recorded in the three outer xylem layers that were damaged (Fig. 14.15). However, at the same time, flow increased in deeper non-damaged xylem layers indicating an ability of the big oak root to compensate for partial root damage, similarly to what was observed in the spruce root.

14.3 Final Remarks

In this chapter, we report data from sap flow measuring experiments carried out in 11 sites across Europe that included 8 tree genera. Obtained results demonstrate well that sap flow measurements are a good tool to study in situ root behaviour. Long-term sap flow measurements are non-destructive, automatic and extremely informative. Seasonal variation of sap flow in roots and stem xylem directly show

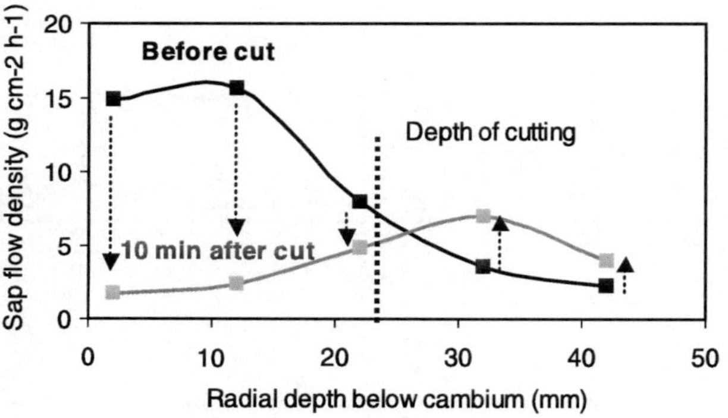

Fig. 14.15 Sap flow radial profile in a big oak tree root, before and after partial root cutting by saw

changes in the water uptake pattern from different soil horizons. The outer trunk xylem was mostly responsible for the water movement from surface soil horizons, whereas deep roots supplied water to the whole stem sapwood area, which is particularly important during drought. These findings were consistent for all studied genera. When sap flow substantially decreased both in shallow roots and in the outer stem xylem during drought, water flow was maintained (shallow rooted trees) or increased (deep rooted species) in the inner sapwood, due to its direct connection to sinker roots. Shortage in soil water caused hydraulic redistribution (HR) of different nature (from the location of higher water availability to the dry soil). Root sap flow has been shown to be a good, species-specific, stress indicator. Our results also highlighted that a compensation mechanism in sap flow may develop between different xylem layers in the same root or between different roots in response to sudden root (water source) destruction.

Overall, results on root functioning obtained through sap flow research help to deepen the understanding of the physical and biological background of the hydraulic movement of water within the soil–leaf pathway. Therefore, results from this type of research are of extreme value for the conceptual modelling in this field.

The experimental studies reported in this chapter were partially supported by the projects:

- Czech national projects: MSM 6215648902; GACR 526/08/1050; GACR 205/08/1174; MECR SP/1a6/151/07; IGA LDF 12/2010
- Portuguese Foundation for Science and Technology projects: POCI-PPCDT/AGR/59152/2004 and PTDC/AAC-AMB/100635/2008 WUSSIAAME
- European projects "WATERUSE" (EVK1-2000-00079EU) and "Sustman" (QLK5-CT-2002-00851)
- Bilateral scientific exchange projects between the University of Antwerp (Flemish Community) and the Mendel University of Brno (Czech Republic), 2004–2005
- Internal project of the Federal Research and Training Centre for Forests, Natural Hazards and Landscape (BFW), Vienna

References

Brooks JR, Meinzer FC, Coulombe R, Gregg J (2002) Hydraulic redistribution of soil water during summer drought in two contrasting Pacific Northwest coniferous forests. Tree Physiol 22:1107–1117

Burgess SSO, Bleby TM (2006) Redistribution of soil water by lateral roots mediated by stem tissues. J Exp Bot 57:3283–3291

Burgess SSO, Adams MA, Turner NC, Ong CK (1998) The redistribution of soil water by tree root systems. Oecologia 115:306–311

Burgess SSO, Adams MA, Turner NC, Beverly CR, Ong CK, Khan AAH, Bleby TM (2001) An improved heat pulse method to measure low and reverse rates of sap flow in woody plants. Tree Physiol 21:589–598

Cabibel B, Do F (1991) Thermal measurement of sap flow and hydric behaviour of trees. 2. Sap flow evolution and hydric behaviour of irrigated and non-irrigated trees under trickle irrigation. Agronomie 11:757–766

Cermák J (1986) Short- and long-term response of transpiration flow rate in full-grown trees to water stress. In: Proc.18th IUFRO World Congress, Whole-Plant Physiology Working Party, Plesko Ljubljana, p 187–193

Cohen M, Ameglio T, Cruiziat P, Archer P, Valancogne C, Dayau S (1997) Yield and physiological responses of walnut trees in semiarid conditions: application to irrigation scheduling. Acta Hort 449:273–280

Fernandez JE, Palomo MJ, Diaz-Espejo A, Clothier BE, Green SR, Giron IF, Moreno F (2001) Heat-pulse measurements of sap flow in olives for automating irrigation: tests, root flow and diagnostics of water stress. Agr Wat Manage 51:99–123

Ginestar C, Eastham J, Gray S, Iland P (1998) Use of sap-flow sensors to schedule vineyard irrigation. I. Effects of post-veraison water deficits on water relations, vine growth, and yield of Shiraz grapevines. Am J Enol Vitic 49:413–420

Hultine KR, Scott RL, Cable WL, Goodrich DC, Williams DG (2004) Hydraulic redistribution by a dominant, warm-desert phreatophyte: seasonal patterns and response to precipitation pulses. Funct Ecol 18:530–538

Kubát K, Hrouda L, Chrtek jun J, Kaplan Z, Kirschner J, Štěpánek J (2002) Klíč ke květeně České republiky (Kye to the Flora of the Czech Republic). Academia, Praha

Massai R, Ferreira MI, Paco TA, Remorini D (2000) Sap flow in peach trees during water stress and recovery in two environmental conditions. Acta Hort 537:351–358

Meinzer FC, Brooks JR, Bucci S, Goldstein G, Scholz FG, Warren JM (2004) Converging patterns of uptake and hydraulic redistribution of soil water in contrasting woody vegetation types. Tree Physiol 24:919–928

Moreira MZ, Scholz FG, Bucci SJ, Sternberg LS, Goldstein G, Meinzer FC, Franco AC (2003) Hydraulic lift in a neotropical savanna. Funct Ecol 17:573–581

Nadezhdina N (1988) Apple tree water relations and their optimization under conditions of southern Ukraine (in Russian). Ph.D. Thesis. Kiev

Nadezhdina N (1989) A physiological algorithm of woody plant irrigation control under air drought (in Russian). Fiziologiya Rastenii 36:972–979

Nadezhdina NE (1992) A method of estimating negative temperature resistance of fruit trees (in Russian). Certificate of authorship 1702937 USSR AO1G 7/00. BI 1

Nadezhdina N (1999a) Sap flow index as an indicator of plant water status. Tree Physiol 19:885–891

Nadezhdina N (1999b) Woody plant behavior and stress assessment based on sap flow measurement. Application in forestry and horticulture. Assoc. Prof. Thesis at the Mendel University of Agriculture and Forestry in Brno, Czech Rep

Nadezhdina N (2010) Integration of water transport pathways in a maple tree: responses of sap flow to branch severing. Ann For Sci 67(107)

Nadezhdina N, Čermák J (2000a) Responses of sap flow in spruce roots to mechanical injury. In: Klimo E, Hager H, Kulhavy J (eds) Spruce monocultures in Central Europe: problems and prospects, EFI Proceedings No 33. European Forest Institute, Joensuu, pp 167–175

Nadezhdina N, Čermák J (2000b) Responses of sap flow rate along tree stem and coarse root radii to changes of water supply. In: Stokes A (ed) Proceedings of the supporting roots of trees and woody plants: form, function and physiology. Klouwen, Dordrecht-Boston-London, pp 227–238

Nadezhdina N, Cermak J (2003) Instrumental methods for studies of structure and function of root systems in large trees. J Exp Bot 54:1511–1521

Nadezhdina N, Cermak J, Nadezhdin V (1998) Heat field deformation method for sap flow measurements. In: Čermak J, Nadezhdina N (eds) Measuring sap flow in intact plants, Proceedings of 4th International Workshop, Zidlochovice, Czech Republic. IUFRO Publications, Brno, pp 72–92

Nadezhdina N, Tributsch H, Čermák J (2004) Infra-red images of heat field around a linear heater and sap flow in stems of lime trees under natural and experimental conditions. Ann For Sci 61:203–213

Nadezhdina N, Čermak J, Gasparek J, Nadezhdin V, Prax A (2006a) Vertical and horizontal water redistribution within Norway spruce (*Picea abies*) roots in the Moravian Upland. Tree Physiol 26:1277–1288

Nadezhdina N, Cermak J, Neruda J, Prax A, Ulrich R, Nadezhdin V, Gasparek J, Pokorny E (2006b) Roots under the load of heavy machinery in spruce trees. Eur J For Res 125:111–128

Nadezhdina N, Cermak J, Meiresonne L, Ceulemans R (2007) Transpiration of Scots pine in Flanders growing on soil with irregular substratum. For Ecol Manage 243:1–9

Nadezhdina N, Ferreira MI, Silva R, Pacheco CA (2008) Seasonal variation of water uptake of a *Quercus suber* tree in Central Portugal. Plant Soil 305:105–119

Nadezhdina N, Steppe K, De Pauw DJW, Bequet R, Cermak J, Ceulemans R (2009) Stem-mediated hydraulic redistribution in large roots on opposing sides of a Douglas-fir tree following localized irrigation. New Phytol 184:932–943

Nadezhdina N, David TS, David JS, Ferreira I, Dohnal M, Tesar M, Gartner K, Leitgeb E, Nadyezhdin V, Čermák J, Jimenez MS, Morales D (2010) Trees never rest: the multiple facets of hydraulic redistribution. Ecohydrology 3:431–444

Remorini D, Massai R (2003) Comparison of water statud indicators for young peach trees. Irrig Sci 22:39–46

Ryel RJ (2004) Hydraulic redistribution. In: Esser K, Lüttge U, Beyschlag W, Murata J (eds) Progress in Botany, vol 65. Springer, Berlin, Heidelberg, New York, pp 413–435

Stokes A, Berthier S, Nadezhdina N, Cermák J, Loustau D (2000) Sap flow in trees is influenced by stem movement. In: Spatz HCh, Speck T (eds) Proceedings of 3 rd plant biomechanics conference, Freiburg-Badenweiler. Georg Thieme Verlag Stuttgart, New York, pp 272–277

Warren JM, Meinzer FC, Brooks JR, Domec JC, Coulombe R (2007) Hydraulic redistribution of soil water in two old-growth coniferous forests: quantifying patterns and controls. New Phytol 173:753–765

Chapter 15
Fine Root Dynamics and Root Respiration

Karibu Fukuzawa, Masako Dannoura, and Hideaki Shibata

Abstract Studies of fine root phenology and respiration in forest ecosystems are reviewed. Direct, nondestructive observation methods, such as the minirhizotron imaging system and the root window, provide simultaneous quantitative measurements of root production and mortality. Temporal patterns are discussed, as well as endogenous and exogenous factors (e.g., soil temperature and moisture) controlling fine root dynamics. Both root respiration and microbial respiration release CO_2 from the soil to the atmosphere; methods that distinguish them are evaluated. Finally, factors controlling root respiration, such as root age and diameter and soil nitrogen concentration, are reviewed, and approaches for scaling up root respiration to stand level are discussed.

15.1 Introduction

Fine roots are usually defined as <2–5 mm and sometimes up to 10 mm in diameter. The processes of production, death, and decomposition are referred to as "fine root dynamics." Understanding temporal patterns of fine root dynamics is crucial for estimating turnover and productivity, which are important to carbon and nutrient cycling in forest ecosystems (Hendrick and Pregitzer 1993a; Jackson et al. 1997; Gill and Jackson 2000). Various destructive and nondestructive research methods have been developed during the last several decades to determine fine root dynamics (Vogt et al. 1998). Some studies using nondestructive methods found short lifespans,

K. Fukuzawa (✉) • H. Shibata
Field Science Center for Northern Biosphere, Hokkaido University, 250 Tokuda, Nayoro, 096-0071, Japan
e-mail: caribu@fsc.hokudai.ac.jp

M. Dannoura
Graduate School of Agriculture, Kyoto University, Kitashirakawa Oiwake-cho, Sakyo-ku, Kyoto 606-8502 Japan

S. Mancuso (ed.), *Measuring Roots*,
DOI 10.1007/978-3-642-22067-8_15,

indicating rapid turnover of fine roots (Jackson et al. 1997). Many studies using nondestructive methods reported that fine root production and mortality occurred simultaneously and fluctuated widely within a year. In contrast, destructive methods, such as sequential coring, do not detect temporal changes in fine root production and mortality. Thus, measurements using nondestructive methods are necessary to determine the contribution of fine root production to total production in terrestrial ecosystems (see Chap. 16).

Temporal patterns of fine root dynamics also influence the seasonal movement of carbon and nutrients, by affecting such processes as soil respiration. Soil respiration is defined as the sum of root respiration and microbial respiration (decomposition). Hanson et al. (2000) reviewed methods of distinguishing root and soil respiration in forest ecosystems, including direct and indirect methods and the use of isotopes. More recently, Kuzyakov (2006) reviewed the technical aspects, and Subke et al. (2006) published a meta-analysis of partitioning of soil respiration. Luo and Zhou (2006) synthesized the results of many studies of soil respiration. These researchers partitioned soil respiration into five or six separate processes, including root respiration. In the strict sense, autotrophic respiration refers only to root respiration, and does not include CO_2 derived from microbes associated with roots; that is, root-derived respiration separates root respiration and rhizo-microbial respiration. A review of these two processes concluded that, on average, root respiration accounts for 48% of total root-derived CO_2 (Kuzyakov 2006). Moreover, Boone et al. (1998) reported that root respiration (including rhizo-microbial respiration) and microbial respiration from bulk soil had different temperature sensitivities, and Burton et al. (2008) found that the effect of temperature on root respiration among ecosystems with different temperature regimes was smaller than the effect of temperature changes within ecosystem on root respiration. Overall, these results demonstrate that metabolic analyses should consider root and rhizo-microbial respiration separately and confirm the importance of using measurement methods that are appropriate to the purpose of each study.

In this chapter, we review patterns of fine root dynamics and respiration, including (1) measurement methods, (2) temporal and spatial variation, (3) modeling approaches, and (4) methods for scaling results up to the stand level.

15.2 Fine Root Dynamics

15.2.1 Methods of Measuring Fine Root Dynamics

15.2.1.1 Destructive Methods

Sequential soil coring is a common method of measuring biomass (Fukuzawa et al. 2007; Table 15.1) and net primary production (NPP) of fine roots (Vogt et al. 1998). This method determines temporal changes by measuring fine root biomass several times a year using an auger. NPP is calculated as (1) the difference between maximum and minimum biomass, (2) the sum of significant changes in live and dead roots and root decay during a specific interval (Santantonio and Grace 1987), or (3) the sum of significant differences in fine root biomass between sampling periods (Persson 1978). In general, fine root biomass has been reported to peak in autumn (McClaugherty et al. 1982). In some cases, however, a temporal change in fine root biomass is small, and it cannot be determined whether the same biomass is being remeasured or the biomass has been replaced by death and regrowth (Vogt et al. 1998). When temporal variation is smaller than variation between spatial replicates, sequential coring is not an appropriate method.

In the in-growth core method, another destructive method of estimating fine root NPP, soil cores are collected, the roots are removed, and the root-free soil is replaced in the hole. After a certain period, roots that have invaded the root-free soil are collected and quantified. The in-growth core method is powerful for comparing effects of treatment on fine root dynamics. However, it is not appropriate for measuring temporal changes in fine root growth, because it takes a relatively long time (e.g., six months) to detect the recolonized roots (Vogt and Persson 1991).

15.2.1.2 Nondestructive Methods

Two methods developed during the last 20 years – the root window and the minirhizotron imaging system – allow nondestructive observation of the same root over time. The minirhizotron imaging system, in particular, enables researchers to follow an individual root from emergence to disappearance. Many studies using these methods reported that the lifespan of fine roots is short, and that roots are replaced by simultaneous production and mortality. The definition of mortality is somewhat vague, because it includes both death and decomposition. In image analysis, it is difficult to distinguish between live and dead roots, so mortality is defined in most studies as disappearance. The minirhizotron imaging system provides information only on root length, so the method must be combined with measurements of root biomass in order to estimate root production. Observations of roots in rhizoboxes using scanners are also available for individual plant-based studies, although minirhizotron system and root windows are used in field studies.

The rest of this chapter reviews growth and mortality patterns determined using nondestructive methods. See Chap. 16 for details of the minirhizotron system and Chap. 18 for a discussion of root turnover.

15.2.2 Root Dynamics and Effects of Exogenous and Endogenous Factors

15.2.2.1 Root Growth Patterns

Compared with patterns of leaf behavior and phenology, little is known about seasonal patterns of root dynamics, despite their importance to the overall carbon balance of vegetation (Steinaker and Wilson 2008). Previous studies using minirhizotron imaging systems or root windows produced contradictory results regarding seasonal patterns. Patterns are more difficult to identify in complex field ecosystems than in controlled environments. In boreal and cool temperate forests where water was not limiting to production, root growth peaked in mid to late summer (Burke and Raynal 1994; Steele et al. 1997; Ruess et al. 1998; Tierney et al. 2003; Fukuzawa et al. 2007). In temperate forests that experienced intense drought during late summer, root growth was maximum during spring or early summer, immediately following the completion of leaf expansion, and decreased during the remainder of the summer (Lyr and Hoffman 1967; Teskey and Hinckley 1981; Hendrick and Pregitzer 1997; Joslin et al. 2001). In Mediterranean evergreen oak forest in Spain, which is characterized by severe summer drought and a relatively mild, wet winter, fine root growth peaked during winter (Lopez et al. 2001). In tropical moist forest in Panama, fine root growth rate was high early in the wet season (around May–June). These different seasonal patterns are due to site-specific differences in climate and species (Tierney et al. 2003).

To predict fine root production in various areas and to determine the cause of differences among sites, factors controlling seasonal patterns of fine root growth must be understood, including a number of exogenous (environmental) and endogenous factors (Tierney et al. 2003). In northern cool temperate and boreal forests without water limitation, soil temperature is thought to be the primary factor controlling root growth, because cold soil inhibits root growth during early summer, despite warmer air (Tryon and Chapin 1983; Steele et al. 1997; Pregitzer et al. 2000; Tierney et al. 2003). However, Tierney et al. (2003) found that air temperature was a better predictor of fine root growth rate than soil temperature, suggesting that the photosynthate availability, an endogenous factor, was also important. However, it is not clear whether root growth during spring is influenced by the same environmental cues that drive shoot development, because endogenous control, such as photosynthate distribution, might be influenced by exogenous controls, including temperature and light (Pregitzer et al. 2000; Tierney et al. 2003).

In temperate areas that experience intense drought during summer, fine root growth is affected by soil moisture, as well as temperature. Teskey and Hinckley (1981) reported that in white oak (*Quercus alba* L.) stands, temperature was the dominant factor affecting growth rate at soil temperatures <17°C, but that soil water potential became the most important factor at temperatures >17°C. In tropical forest, where temperature is almost constant, low soil moisture inhibits fine root growth during the dry season (Yavitt and Wright 2001).

Some studies reported that root growth rate was not related to any environmental factors (Hendrick and Pregitzer 1997; Joslin et al. 2001). Joslin et al. (2001) found that root growth rate did not return to high early summer rates during late summer and early autumn, even though soil moisture and temperature conditions were favorable. They concluded that root growth is phenologically programmed to produce an extensive fine root system when soil moisture and temperature conditions are favorable and the supply of carbohydrates produced by the newly expanded canopy is adequate. Although there have been few relevant studies, if root growth pattern is phenologically programmed, the dynamics of shoot and root would be linked. Reich et al. (1980) showed that root and shoot growth were not synchronized in seedlings and trees of oak species, suggesting competition for carbon between roots and shoots. Steinaker and Wilson (2008) reported that synchronization of root and shoot growth was rarer in aspen (*Populus tremuloides* Michaux) forest than in grassland, because greater reserves of photosynthate in spring allowed earlier shoot development in forest. Reduction of available photosynthate by shading reduced fine root production in grassland, but increased soil temperature did not, indicating endogenous control (Edwards et al. 2004). Different patterns of fine root growth and different optimum temperatures for root growth have been reported for various tree species under similar conditions (Kozlowski and Pallardy 1997), further supporting the importance of species-specific phenology. Joslin et al. (2001) suggested that growth patterns might differ among functional groups, such as fixed-shoot-growth species vs. free-shoot-growth species.

Determining the exogenous and endogenous factors controlling fine root dynamics is difficult, because phenological cues would be expected to maximize root growth when environmental conditions are normally favorable, and the relative importance of each factor would depend on the environmental conditions (Tierney et al. 2003). In mixed forests, the root growth pattern of each functional group must be evaluated for precise estimation and prediction of fine root production and turnover. For example, Fukuzawa et al. (2010) showed different root growth patterns of oak seedling and understory dwarf bamboo, which coexist in cool-temperate forests in Japan, by growing them in separate rhizoboxes. Finally, long-term observations, as well as measurements of aboveground phenology and environmental factors, are necessary to detect interannual variation and to separate the effects of exogenous and endogenous factors on fine root dynamics.

15.2.2.2 Root Mortality Patterns

Compared with root growth, root mortality has been the subject of relatively few studies and is not well understood. A limited number of studies using direct observations, such as the minirhizotron imaging system, showed that growth and mortality occur simultaneously, and that mortality has seasonal patterns. In northern forests, fine root mortality was highest in late summer to autumn (Tierney et al. 2003; Tingey et al. 2005), after the peak period of growth, resulting in an inverse relationship between growth and mortality within a year (Ruess et al. 1998). In contrast, evergreen oaks in the Mediterranean region of Spain had high root mortality in spring, following the winter increase in root production (Lopez et al. 2001). Substantial mortality of fine roots in autumn is thought to be controlled by endogenous factors related to defoliation (Hendrick and Pregitzer 1992). Eissenstat and Duncan (1992) reported large losses of fine roots 4 weeks after removal of the top third of the canopy of citrus trees in FL, USA. Lopez et al. (2001) suggested that high root mortality in spring could be related to carbon limitation when carbon is being allocated to new leaves. Thus, carbon distribution within vegetation appears to influence fine root mortality.

Eissenstat and Yanai (1997) used a cost (construction, maintenance, and ionic uptake) and benefit (nutrients acquired) approach to predict the optimum lifespan of fine roots, and concluded that the lower the cost of respiration, the longer the lifespan. If so, temperature might be an important factor in determining root mortality, although previous studies found weak relationships between fine root mortality and environmental factors. However, results of some research did show increased maintenance respiration with increasing temperature (Eissenstat and Yanai 1997), as well as shorter lifespan of fine roots at sites with higher soil temperatures (Hendrick and Pregitzer 1993b). Furthermore, soil temperature was selected as an explanatory variable of temporal variation in fine root mortality (Tierney et al. 2003; Fitter et al. 1999). Although results showing overwinter mortality are not common, some studies pointed out that soil freezing causes fine root death (Pregitzer et al. 2000; Tierney et al. 2001). Temperature and other environmental factors could influence fine root mortality not only via direct physiological effects, but also through changes in indirect processes, such as herbivory, parasitism, and nutrient availability.

15.3 Root Respiration

15.3.1 Methods of Measuring Root Respiration

15.3.1.1 Root Excavation

The simplest method of measuring root respiration relies on root excavation. Excavated root samples are set in a chamber, and the increase in CO_2 concentration is measured by absorption in KOH in a closed system (Mori and Hagihara 1991), or by infrared gas analyzer, in a closed system (Dannoura et al. 2005, 2006b) or an open system (Ryan et al. 1996). One advantage of the root excavation method is that root respiration can be compared with respiration rates of other tissues, such as stem (Zach et al. 2008). Another advantage is that other root characteristics, such as length, diameter, and weight, can be measured and compared with root respiration. The results can then be scaled up to the forest stand level by using allometric relationships, such as between root diameter and respiration rate (Dannoura et al. 2006b; Marsden et al. 2008b), or between tree diameter at breast height (DBH) and root respiration rate (Mori et al. 2010).

The root excavation method has some disadvantages and limitations. Cutting and wounding roots could affect respiration, especially in coarse roots. Ohata et al. (1976) found little effect of root cutting on fine root respiration, but a greater effect of branch cutting, due to the relatively large cut area. Dannoura et al. (2006b) put silicone on the cut surfaces to avoid an artifactual increase in CO_2 production. Organic matter on the root surface should be removed, because respiration by fungi and bacteria should not be included in root respiration: Most studies wash the roots with water before respiration measurements (Burton et al. 1997, 2002; George et al. 2003; Marsden et al. 2008a, b; Makita et al. 2009), or just brush the

Table 15.1 Vertital distribution of fine root biomass of dwarf bamboo (Sasa) and trees (in grams per square meter) at different root diameters in a cool-temperate forest. The values represent the average of data from six sampling points, with SE in parentheses

Soil depth (cm)	*Sasa* (g m^{-2})		Trees (g m^{-2})		Total roots (g m^{-2})
	<0.5 mm	0.5–2 mm	<0.5 mm	0.5–2 mm	
0–15	181 (38)^{a}	128 (24)^{a}	100 (28)^{a}	57 (20)^{a}	466 (39)^{a}
15–30	87 (23)^{a}	77 (28)^{ab}	13 (3.4)^{b}	16 (11)^{a}	193 (46)^{b}
30–45	26 (8.0)^{ab}	32 (18)^{bc}	8 (3.0)^{b}	28 (22)^{a}	94 (26)^{b}
45–60	9 (0.2)^{b}	6 (1.9)^{c}	6 (2.0)^{b}	0^{a}	21 (2.0)^{c}
Total	303	243	127	101	774 (100)

Values in a column followed by different superscript letters differ significantly among soil depth ($P < 0.05$). The Tukey–Kramer HSD test was used for *Sasa* 0.5–2 mm, trees <0.5 mm and total roots; Kruskal–Wallis all-pairwise comparisons test was used for *Sasa* < 0.5 mm, trees 0.5–2 mm. Data is from Fukuzawa et al. (2007)

soil off (Burton et al. 2002; Burton and Pregitzer 2003; Gough and Seiler 2004). Anyhow, respiration of excavated roots should be measured before roots transubstantiate. Pregitzer et al. (1998) kept roots in cool deionized water (1°C) before laboratory measurements. Makita et al. (2009) measured respiration immediately after the excavation of roots in the field.

15.3.1.2 The Intact Root Method

Some studies measured respiration by enclosing intact roots in a chamber (Marsden et al. 2008a, b; Zach et al. 2008). This method involves some disturbance because condition in a chamber is different from ambient soil condition, but provides continuous, direct measurements of root respiration. The advantage of intact root method is that continuous measurement of respiration could be compared with environmental factors. Respiration rates of intact roots have been measured in the laboratory, by placing plant pots in chambers (Dannoura et al. 2005; Edwards 1991), as well as in the field (Dannoura et al. 2006a; Chen et al. 2009).

The intact root method has been used to examine effects of environmental factors on root respiration. Bouma et al. (2001) found that soil water content and temperature were controlling factors (Bouma et al. 1997). Colpaert et al. (1996) compared root respiration in plants with and without mycorrhizae and found that mycorrhizae increased root respiration. Espeleta and Eissenstat (1998) compared respiration of fine roots in chambers attached to fine roots of mature trees and seedlings, to determine effects of soil drying. The results suggested that soil drying results in significantly reduced root respiration in mature trees.

15.3.2 Root Respiration Patterns and Effects of Environmental Factors

15.3.2.1 Relationships with Root Morphology and Chemical Property

Results of root excavation showed that finer roots usually have higher respiration rates than larger roots, suggesting that roots with different diameters have different functions (Makita et al. 2009; Marsden et al. 2008a, b). Similarly, Bidel et al. (2001) found a significant relationship between respiration rate and meristem radius at the root tip.

A number of studies found strong relationships between root respiration rate and N concentration (Ryan et al. 1996; Pregitzer et al. 1998; Burton et al. 2002; Makita et al. 2009). Bidel et al. (2001) found that meristem of tap roots has both a higher N concentration and a higher respiration rate than meristem of other roots. Gough and Seiler (2004) reported that increasing nutrient availability through fertilization increased root respiration. These results are not surprising, as N concentration is closely correlated with growth and physiological factors, such as protein concentration, nutrient assimilation, and maintenance of ionic gradients. The strong correlations between fine root respiration and both root structure and N concentration support the hypothesis that root morphology reflects physiological function (Eissenstat et al. 2000; Hishi 2007). The root excavation method also demonstrated that root age influences respiration: younger roots have higher respiration rates than older roots (Bouma et al. 2001). This relationship was confirmed with roots grown in pots with organic-free soil (Dyckmans and Flessa 2002). Both studies attributed the age effect to differences in N uptake capacity.

15.3.2.2 Effects of Environmental Factors

Continuous measurements of root respiration using the intact root method has enables researchers to examine relationships between respiration rate and environmental factors. Root respiration rates are generally high during the periods of high soil temperature. Q10 values, which indicates the temperature sensitivity, differed between root and soil respiration, in both pot measurements and field measurements (Edwards 1991; Boone et al. 1998) However, Bhupinderpal-Singh et al. (2003) found large seasonal variation in root respiration during a period of a small variation in soil temperature. They concluded that above-ground photosynthetic activity and C supply, as well as environmental factors, influence the seasonal changes in root respiration.

Finally, results of root respiration measurements using the intact root method have been combined or compared with other field measurements. For example, Coleman et al. (1996) showed that periods of high root respiration rates coincided with periods of high root growth rates. Bryla et al. (2001) used a chamber to measure CO_2 from single intact root and combined the results with fine root distribution to estimate root respiration on an area basis. Dannoura et al. (2006a) developed new techniques for comparing respiration of intact roots and soil respiration. They found that root and soil respiration responded differently to soil water content and soil temperature, and that the ratio of root respiration to soil respiration varied seasonally. As the phenological phase of plant affect root activity, studies such as these, which measured root respiration, as well as temporal changes in environmental factors and root biomass, could provide precise evaluation of root respiration.

15.4 Conclusions

To understand the effects of a changing global environment on fine root dynamics, it is necessary to determine temporal patterns of fine root growth and mortality and of root respiration through long-term observations. Nondestructive methods, including the minirhizotron imaging system and the root window, are particularly useful for continuous observation. Root dynamics can then be related to exogenous factors, such as temperature and soil moisture, and endogenous factors, such as carbon allocation. Patterns of root mortality include both death and decomposition, making it difficult to determine the relationships with controlling factors, and new techniques are needed to improve mortality measurements. Methods of measuring root respiration include destructive (root excavation) and nondestructive (intact root) methods. The root excavation method allows measurements of root structure, such as diameter, which are important for relating respiration to function. On the other hand, the intact root method provides continuous measurements of root respiration and allows examination of the relationships between root respiration, root growth, and environmental factors in the field. Overall, integrated studies of root dynamics and root respiration would improve our understanding of carbon and nutrient cycling in terrestrial ecosystems.

References

Bidel LPR, Renault P, Pagès L, Rivière LM (2001) An improved method to measure spatial variation in root respiration: application to the taproot of a young peach tree *Prunus persica*. Agron Sustain Dev 21:179–192

Bhupinderpal-Singh NA, Lofvenius MO, Hogberg MN, Mellander PE, Hogberg P (2003) Tree root and soil heterotrophic respiration as revealed by girdling of boreal Scots pine forest: extending observations beyond the first year. Plant Cell Environ 26:1287–1296

Boone RD, Nadelhoffer KJ, Canary JD, Kaye JP (1998) Roots exert a strong influence on the temperature sensitivity of soil respiration. Nature 396:570–572

Bouma TJ, Nielsen KL, Eissenstat DM, Lynch JP (1997) Soil CO_2 concentration does not affect growth or root respiration in bean or citrus. Plant, Cell Environ 20:1495–1505

Bouma TJ, Yanai RD, Elkin AD, Hartmond U, Flores-Alva DE, Eissenstat DM (2001) Estimating age-dependent costs and benefits of roots with contrasting life span: comparing apples and oranges. New Phytol 150:685–695

Bryla DR, Bouma TJ, Hartmond U, Eissenstat DM (2001) Influence of temperature and soil drying on respiration of individual roots in citrus: integrating greenhouse observations into a predictive model for the field. Plant Cell Environ 24:781–790

Burke MK, Raynal DJ (1994) Fine root growth phenology, production, and turnover in a northern hardwood forest ecosystem. Plant Soil 162:135–146

Burton AJ, Zogg GP, Pregitzer KS, Zak DR (1997) Effect of measurement CO_2 concentration on sugar maple root respiration. Tree Physiol 17:421–427

Burton AJ, Pregitzer KS, Ruess RW, Hendrick RL, Allen MF (2002) Root respiration in North American forests: effects of nitrogen concentration and temperature across biomes. Oecologia 131:559–568

Burton AJ, Pregitzer KS (2003) Field measurements of root respiration indicate little to no seasonal temperature acclimation for sugar maple and red pine. Tree Physiol 23:273–280

Burton AJ, Melillo JM, Frey SD (2008) Adjustment of forest ecosystem root respiration as temperature warms. J Integr Plant Biol 50:1467–1483

Coleman MD, Dickson RE, Isebrands JG, Karnosky DF (1996) Root growth and physiology of potted and field-grown trembling aspen exposed to tropospheric ozone. Tree Physiol 16:145–152

Chen DM, Zhang Y, Lin YB, Chen H, Fu SL (2009) Stand level estimation of root respiration for two subtropical plantations based on in situ measurement of specific root respiration. For Ecol Manage 257:2088–2097

Colpaert JV, van Laere A, van Assche JA (1996) Carbon and nitrogen allocation in ectomycorrhizal and non-mycorrhizal *Pinus sylvestris* L. seedlings. Tree Physiol 16:787–793

Dannoura M, Kominami Y, Hirano Y, Kanazawa Y, Goto Y (2005) Mem Grad School Sci Technol Kobe Univ 23-A:67–73

Dannoura M, Kominami Y, Tamai K, Jomura M, Miyama T, Goto Y, Kanazawa Y (2006a) Development of an automatic chamber system for long-term measurements of CO_2 flux from roots. Tellus 58B:502–512

Dannoura M, Kominami Y, Tamai K, Jomura M, Kanazawa Y (2006b) Short term evaluation of the contribution of root respiration to soil respiration in a broad-leaved secondary forest in the southern part of Kyoto prefecture. J Agric Meteorol 62:15–21 (In Japanese with English Summary)

Dyckmans J, Flessa H (2002) Influence of tree internal nitrogen reserves on the response of beech (*Fagus sylvatica*) trees to elevated atmospheric carbon dioxide concentration. Tree Physiol 22:41–49

Edwards NT (1991) Root and soil respiration response to ozone in *Pinus taeda* L. seedlings. New Phytol 118:315–321

Edwards EJ, Benham DG, Marland LA, Fitter AH (2004) Root production is determined by radiation flux in a temperate grassland community. Global Change Biol 10:209–227

Eissenstat DM, Duncan LW (1992) Root-growth and carbohydrate responses in bearing citrus trees following partial canopy removal. Tree Physiol 10:245–257

Eissenstat DM, Yanai RD (1997) The ecology of root lifespan. Adv Ecol Res 27:1–60

Eissenstat DM, Wells CE, Yanai RD, Whitbeck JL (2000) Building roots in a changing environment: implications for root longevity. New Phytol 147:33–42

Espeleta JF, Eissenstat DM (1998) Responses of citrus fine roots to localized soil drying: a comparison of seedlings with adult fruiting trees. Tree Physiol 18:113–119

Fitter AH, Graves JD, Self GK, Brown TK, Bogie DS, Taylor K (1999) Root production, turnover and respiration under grassland types along an altitudinal gradient: influence of temperature and solar radiation. Oecologia 114:20–30

Fukuzawa K, Shibata H, Takagi K, Satoh F, Koike T, Sasa K (2007) Vertical distribution and seasonal pattern of fine root dynamics in a cool-temperate forest in northern Japan: implication of the understory vegetation, *Sasa* dwarf bamboo. Ecol Res 22:485–495

Fukuzawa K, Dannoura M, Kanemitsu S, Kosugi Y (2010) Seasonal patterns of root production of Japanese oak seedlings and dwarf bamboo grown in the rhizoboxes. Plant Biosystems 144:434–439

Gill RA, Jackson RB (2000) Global patterns of fine root turnover for terrestrial ecosystems. New Phytol 147:13–31

George K, Norby RJ, Hamilton JG, DeLucia EH (2003) Fine-root respiration in a loblolly pine and sweetgum forest growing in elevated CO_2. New Phytol 160:511–522

Gough CM, Seiler JR (2004) Belowground carbon dynamics in loblolly pine (*Pinus taeda*) immediately following diammonium phosphate fertilization. Tree Physiol 24:845–851

Hanson PJ, Edwards NT, Garten CT, Andrews JA (2000) Separating root and microbial contributions to soil respiration: a review of methods and observations. Biogeochem 48:115–146

Hendrick RL, Pregitzer KS (1992) The demography of fine roots in a northern hardwood forest. Ecology 73:1094–1104

Hendrick RL, Pregitzer KS (1993a) The dynamics of fine root length, biomass, and nitrogen content in two northern hardwood ecosystems. Can J For Res 23:2507–2520

Hendrick RL, Pregitzer KS (1993b) Patterns of fine root mortality in two sugar maple forests. Nature 361:59–61

Hendrick RL, Pregitzer KS (1997) The relationship between fine root demography and the soil environment in northern hardwood forests. Ecoscience 4:99–105

Hishi T (2007) Heterogeneity of individual roots within the fine root architecture: causal links between physiological and ecosystem functions. J For Res 12:126–133

Jackson RB, Mooney HA, Schulze ED (1997) A global budget for fine root biomass, surface area and nutrient contents. Proc Natl Acad Sci USA 94:7362–7366

Joslin JD, Wolfe MH, Hanson PJ (2001) Factors controlling the timing of root elongation intensity in a mature upland oak stand. Plant Soil 228:201–212

Kuzyakov Y (2006) Sources of CO_2 efflux from soil and review of partitioning methods. Soil Biol Biochem 38:425–448

Kozlowski TT, Pallardy SG (1997) Physiology of woody plants, 2nd edn. Academic, San Diego, CA

Lopez B, Sabate S, Gracia CA (2001) Annual and seasonal changes in fine root biomass of a *Quercus ilex* L. forest. Plant Soil 230:125–134

Luo Y, Zhou X (2006) Soil respiration and the environment. Academic Press/Elsevier, San Diego, CA

Lyr H, Hoffman G (1967) Growth rates and growth periodicity of tree roots. Int Rev For Res 2:181–236

Makita N, Hirano Y, Dannoura M, Kominami Y, Mizoguchi T, Ishii H, Kanazawa Y (2009) Fine root morphological traits determine variation in root respiration of *Quercus serrata*. Tree Physiol 29:579–585

Marsden C, Nouvellon Y, Epron D (2008a) Relating coarse root respiration to root diameter in clonal Eucalyptus stands in the Republic of the Congo. Tree Physiol 28:1245–1254

Marsden C, Nouvellon Y, M'Bou AT, Saint-Andre L, Jourdan C, Kinana A, Epron D (2008b) Two independent estimations of stand-level root respiration on clonal Eucalyptus stands in Congo: up scaling of direct measurements on roots versus the trenched-plot technique. New Phytologist 177:676–687

McClaugherty CA, Aber JD, Melillo JM (1982) The role of fine roots in the organic matter and nitrogen budgets of two forested ecosystems. Ecology 63:1481–1490

Mori S, Hagihara A (1991) Root respiration in *Chamaecyparis obtusa* trees. Tree Physiol 8:217–225

Mori S et al (2010) Mixed-power scaling of whole-plant respiration from seedlings to giant trees. Proc Natl Acad Sci USA 107:1447–1451

Ohata S, Shidei T, Tsuji H, Hatayama I (1976) Changes in respiratory rate of excised tree organs. Bull Kyoto Univ For 39:100–109

Persson H (1978) Root dynamics in a young Scots pine stand in central Sweden. Oikos 30:508–519

Pregitzer KS, Laskowski MJ, Burton AJ, Lessard VC, Zak DR (1998) Variation in sugar maple root respiration with root diameter and soil depth. Tree Physiol 18:665–670

Pregitzer KS, King JS, Burton AJ, Brown SE (2000) Response of tree fine roots to temperature. New Phytol 147:105–115

Reich PB, Teskey RO, Johnson PS, Hinckley TM (1980) Periodic root and shoot growth in oak. For Sci 26:590–598

Ruess RW, Hendrick RL, Bryant JP (1998) Regulation of fine root dynamics by mammalian browsers in early successional Alaskan taiga forests. Ecology 79:2706–2720

Ryan MG, Hubbard RM, Pongracic S, Raison RJ, McMurtrie RE (1996) Foliage, fine-root, woody-tissue and stand respiration in *Pinus radiata* in relation to nitrogen status. Tree Physiol 16:333–343

Santantonio D, Grace JC (1987) Estimating fine-root production and turnover from biomass and decomposition data: a compartment-flow model. Can J For Res 17:900–908

Steele SJ, Gower ST, Vogel JG, Morman JM (1997) Root mass, net primary production and turnover in aspen, jack pine and black spruce forests in Saskatchewan and Manitoba, Canada. Tree Physiol 17:577–587

Steinaker DF, Wilson SD (2008) Phenology of fine roots and leaves in forest and grassland. J Ecol 96:1222–1229

Subke JA, Inglima I, Cotrufo MF (2006) Trends and methodological impacts in soil CO_2 efflux partitioning: a metaanalytical review. Global Change Biol 12:921–943

Teskey RO, Hinckley TM (1981) Influence of temperature and water potential on root growth of white oak. Physiol Plant 52:363–369

Tierney GL, Fahey TJ, Groffman PM, Hardy JP, Fitzhugh RD, Driscoll CT (2001) Soil freezing alters fine root dynamics in a northern hardwood forest. Biogeochem 56:175–190

Tierney GL, Fahey TJ, Groffman PM, Hardy JP, Fitzhugh RD, Driscoll CT, Yavitt JB (2003) Environmental control of fine root dynamics in a northern hardwood forest. Global Change Biol 9:670–679

Tingey DT, Phillips DL, Johnson MG, Rygiewicz PT, Beedlow PA, Hogsett WE (2005) Estimates of Douglas-fir fine root production and mortality from minirhizotrons. For Ecol Manage 204:359–370

Tryon PR, Chapin FS III (1983) Temperature control over root growth and root biomass in taiga forest trees. Can J For Res 13:827–833

Vogt KA, Persson H (1991) Measuring growth and development of roots. In: Lassoie JP, Hinckley TM (eds) Techniques and approaches in forest tree ecophysiology. CRC, Boca Raton, FL, pp 407–501

Vogt KA, Vogt DJ, Bloomfield J (1998) Analysis of some direct and indirect methods for estimating root biomass and production of forests at an ecosystem level. Plant Soil 200:71–89

Yavitt JB, Wright SJ (2001) Drought and irrigation effects on fine root dynamics in a tropical moist forest, Panama. Biotropica 33:421–434

Zach A, Horna V, Leuschner C (2008) Elevational change in woody tissue CO_2 efflux in a tropical mountain rain forest in southern Ecuador. Tree Physiol 28:67–74

Chapter 16
Biases and Errors Associated with Different Root Production Methods and Their Effects on Field Estimates of Belowground Net Primary Production

Daniel G. Milchunas

Abstract Many estimates of root production in the literature were based on sequential coring of biomass methods now considered unreliable. New methods such as minirhizotron and isotope labeling were developed to overcome known biases of old methods, but also have different assumptions and biases that can have large effects on estimates. Variations on the old root ingrowth method are now widely used because it is easy, straightforward, and does not require expensive specialized materials. Only recently have there been sufficient studies of new methods to better define the importance of various potential biases and to compare among ingrowth, minirhizotron, and isotope labeling studies. How these three methods are executed in the field and/or in how data are used to calculate root production can have large effects on mitigating some of the potential biases of each of these methods. There are three main objectives of this chapter. First, known biases of each method are described and suggestions of procedures that reduce problems leading to over- and underestimation of absolute values of root production are made. Second, a comparative approach of methods is taken to assess the importance of biases, because the correct absolute answer is never known. Third, how method biases may manifest differently in annual, short-lived perennial, and long-lived perennial plant root systems are assessed. Comments on root biomass separation from soil and how this affects estimates of root production are also made, because all three methods rely either directly or indirectly on this procedure to derive an estimate of root production.

D.G. Milchunas (✉)
Long-Term Ecological Research program, Colorado State University, Ft. Collins, CO 80523-1499, USA
e-mail: Daniel.Milchunas@Colostate.edu

S. Mancuso (ed.), *Measuring Roots*,
DOI 10.1007/978-3-642-22067-8_16, © Springer-Verlag Berlin Heidelberg 2012

16.1 Introduction

Net primary production is one of the most basic agronomical and ecological processes. While there are methodological problems associated with estimating aboveground net primary production (ANPP) (Sala and Austin 2000; McNaughton et al. 1996), the numbers and magnitudes of the biases and errors in estimating belowground net primary production (BNPP) under field conditions are much greater (Fahey et al. 1999; Milchunas and Lauenroth 2001). Differences among methods in basic assumptions and/or in how samples are collected, processed, and data calculated within a method can determine the estimate obtained (Swinnen et al. 1995; Tierney and Fahey 2001; Milchunas et al. 2005b, Milchunas 2009) to the degree that syntheses across studies concerning issues ranging from effects of climate change to carbon allocation are severely hampered (Hendricks et al. 1993, 2006; Nadelhoffer 2000; Norby and Jackson 2000; Arnone et al. 2000). Much of the historic literature on BNPP is based on sequential coring of root biomass (Singh et al. 1975) which has been shown to provide unreliable estimates (Singh et al. 1984; Lauenroth 2000; Milchunas and Lauenroth 2001). New methods such as isotope labeling and minirhizotron offer promising alternatives but also have numerous biases and errors in data collection, processing, and calculation, and are only now becoming sufficiently used and studied that critical attempts at evaluations are possible. Old methods such as root ingrowth are still widely used and also have a large number of potential biases and errors. Even though there are problems and assumptions in all methods of estimating BNPP, and the true absolute answer is never known, comparative approaches to reconcile differences among methods and establish which potential biases are problematic are rare (Hendricks et al. 2006; Milchunas 2009).

Root-to-shoot ratios vary widely among native plant biomes and croplands (Jackson et al. 1996) and can be as high as 90% in semiarid grassland where BNPP is two-thirds of total primary production (Milchunas and Lauenroth 2001). Aboveground to belowground biomass ratios increase with increasing precipitation in perennial grasslands (Milchunas et al. 2008) and are greater in annual than perennial communities (Schenk and Jackson 2002). Relatively lowest average root-to-shoot ratios were found for croplands (0.10), coniferous forests (0.18), and tropical evergreen forests (0.19) and highest for tundra (6.6), cold deserts (4.5), and temperate grasslands (3.7) (Jackson et al. 1996). Root derived carbon inputs to soils can be large even when root-to-shoot ratios are low. Balesdent and Balabane (1996) observed root-to-shoot ratios of 0.5 in maize and root–carbon inputs 1.5 times that of stalks and leaves. Factors contributing to greater carbon input from roots can be a high turnover and/or exudation and sloughing. Further, the relatively lower tissue quality of roots compared with shoots (Fig. 16.1) and the potential for UV-induced photo-oxidation (photo-degradation) of aboveground exposed litter (Austin and Vivanco 2006; Brandt et al. 2007, 2010) are additional factors in potentially greater soil organic matter inputs and carbon sequestration from roots than shoots. Understanding root turnover and production has always been important in terms of influences on ecosystem nutrient and water dynamics but

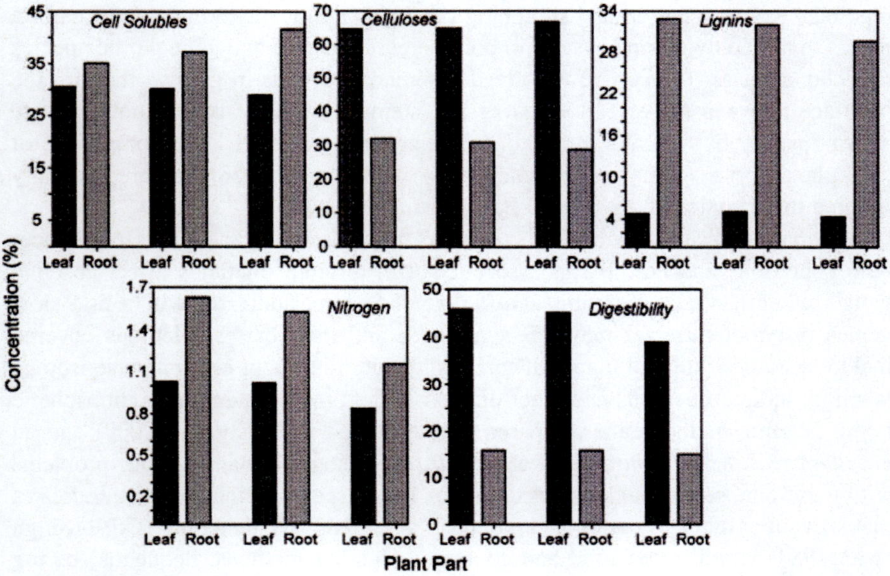

Fig. 16.1 Tissue quality for concentrations (%) of solubles, celluloses (hemicelluloses plus cellulose), lignins, nitrogen, and digestibility for aboveground and belowground biomass from a shortgrass steppe plant community averaged over three CO_2 treatments and four years. "Leaf" represents all aboveground leaf, stem, and seed clipped above crown level and "root" represents roots and rhizomes sampled in late autumn after aboveground senescence and both are for current annual growth only (aboveground previous year growth sorted out and roots from annually sampled ingrowth donuts). Data from Milchunas et al. (2005a, 2005c, unpublished data)

has become increasingly important in the context of climate change research (Norby 1994; Curtis et al. 1994; Norby and Jackson 2000).

Annual and perennial crop species have root longevities generally ranging from weeks to months (Pritchard and Rogers 2000) while native perennial systems can have highly variable longevities with some fine roots surviving many years in some plant communities (Eissenstat and Yanai 1997; Gaudinski et al. 2001; Milchunas et al. 2005b). A simple estimate of standing root biomass at the end of the season may adequately estimate root carbon inputs in some annual crop species (Fitter et al. 1996; Pritchard and Rogers 2000; Rees et al. 2005), but longevity of the roots in the soil can still vary widely. In other annual systems there can be considerable root mortality prior to plant senescence (Huck et al. 1987; Rees et al. 2005), and this is the most common pattern for annual crop species (Pritchard and Rogers 2000). Swinnen (1994) reported that about one-half of barley and wheat roots had decayed by the end of the growth period. In perennial native plant communities, root growth and decomposition can occur simultaneously with no net change in seasonal root biomass in the majority of years (Milchunas and Lauenroth 2001). The method used to estimate BNPP (Pritchard and Rogers 2000) or the means of calculating turnover within a method (Milchunas 2009) needs to consider these differences in root life histories and dynamics of turnover. Other factors may also influence choice of methods. Root exudation and sloughing can be a significant component of carbon inputs to the belowground system in both perennial native and annual crop species (Milchunas et al. 1985). Crowns are the perennial organs in grasses that are the interface between aboveground leaves and stems and belowground roots and are approximately half above- and half below-ground, can be a large component of total plant biomass in perennial bunchgrasses (Sims et al. 1978), and are generally ignored in estimates of ANPP and BNPP (Milchunas 2009).

The main focus of this chapter is on direct methods of estimating of BNPP under field conditions and how biases associated with different methods affect absolute values rather than relative treatment differences. An issue addressed will be how slow versus fast root turnover may affect methods and their biases. Methods covered include variations on root ingrowth core and donut, minirhizotron, and pulse-isotope labeling. Indirect methods and other direct, old field methods and new atmospheric bomb labeling methods are compared and discussed in Milchunas (2009) and an excellent review by Hendricks et al. (2006), including the many serious problems with traditional sequential biomass coring. A recent search of the literature indicates that sequential biomass coring and the many variations on calculating BNPP (Singh et al. 1984) is still often used and compared to other methods. Sequential coring methods may provide nonzero estimates in systems with highly seasonal growth patterns or in some systems where production and decomposition are decoupled, but the high possibility of zero or very low estimates as well as artificially very high estimates makes this method unreliable (Lauenroth 2000; Milchunas 2009) and will not be discussed here. BNPP rather than root production is considered the bottom-line variable of interest because crowns, rhizomes, exudation, and sloughing are also components of BNPP. All methods covered here either require a direct separation of roots from soil (ingrowth and isotope) or indirectly rely on root separation from soil to convert from turnover to a weight per area per time estimate of production (minirhizotron). Therefore, some comments on this often ignored but critical and problematic separation are also included.

For each of the three primary methods, we first describe very briefly the method and variations on the method, then list potential biases, problems, and advantages, and finally summarize what can be deduced about estimate values relative to other methods. This method-comparative approach, when viewed in conjunction with directions that biases may skew estimates, provides a means of evaluating not only which biases are most critical to a method, but also can help place methods in a common context and provide some insight to where absolute values may lie when the absolute is not really known (Hendricks et al. 2006, Milchunas 2009).

16.2 Root Ingrowth

An electronic search at the time of writing this manuscript revealed that root ingrowth was the most commonly used of the three methods covered here. This may be because it does not require expensive equipment to obtain or process samples, it is straightforward and simple, and generally no equilibration periods are necessary before sampling begins and time steps of BNPP estimates can be short if desired (see constraints on other methods in this latter issue below). However, the tradeoff is that potential biases influencing absolute values of root production may be somewhat more severe than for other methods, although possibly less so for relative values. All ingrowth methods involve placement of sifted, root-free soil in the ground and allowing new roots to grow in for a specific time period, and then sampling that soil for roots (Fig. 16.2).

16.2.1 Root Ingrowth Methods

There are three minor and two major variations on the traditional root ingrowth core method. The original method dates back to Hendrickson and Veihmeyer (1931), with additional development by Lund et al. (1970), Persson (1978, 1979), and Steen (1983, 1991). A mesh "bag" that lines a core or augured hole is filled with root-free soil and pulled or dug from the ground after some period of time has allowed for new root growth into the "stocking." Pulling bags out of some very heavily compacted untilled soils can be impossible without breakage and digging bags out very labor intensive, especially if root profiles are deep. Re-coring out the ingrowth area with or without a mesh bag or liner, or a concentric core method is used in soils or treatments not conducive to pulling or digging. The concentric core method uses a larger diameter core in which sifted soil is placed and a subsequent smaller diameter core that is sampled after a period of ingrowth of new roots. The concentric core method allows some leeway for error in missing the sifted soil and going into root-dense surrounding soil, and is often used without a mesh liner. However, it is difficult to core deeply without deviating from the original cored area even with the use of guides or levels, and practically impossible to drive the second core exactly down the middle of the slightly larger original cored hole.

The ingrowth donut method was developed for use in experimental setups that are highly space restricted, such as field CO_2 fumigated plots, where multiple coring would be too destructive (see Milchunas et al. 2005a for details). A single, but large, hole is installed and sampled and refilled with sifted soil repeatedly (Fig. 16.2). The large diameter outer donut ingrowth area lowers variability of smaller point-in-space cores, but keeps the area and volume sampled small relative to the outer circumference surface area that is exposed to in-growing roots (see implications of this in Sect. 16.2.2). Installing ingrowth donuts is much more laborious than ingrowth-core methods, but sampling and refilling afterwards is much less labor intensive as no further coring is necessary. Plastic mesh material, as described in the original paper, is no longer available in sufficient size for most deep profile applications, but wire mesh materials of various sizes are available from specialized suppliers (http://twpinc.com/galvanized). This author has recently also used tractor augering instead of steel tubing (steel pipe is too thick to drive into

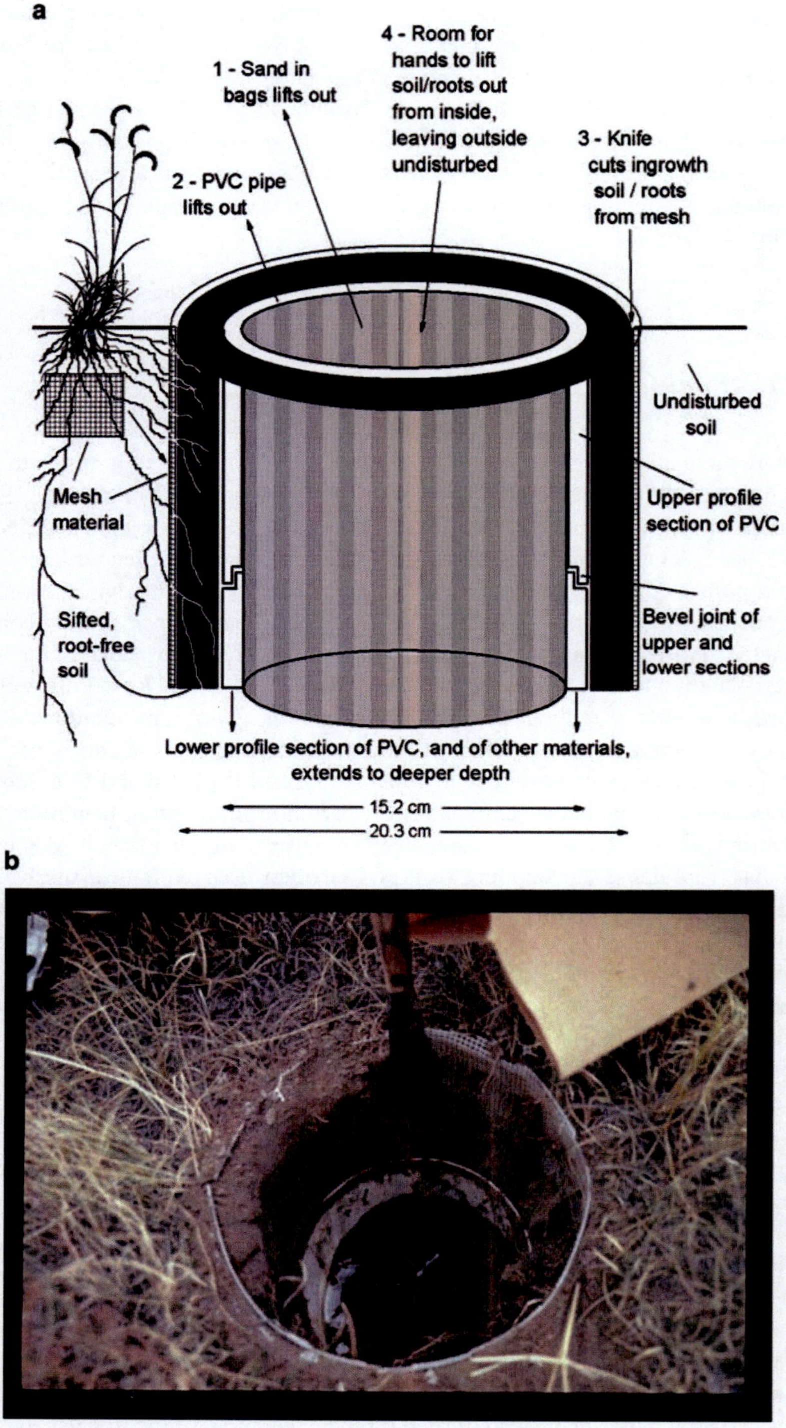

Fig. 16.2 Root ingrowth donut (**a**) schematic drawing, and (**b**) field sampling one year after filling donut area with sifted soil. (**a**) is from Milchunas et al. (2005a)

compact soil), if auger bits can be ground-down to adjust for a sufficiently small donut between the outer mesh and the inner filled PVC pipe (Fig. 16.2a).

Another variation on the ingrowth core method is the ingrowth screen. Screen planes are inserted into the soil using a steel plate to cut and slightly open a slot, root intercepts and diameters recorded upon peeling the screens and soil away on one side, and relationships between intercepts and root length or biomass per unit area used to calculate production (Fahey and Hughes 1994). Lucak and Godbold (2010) further modified this method to arrive at an inclusion net method. Screens are removed along with a slab of soil on each side, somewhat like a mini root monolith. Soil is removed, root lengths trimmed to a uniform length on each side of the screen, and roots then removed, dried, and processed. In contrast to ingrowth cores, screen methods have been used and studied in only a few reported cases.

16.2.2 Potential Biases, Problems, and Advantages of Root Ingrowth

Sampling frequencies for ingrowth methods are constrained by temporal dynamics of root lifespan plus rates of decomposition/herbivory. New roots growing into the space that die and decompose or are consumed before samples are obtained represent an underestimation bias. Either some understanding of lifespan or decomposition rates needs to be known prior to establishing an experimental design, or differences between two short periods and one equally long period can be compared to establish appropriate time steps for sampling. Loss occurred if biomass from the one long period is less than that from the two short periods for ingrowth cores or donuts, but growth into the space between concentric core diameters represents an additional uncertainty in that particular core method.

Other potential biases that can cause underestimation of absolute values include ingrowth area outside spatial range of roots, soil volume not at equilibrium with outside in terms of new root growth, for the concentric core method there is space that new roots grew in to but is not cored, and mesh size constraint on root entry (Milchunas 2009). The former two potential biases both represent a growth rate versus size of ingrowth area factor. New incoming roots may never fully exploit the soil volume available and a small size may allow as much new root growth as a larger size, but the same quantity of production is calculated on the small or the large area and the larger area calculation results in a lower estimate. Hertel and Leuschner (2002) observed that smaller ingrowth areas produced greater estimates of production than larger areas. This is an important but often overlooked bias of root ingrowth (see 16.2.4).

Root proliferation in a nutrient and water rich space due to low competition from existing roots is more often expressed as a possible important potential bias that may act in a positive direction and cause overestimates (Steingrobe et al. 2000). Other potential biases that may cause overestimation include root proliferation after severing, and accidentally coring into outside nonsifted soil. The concentric core method minimizes this latter potential problem in place of the potential underestimate mentioned earlier due to size. Root proliferation after severing has been primarily studied and observed in woody species (Geisler and Ferree 1984; Joslin and Wolfe 1999), no altered root patterns have been observed for wheat or barley (Steingrobe et al. 2001), and inhibition of proliferation may have occurred in a marsh grass study where production estimates from a single long-term ingrowth were more than from the sum of several short-term ingrowths (Neill 1992). The effect of multiple root cutting at one location in the root ingrowth donut method has not been examined, but relative treatment effects based on donuts and minirhizotrons were maintained over several years of comparative sampling (Milchunas 2009).

Other potential biases that may act in either positive or negative or unknown directions include an altered soil profile during refilling with sifted soil, different soil bulk densities of sifted soil inside compared with outside soil and the effects on root penetration and soil water dynamics, sifting effects on mineralization rates and microbial communities, and root architecture can interact with the angle of insertion of the ingrowth area to affect possibilities of interception that most likely vary with depth in the profile (see Fig. 2 in Milchunas 2009). Root ingrowth donuts generally must be inserted vertically, but cores can be inserted at an angle that optimizes probability of interceptions, particularly if the species' architecture is known. Packing of soil and source of sifted soil can be done in a manner that minimizes the former two potential biases, and the former two are less of a factor in tilled crop systems where the soil profile is mixed and loosened anyway.

Use of standard soil sampling cores for installation of ingrowth cores or donuts can also result in biases and the method of root separation from soil will have a large effect on estimate values (see Sect. 16.5). When installing ingrowth holes it helps to use a reverse taper bit after taking a smaller regular core (Fig. 16.3) in order to avoid compacting the sides of the hole and restricting root entry, or use an auger with a sharp blade if uniform sides can be cut.

Another potential bias in root ingrowth methods concerns the root separation from the sifted "root-free" soil used to fill the structures. This supposedly root-free soil must be dry sieved and hand picked of roots even though final soil with ingrown roots may either be wash floated, wet sieved, and hand picked or dry sieved and hand picked. Dry processing never get as many roots out as flotation methods (see Sect. 19.5) and "root-free" soil is therefore practically impossible to obtain. The potential for root growth estimates to partly contain root contamination is particularly a problem when flotation wet processing is the final means of root separation from soil. This potential bias can be to some extent corrected for by including a number of "blanks" in each treatment replicate whereby solid sheet liner rather than mesh or no liners are used. Decomposition of roots from the sifted soil can thus partly be accounted for and subtracted from final ingrowth weights, although the quantities will probably be less than in soil that also had live-root effects on dead root decomposition.

There are very few studies using root screen methods, which will not be discussed in detail in this section. However, a potential bias not associated with

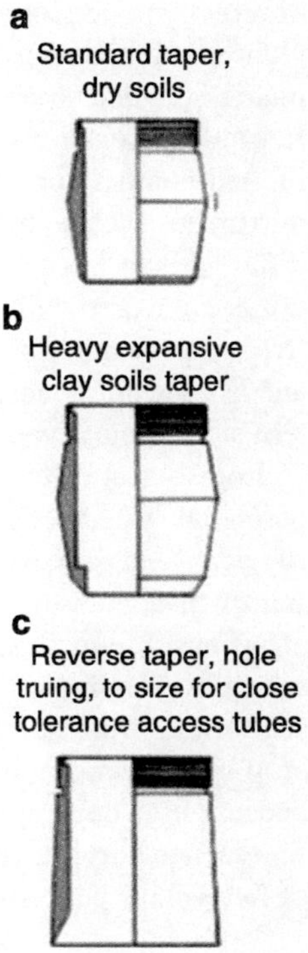

Fig. 16.3 Soil core and reamer bit designs. Note differences in the direction of potential compaction for cores versus reamers. Cores force soil to outside of tube for easier extraction, and allow for expansion of sample into inner tube for easier extraction. Reamers do not compress soil to outside of tube, but reverse the compression to the inside, leaving outside wall at a more natural and uncompressed state conducive to normal penetration and root growth for both root ingrowth core or donut and minirhizotron tube installations

these methods, which is a component of other ingrowth methods, is that the soil profile is undisturbed, although cutting of roots during installation and some soil disturbance occurs when making the insertion slot. Surface samples may be relatively easy to obtain compared with the excavations required for deep sampling.

Root ingrowth methods can account for rhizome production if they are not sorted out or difficult to distinguish from aboveground contamination that is common in tilled systems and if their size is not larger than the mesh size used as the liner. Root ingrowth will not provide estimates of crown production or of sloughing and exudation components of BNPP.

16.2.3 Application of Root Ingrowth Methods in Different Ecosystems

There are no differences in ingrowth methods or calculations that are more or less appropriate for fast-growth/decomposition annual crop systems versus slow growing long-lived perennial systems, other than for different temporal and spatial sampling protocols. This is in contrast to minirhizotron and isotope pulse labeling methods discussed below. However, ingrowth methods may manifest biases differently in agricultural annual crop systems where plowing mixes soil profiles and reduces soil bulk densities whereby ingrowth mediums are not as different as outside matrix soil, and where fertilization/irrigation lessen potential effects of competition-free space. Annual seeded agricultural systems start the season or growth period with limited competition, and weed control and spaced crop plantings are designed to maximize yields by limiting inter- and intra-specific competition, whereby differences in "root-free" space inside versus outside ingrowth volumes are relatively minor compared with late-seral native communities. There are insufficient comparative studies among ingrowth and nonsequential coring methods like minirhizotron and isotope pulse labeling to suggest how these differences in biases may manifest in terms of absolute values of root production in fast versus slow moving systems.

Logistics of different ingrowth methods may vary with annual versus long-lived perennial systems. Excavation of ingrowth liners is much more feasible in shallow rooted, tilled soft soils, possibly reducing necessity of concentric core method. Ingrowth donuts that require relatively more effort in initial installation but less effort in subsequent sampling may not be advantageous in systems that are plowed annually (requiring re-installation) and that do not require many samplings per season to minimize decomposition losses between intervals. This may depend on crop specific root mortality rates and decomposition rates once mortality occurs, because high decomposition rates would be irrelevant if there is low mortality up to senescence/harvest. How roots are processed can have a large effect on proportions of live versus partially decomposed root ratios (see Sect. 19.5).

16.2.4 Root Ingrowth Estimate Values Relative to Other Methods

Although limited in number, a review of comparisons of methods at the same study sites found that ingrowth methods generally underestimated root production relative to minirhizotron and relative to isotope labeling (Milchunas 2009). A comparison of ingrowth donuts, minirhizotron, and isotope labeling at the same shortgrass steppe grassland showed estimate orders of ingrowth < minirhizotron < isotope pulse labeling. This means that, under the conditions and plant systems compared, positive biases listed earlier are either not acting or are overcompensated for by negative direction biases that result in underestimation. Based on evidence in the detailed review, it appeared that size of the ingrowth area and some degree of rapid turnover of roots that is missed between sampling are the most commonly reported causes of the net underestimation. Positive and other negative biases may be and probably are in operation, and how these weigh out will be site specific.

Ingrowth screens were not evaluated in the Milchunas (2009) review, and these methods are the most different from variations on the traditional ingrowth method. There are, however, few studies comparing ingrowth screens with either minirhizotron or isotope labeling. Lucak and Godbold (2010) found no difference between root production estimates obtained by root inclusion nets and ingrowth cores sampled the traditional way (mesh lined and roots cut from outside of mesh). Fahey and Hughes (1994) found root production estimates 50–70% of those obtained by ingrowth concentric core method (outside 5 cm core with sifted soil, 1.9 cm core sampled for ingrowth, no mesh liner).

Mortality by death or herbivory and the decomposition of new root growth between sampling periods and the size-associated bias of underexploitation of the soil volume are two of the more major problems of root ingrowth that to some extent can be minimized by adjusting timing of sampling and size of ingrowth volume. These are of course ecosystem-specific adjustments requiring some prior understanding or testing in the system of study. To some extent, however, the solutions are mutually exclusive. Long enough periods for roots to more fully exploit the ingrowth volume may overlap with greater opportunity for decomposition to become a factor. Root ingrowth methods in many systems may always underestimate absolute values compared with minirhizotron and isotope pulse labeling, assuming isotope methods are considered to possess the least number and magnitude of potential biases (see Sect. 16.4 and Milchunas 2009). However, relative treatment differences in root production between methods may be similar even when those from ingrowth are lower (Milchunas 2009). Unbiased relative treatment differences cannot necessarily be assumed if species composition changes result in different root architectures that affect insertion-plain interception probabilities or if treatments influence decomposition rates between sampling periods, herbivore composition or abundance, or other factors of bias.

16.3 Minirhizotrons

There are no major variations on the basic minirhizotron method, but substantial variation in all aspects of execution during installation, image collection and processing, and data calculation. There can be biases associated with details of execution as well as inherent in the basic method, and these are discussed in the section below. Minirhizotrons are a much improved version of large rhizotron windows that either have additional biases or accentuate those of the minirhizotron and are much more automated and permit greater spatial and/or treatment coverage (Fahey et al. 1999; Milchunas 2009). Minirhizotrons require substantial financial investment in equipment and labor investment in sample processing while field work is minimal after installation.

16.3.1 Minirhizotron Methods

Basically, clear plastic tubes (cellulose acetate butyrate or acrylic) are inserted into the ground at an angle and photographic images taken along the top of the tube at known distances down the tube (Fig. 16.4a). The images are captured on a computer a

b

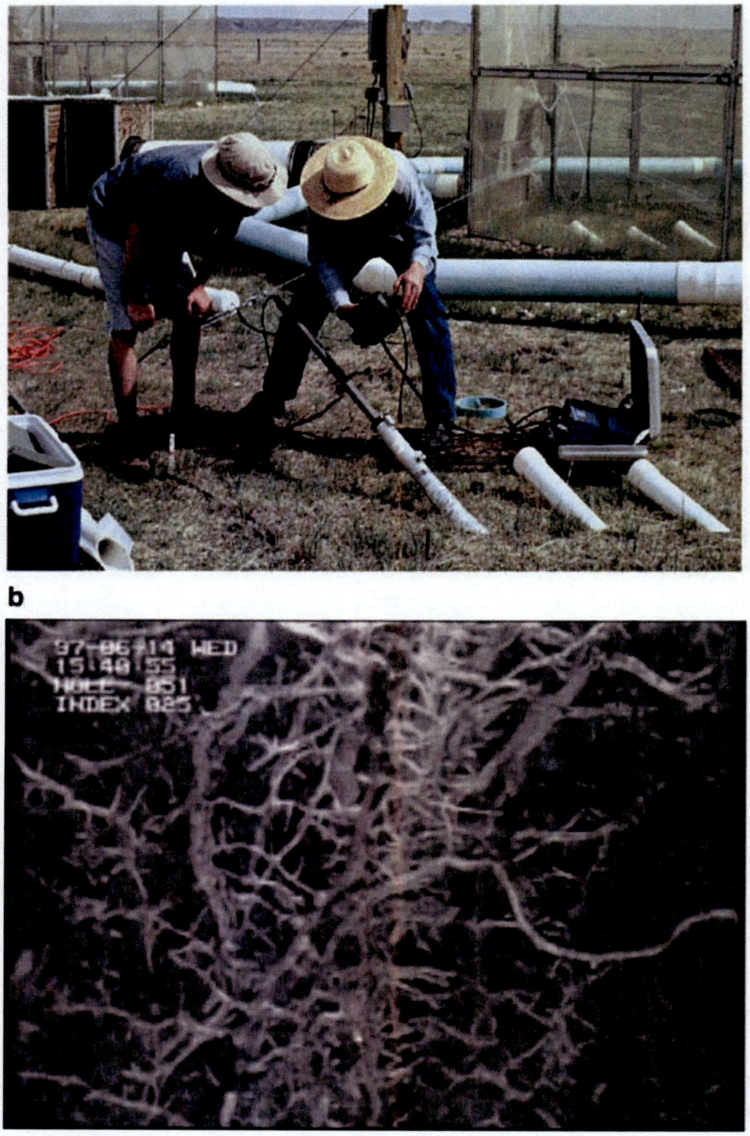

Fig. 16.4 Minirhizotron (**a**) image collection sampling and (**b**) example image from shortgrass steppe illustrating the very dense root structure in some ecosystems and the long and difficult process of digitizing such images

and the roots digitized according to length and diameter, with each individual root uniquely numbered. A calibration frame image of a grid of known distances is also captured to convert root measurements to known lengths, and to adjust for any differences between sampling dates in camera distance from the tube wall that would affect consistency of image dimensions. Only a portion of the whole picture area is used to allow adjustment of images for small differences in vertical and horizontal location between sample dates. Software developed to convert image pixels to millimeter lengths produces a file which identifies the tube, frame, root number, and its length/diameter and a number of other chosen observation categories concerning color branching order, etc. Another set of images is collected at a later date at the same location in the tube. The previous digitized "stick-drawings" are then projected on top of the new picture image. Additional length/diameter due to growth or loss of length/diameter due to decomposition/herbivory to existing roots in the stick drawing is indicated by redrawing the root to the new dimensions. Roots not in the stick-drawing are added by digitizing and assigning a new number, and roots no longer in the new image are deleted. Total addition and loss of length between the two dates can then be calculated on an individual root basis along with total length/frame/date. Root demographic information on longevity, birth and death rates of different cohorts, etc., can thus be collected over many consecutive sample dates.

It is critical that the same location in the tube and distance of the camera from the wall of the tube be maintained in order that images overlay properly and maintain the same pixel to known length relationship. Location down the tube can be established by specially designed indexing handles that the camera attaches to and fixed locations for attachment of the apparatus, available from manufacturer, or by etching the tubes prior to installation. The digitizing software uses changes in the calibration image to adjust picture size in order to maintain the same image size as that of the previous date based on known distance (but see problems in section below).

Minirhizotron hardware is available from Bartz Technology Corporation (Carpinteria, CA, USA) and two of the most commonly used software packages are RooTracker (David Tremmel, Duke University, http://www.biology.duke.edu/rootracker/index.html) and MR-RIPL (Alvin Smucker, Michigan State University, http://rootimage.msu.edu/MR-RIPL/index.html).

16.3.2 Potential Biases, Problems, and Advantages of Minirhizotron

This section is organized in the temporal manner that would be encountered when first using minirhizotrons: field installation, image collection, data processing, and conversion of length growth and decomposition to turnover or biomass and then root production.

The first question one may have when designing a minirhizotron study is what number of tubes will be necessary. Although this will obviously depend on the particular plant community, a comparative approach can give some guidance. Most root researchers will have some data on root biomass, and if not it is easy to obtain.

Table 16.1 Variance (coefficient of variation) associated with estimates of various root parameters based on biomass sampling by cores compared to parameters from minirhizotron images collected in a shortgrass steppe semiarid grassland. Values are calculated from data in Milchunas and Lauenroth (2001) and Milchunas et al. (2005b) for a representative sampling date in the middle of the growing season, or 3 years after minirhizotron tube installation

Parameter measured	Coefficient of variation	Mean
Roots from cores (40 cores, 40 cm deep, 66.5 cm inside diameter)		
Dry weight (g/core not ash corrected)	0.32	6.8
Organic matter coefficient (ash correction)	0.16	59
Organic matter (g/m^2)	0.26	1239
Minirhizotron root length (18 tubes, 50 images/tube of 12.5 by 18 mm, 40 cm deep, 23° angle)		
Total length (mm/tube)	0.34	1354
New growth (mm/tube)	0.55	331
Disappear/decomposition (mm/tube)	0.81	130

For a grassland example, 18 minirhizotron tubes were similar to 40 root cores in terms of the coefficient of variation for total root length compared to raw total biomass, respectively (Table 16.1). Ash correcting root biomass to an organic matter basis reduced the variance. New root growth and disappearance or decomposition of roots from minirhizotron tubes was, however, substantially noisier that total root length. In general, the narrow plane of images sampled down the top of a minirhizotron tube makes measures of interest for calculation of turnover relatively more variable than biomass core data.

The artificial curved plane (tube top when inserted at an angle to the surface) that roots grow across is a critical aspect of two potential bias factors, and relationships of distributions through the soil profile with depth between total root length from minirhizotrons and biomass from cores generally are poor (Fahey et al. 1999). First, roots hitting the surface at an angle will tend to grow all along the surface rather than penetrating through if there had been no obstruction, and these result in elevated estimates. Second, this will interact with the architectural bias that arises due to differential probabilities of hitting the surface when growing at different angles to the ground surface (see Fig. 2 in Milchunas 2009).

Different tube insertion angles can result in different estimates of total, new, and disappearing root lengths, and this can create bias for some methods of calculating turnover and then root production from minirhizotrons. The distribution of angles of root growth (the pattern of horizontal, vertical, and in-between angles) and the angle of tube insertion into the ground can affect probabilities of intercepting the tube-top surface. Insertion at an angle, often from 20° to 45° to the surface, generally improves the possibility of a growing root contacting the surface. The angle chosen should depend on root architecture of the particular system. If all roots grew straight down then horizontal tube placements through the profile would be most efficient, and the vertical inverse if all roots only grew horizontal. However, neither of the two is true, there is some optimal angle, and that most likely varies with depth in the profile. Implications of this bias are discussed below in reference to methods of calculating data, because some methods of calculating turnover suffer from this problem while in others it can be minimized.

Tube installation at any angle should attempt to minimize several other tube related potential biases. First, cores fitted with bits that result in compaction outside the core should not be the final diameter size used to fit the tube. Standard soil and root sampling cores utilize a design that minimizes side compaction of the sample collected inside the core, and allows for more easy removal of the sample from the core and the ground (Fig. 16.3). However, these cores force soil to the outside of the core, causing compaction, and this can inhibit root penetration to soil around the minirhizotron tube that will subsequently be placed in the hole. Reamer, or reverse taper, bits should be used to size the final hole, after coring with a smaller diameter regular taper core and bit. The reverse taper bit forces compaction to the inside rather than the outside of the core tube. This procedure should be used for installing root-ingrowth spaces as well. Root injury during tube or core installation has also been mentioned as a potential problem in minirhizotron and ingrowth methods, and this is discussed earlier (Sect. 16.2.2).

Minirhizotron tubes must fit snug into the hole created with the reverse taper bit or (1) root growth can be influenced by air spaces (possibly positively), (2) water can flow directly down the space, and (3) tubes can move in the hole from date to date and result in images that do not overlap. This problem can occur in expansive soils that are drier after installation. Inflatable tubes have been designed to overcome loss of soil–tube contact in expansive and stony soils (Gijsman et al. 1991; Lopez et al. 1996), or anchoring systems that hold the tube in place can be used if tubes are prone to becoming loose but not so severely as to create space between the soil and tube. This type of tube movement is not a problem in many soils, but should be checked during an equilibration period (see below) if not working in an annual system. The type of material tubes are made of can affect root life spans (Withington et al. 2003) and flexibility of the tubes. Tube flexibility can be a different type of tube movement problem (see below).

Minirhizotron tubes need to be sealed at the bottom, and top when not in use, to prevent moisture from accumulating in the tube. Even when sealed with rubber lab stoppers, moisture in warm ambient air entering during sampling can condense on surfaces below that are cooler after sealing. Images should be viewed while taking them to avoid foggy images later. Missed data for a time interval can result in positive or negative bias. Tube mops can be made of soft absorbent materials and used prior to placement of the camera.

Tubes need to also be insulated in the part that protrudes aboveground to minimize inside–outside temperature differentials with depth, as well as to prevent light from influencing root growth. Insulation tape wrapped in reflective foil or closed-cell polyethylene sleeves can be used (Milchunas et al. 2005b; Pritchard et al. 2006), but large PVC caps can also help in harsh environments (Fig. 16.4a).

Calibration frames on tubes are necessary to provide data-processing software known reference lengths in order to adjust two different sample date image sizes to a constant. Pregitzer et al. (2000) robotically etched each window frame outline onto the plastic tubes. Milchunas et al. (2005b) used 1-mm-square gridded graph paper taped to the top of tubes just aboveground surface, and averaged 6–8 digitizings of vertical, horizontal, and diagonal distances. Some software used to digitize roots permits only one calibration average per tube per sample date. Close inspection of data revealed a systematic error in root lengths for part of some tubes segments during the fourth and fifth year of sampling, while data for other parts of the tubes looked normal (usually frames near the surface). Many of the roots in a series of sequential frames appeared to grow or shrink a certain very small percentage, based on subtracting previous lengths from current lengths columns in the data files, and the common percentage would gradually change with distance down the tube. Concomitantly, passing the camera housing down the tube became difficult for many of the tubes due to rubbing on the upper and lower sides of the tubes. The camera housing became rubbed-worn on the leading and trailing ends. There is some play between camera housing diameter and tube diameter, and this had become lost due to tube bending. Although the soils at this site were not heavy, freeze-thaw soil movement during the harsh winters or soil moisture differences through the profile must have caused this problem. Use of automated data calculation software on raw output files or in new versions of software that include manipulations prior to output should be used with caution even though they save time. If close inspection of the raw data was not made, this systematic error could result in artificially increased turnover estimates and increased temporal variability in the data. Because it became clear that some roots had not changed between dates (they spanned the entire grid so could not grow or shrink) the data could be manually adjusted segment by segment for the bend in the tube that could not be adjusted for by the single calibration frame. Raw data from minirhizotrons should always be closely inspected, and multiple calibration frames down a tube employed. Software that will not permit subsequent division of tubes into separate files may want to consider separate files by tube segments at the beginning of long-term studies.

Tube movement not large enough to cause tube bending and calibration problems can still occur and result in an overestimation of turnover. Small amounts of tube or soil movement can result in the loss of roots from view in the image even though it is still present in the soil. The artificial tube surface may also affect the accuracy of estimating root diameters because the surface is a plane and roots are tubular. This can best be illustrated by picturing a very small and a very large coin standing on edge on a table; the difference in the distances the two coins touch the table (what may be seen by the camera as a diameter) may be very different from the actual two root diameters at a distance from the tube (and possibly hidden from camera view by soil).

Root longevities are highly variable from system to system (see 16.2) so there can be months to years of time before a newly installed minirhizotron surface comes into equilibrium between new root growth and root decomposition. This can result in an overestimation bias, minimal bias, or used as a method of calculating turnover depending on the method of calculating turnover, and this is discussed below in reference to calculation methods. Equilibration periods prior to start of sampling have been suggested in the literature, but placing a specific time on this is not recommended unless prior understanding of turnover in the system you are working is known. For example, time for complete root turnover of 8.9 years based on ^{14}C labeling (Milchunas and Lauenroth 2001) and 5.8 years based on minirhizotron (Milchunas et al. 2005b) has been observed for a perennial semiarid grassland where individuals are very long lived, but it would be totally impractical to wait that long after tube installation to initiate an experiment. For example, equilibration may never occur in an annual row-crop system where most loss occurs after senescence when there is no new root growth. Clearly, methods of calculation of minirhizotron data will be the solution to this important issue and will vary with type of system.

Similar to root ingrowth methods, new root growth onto the tube surface after a sample date that dies and decomposes or is consumed by an herbivore before the next sample date will result in an underestimation of production. Sampling intervals ideally are short enough to minimize this loss between dates while long enough to minimize the huge labor demand in digitizing images. Unfortunately, this optimum sampling frequency is difficult to know without some prior knowledge of root longevity in terms of both lifespan and decomposition in any particular system. While arid systems will generally have longer decomposition rates, an evolutionary history of drought can result in survival adaptations to short-term drought (Lauenroth et al. 1987) even though long-term severe drought can cause mortality (Milchunas et al. 2005b) but would also extend decomposition times. Annual row crop systems may have a defined limit on lifespan, but the highly resource enriched environment may produce greater herbivory loss and will speed decomposition of any dead roots, thereby possibly requiring very short sampling intervals. Since it is impossible to predict these dynamics for any particular system, one could start conservatively and adjust sampling intervals down based on data of when losses start to become significant. Alternatively and more sophisticated, Johnson et al. (2001) developed a means of calculating sampling intervals for achieving a set level of statistical accuracy based on demographic data. As a starting point, literature from similar systems could be consulted for turnover times.

Problems concerning equilibration periods between new-root growth or initiation and loss, and the loss between sampling periods, bring up the issue of how root longevity is defined, and on possibilities of using root death rather than loss in estimates of turnover. There are many reports in the minirhizotron literature describing the classification of roots as live or dead. In forest, Hendrick and Pregitzer (1992) report distinguishing four categories of roots: white, tan, brown, and black with black being dead roots. Fahey et al. (1999) recommend following all roots until they disappear, because some can show continued growth. In grassland, Milchunas et al. (2005b) abandoned attempts to categorize even live from dead roots, as the large majority disappeared before appearing gray or black. This was not unexpected, as previous attempts to distinguish live and dead roots from cores taken one month after herbicide application to kill all plants and cores from adjacent live-plant plots could not be told apart in the lab, even after dying with a stain used for that purpose in other systems (unpublished observations). Classifying roots into condition categories is useful if possible, but should not be relied upon, and disappearance will be the only metric comparable to many other systems. There can be some confusion in the minirhizotron literature concerning whether "lifespan" means death or decomposition/loss, and possibly "longevity" could be used for situations of loss and "lifespan" for death.

The method of calculating turnover from minirhizotron data can have a large effect on how some of the above potential biases manifest, and that is discussed in the next section.

16.3.3 Calculation of Turnover and Application of Minirhizotron Methods in Different Ecosystems

The method of calculating root turnover from minirhizotron data can have large influences on the answer derived for BNPP. Tierney and Fahey (2001) observed root production estimates based on life-span methods that varied from 274 to 401 g/m^2/year depending on use of single or multiple cohorts and root numbers, length, or mass. Milchunas (2009) calculated 5-year mean estimates ranging from 160, 191, to 414 g/m^2/year for the same images and data set but using growth to decomposition regressions to equilibrium to obtain turnover times and two methods based on new length growth/total length to obtain turnover coefficients. Importantly, the method of calculating turnover from minirhizotron data can avoid or at least mitigate some of the biases mentioned earlier. However, the type of system in terms of root demography being studied can constrain the method of calculation used. Therefore, the topic of biases is continued in this section on methods of calculating turnover and in the context of type of plant system.

Some of the ways of calculating minirhizotron data have been developed in response to known biases discussed earlier. The most critical biases concern those associated with the artificial surface of the minirhizotron tubes; growth extending along the surface, root architecture growth from various angles to vertical and probabilities of intercepting the plane, equilibration of new growth, decomposition, and total length, and movement of the tube surface and/or what may be visible to the camera. Method of calculation also may vary with root life span and turnover of the plant system; long-lived perennial systems with long life spans of roots, short-lived perennial or systems with relatively short life spans of roots, and short-lived annual systems where the plant and all roots die within one growing season. Turnover is generally calculated and applied to root biomass data from cores. Turnover has been calculated as the amount of new length growth/existing total length for the year (Hendrick and Pregitzer 1993; Tierney and Fahey 2007, Milchunas 2009), based on cohort demographic analyses (see Hendrick and Pregitzer 1992; Eissenstat and Yanai 1997; Tierney and Fahey 2001, 2007; Ruess et al. 2003; Majdi et al. 2005, for details and options of analytical methods), or as a regression of time to equilibrium for new length growth to length loss (Milchunas 2009). This topic is also discussed in more detail in Chap. 18.

Long-lived perennial systems with long root longevities can have complete turnover times of roots of up to 5.8 years or 8.9 years based on minirhizotron

(Milchunas et al. 2005b) or isotope pulse labeling (Milchunas and Lauenroth 2001) methods, respectively. In this type of a system, calculation by new length growth/total existing length that year will result in overestimation of turnover and production because the denominator is artificially low until the equilibrium represented by a complete turnover is reached. For example, 1 unit of new growth/2 total units = 0.5 turnover coefficient and if there is 1,000 g/m^2 roots, the BNPP = 500 g/m^2/year. But, if there is the same 1 unit of new growth but 4 units of total length when growth and decomposition are eventually balanced, then the turnover coefficient is ¼ = 0.25 and only 250 g/m^2/year production. Waiting for roots to equilibrate before starting an experiment or monitoring is impractical in a system like this. Demographic cohort analyses could be used, but if the experiment only lasted a couple years, there would be greater uncertainty for extrapolation predictions for long-lived roots. The longer the study and more complete the estimates of all life spans, the better the estimates obtained from demographic analyses. The long-tailed distribution of root longevities due to the long time for disappearance of some roots is a factor in choosing the specific demographic method of calculation (Tierney and Fahey 2007) and may interact with choice of calculation when a large proportion of roots are not followed until loss (Majdi et al. 2005). Regression of growth:loss to equilibrium time could be used in long-lived perennial systems (Milchunas et al. 2005b). For example, if the growth:loss regression line intercepted the X-axis (time) at 8 years and there were 1,000 g/m^2 roots on average, then 1,000/8 = 125 g/m^2/year production. Differences between years can be adjusted for based on root biomass differences. For systems with very long life spans of roots and only short periods of study, the regression of growth:loss to equilibrium time may be the most appropriate, but comparisons with estimates based on demographic modeling need to be conducted.

Not accounted for in either the demographic or equilibrium regression approach, and still a source of bias, is the problem of moving tube surfaces loosing sight of a still existing root (a loss/decomposition or demographic death, or possibly also "seeing" a root that was already present and did not actually grow into the frame) or of the bending tube problem altering calibration reference down the tube (see above). There is the question in the demographic approach of whether roots growing along the plastic surface behave similarly to bulk roots (Withington et al. 2003; Fahey et al. 1999; Tierney and Fahey 2007) and this would apply to the regression approach as well. For all three methods, root diameter differences may have to be accounted for because thicker roots weigh more than thin roots and turnovers can be different. This will be a concern more in woody systems and greatest in forests, whereas grass roots are generally all fine roots with much less range in diameter classes.

The short-lived perennial or systems with relatively short longevities of roots, but where some roots survive more than 1 year, are possibly the least restricted in choice of method of calculation and all three methods may be options with the limitations imposed by bias mentioned earlier. The most important bias factor may be clearly defining that growth and decomposition have equilibrated before using data for calculation by the new growth/total length proportion method, and that there has been a complete turnover of all roots from multiple cohorts for the demographic approach. Fahey et al. (1999) recommend calculating and presenting estimates based on multiple methods, and this would help when comparing results among studies as well as in evaluating the different methods.

Annual plant systems may represent the most problematic for choosing a method of calculating turnover from minirhizotrons. Root demography in annual plant systems is clearly constrained by whole plant life span and in annual crop systems by harvest in some cases. Rees et al. (2005) review two general scenarios after an initial phase of high root production from the seedling through approximately flowering stage (1) new production ceases and root mortality is low until end of season, or (2) new production decreases and mortality increases with both occurring simultaneously (Fig. 16.5a). Fitter et al. (1996) is an example of the former. Huck et al. (1987) theoretically described the latter, Rees et al. (2005); Pritchard et al. (2006 and many citations therein) provide examples; and Pritchard and Rogers (2000) describe it as the most common for annual crop systems. In the former, end-of-season peak biomass would be a good estimate of annual BNPP, provided old versus current-year roots could be distinguished and there is little herbivory loss. In the latter, peak biomass would not be a good estimate of root production due to the simultaneous new growth/decomposition processes as well as in difficulty determining when peak biomass would occur. These dynamics of new growth and decomposition for annual systems are very different from those observed in perennial systems (Fig. 16.5b). Minirhizotrons can be a method to separate the growth/decomposition processes, but only some methods of calculating data may be appropriate. We focus here on this latter scenario.

The regression method of calculating turnover from minirhizotron data is an important option for systems with root longevities of many years (described earlier), but may not be a good option in annual systems because many sampling dates would be required to establish the regression, some new root growth can occur after the date where production and decomposition intersect (Fig. 16.5a), and root biomass used to multiply by turnover varies dramatically through the season unlike the relatively constant biomass in long-lived perennial systems (Milchunas and Lauenroth 2001). Demographic analyses to determine turnover may also require that a number of root cohorts be sampled in order to provide longevity estimates. The traditional simple method of calculating turnover from new length growth divided by average total length has the problem of invoking many of the biases associated with the artificial tube surface described earlier and in establishing a reasonable estimate of "average" length and biomass when both vary greatly through the season (Fig. 16.5a). Estimation of specific root lengths (length-diameter to mass relationships) and then conversion to total new length growth to a mass production value suffers all the biases associated with the artificial tube surface discussed earlier and is evidenced in the lack of relationships between total length and total biomass with depth in the profile. Therefore, a potential new method for annual systems is proposed in the next section below.

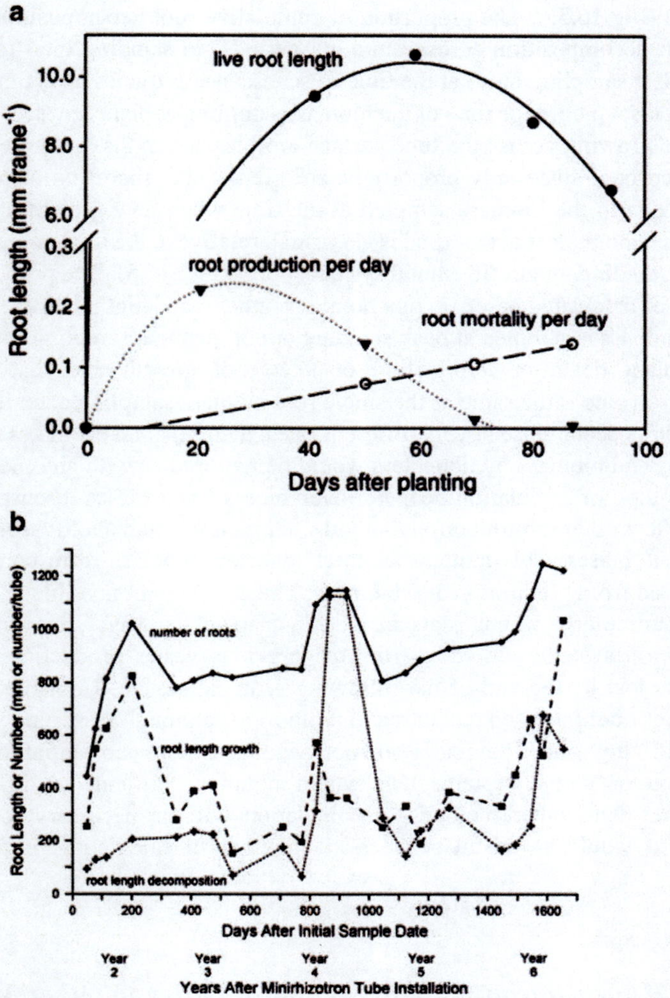

Fig. 16.5 Temporal dynamics of new root growth and decomposition in **(a)** an annual sorghum crop system (from Pritchard et al. 2006), and **(b)** a native shortgrass steppe grassland system with long-lived perennial plants (from Milchunas et al. 2005b)

16.3.4 A New Minirhizotron Method for Estimating Root Production in Annual Crop Systems

Based on the biases and/or labor intensive requirements in estimating root production from minirhizotrons in annual systems discussed earlier, a new but untested method is proposed here. This method uses only the cumulative total root length decomposed as a proportion of the total length at the time of root biomass sampling that coincides with senescence/whole-plant death or the end of new root length production (Fig 16.5a). The proportion of cumulative root decomposition (sum of new length decomposition across sampling periods) to standing total root length (length at last sampling only) at the time of senescence provides an estimate of the proportion lost up until the time of the biomass sampling at plant senescence. While the lengths growing across the tube surface are biased lengths, both are from the same biased pool since only proportions are used. This proportion of biomass is added back in to the biomass sampled at the time when new growth has ceased. Further, the amount lost at that time is very small relative to the total amount of new growth or standing length (in annual systems only) (Fig. 16.5). The proportion lost could also be calculated based on root numbers rather than length. It does not matter whether biomass is sampled at peak standing or not, just that sampling is at earliest time of plant death or termination of new root growth and that the final minirhizotron date is the same as the single root biomass sampling date. If different root diameters decompose at very different rates, then separate estimates for length proportions and biomass by diameters would be required as with any method.

This method of calculation does not overcome several biases discussed earlier that also affect other minirhizotron methods, but is simple and avoids several of the most serious biases. This method assumes that new biomass from cores can be differentiated from previous years' biomass. This is often not possible in perennial systems, particularly where roots can be long-lived. However, in watered and fertilized crop systems old roots from the previous year's production are more likely to be lost by the end of the following growing season. Unplanted "blank" control cores could be used to subtract out old roots, although losses are less in the absence of living roots. Previous year roots would be more substantial in dry-land annual crop or native systems and would be more difficult to differentiate. Comparisons with other methods of calculation will be necessary to test this approach, as would be useful between all methods of calculating minirhizotron data.

16.3.5 Minirhizotron Estimate Values Relative to Other Methods

As mentioned in Sect. 16.2.4, minirhizotrons generally overestimate root production compared with those obtained from root ingrowth. Steele et al. (1997), Fahey and Hughes (1994), Tierney and Fahey (2001, 2002), Hendricks et al. (2006) in three different sites, and Milchunas (2009) observed greater values for root production from minirhizotron than ingrowth and only Higgins et al. (2002) reported the opposite. However, Higgins et al. (2002) estimated root biomass for minirhizotron data from image root lengths and diameters and a root volume to mass relationship, whereby all artificial surface biases would manifest in the calculations of production. The Hendricks et al. (2006) study provides some insight because tube equilibration was explicitly tested and the three sites studied were over a gradient of hydric, mesic, and xeric longleaf pine systems, but sampling intervals were long at two months. The underestimation of production from ingrowth compared with minirhizotron was greater in the hydric compared to xeric site, and mortality rates were also greater in the hydric site. Even though the interval between sampling dates probably missed some production due to turnover between samples the ingrowth method still produced lower values. In a semiarid grassland with very long-lived roots and slow turnover, minirhizotron also produced greater production estimates than ingrowth even when minirhizotron values were calculated based on the regression of growth:loss to equilibrium time (Milchunas 2009). These comparisons suggest that minirhizotron results in greater estimates of root production than ingrowth methods even when equilibration between new growth and loss and root architecture has been accounted for in the minirhizotron estimates. Comparison of minirhizotron estimates of production with the isotope pulse labeling method is discussed below, after considering biases associated with isotope methods.

16.4 Isotope Pulse Labeling

Isotope pulse labeling involves the release of $^{14}CO_2$ or $^{13}CO_2$ for plant uptake in an enclosed environment (such as a closed tent) and using the non-^{12}C isotope as a tracer (Fig. 16.6). ^{14}C is radioactive so involves permits and intense safety and disposal regulations, while ^{13}C is a stable isotope. However, there are two potential problems with using ^{13}C rather than ^{14}C: the quantity and variability of background noise is much greater for ^{13}C than ^{14}C, and therefore the time necessary to get sufficient label in plants from a pulse can also be a problem. Theoretically, ^{13}C should be an option in this method, but we will talk here in terms of ^{14}C because further evaluation of these potential restrictions may be necessary for use of ^{13}C for root production and BNPP purposes.

In perennial plant systems and especially long-lived ones, the time necessary to obtain an estimate of turnover based on isotope pulse labeling can be many years, similar to the turnover time of all roots of various life spans described earlier in the minirhizotron section. Further, there is a requirement that all labile ^{14}C is incorporated into structural tissue, so there is also a time lag necessary before starting sampling that will be used in calculation of turnover. These two points are discussed further below, but the practical aspect of these two time constraints and the handling requirements means that isotope labeling may not be as practical as some other methods for wide use. However, as described below, the relatively few assumptions and potential biases of the ^{14}C pulse labeling method also make it an important reference point for comparing and evaluating the two other methods described earlier. Thus, from root ingrowth to minirhizotron to isotope labeling there is a progression from most to least eases of use, but possibly also the most to least severity of potential biases.

There are two primary and very different methods of calculating root production from isotope pulse labeling: isotope dilution and isotope turnover. Isotope dilution (Caldwell and Camp 1974; Milchunas and Lauenroth 1992) is not discussed in

Fig. 16.6 Pulse labeling native shortgrass steppe with ^{14}C. (**a**) Tents are covered with clear plastic and sealed at soil surface. (**b**) $^{14}CO_2$ is released and fans circulate air in tent

detail here, because it has been shown to possess a very severe bias associated with requiring the assumption of a uniform labeling of all roots and root tissue types be met. The details of why this assumption cannot be met in the isotope dilution method and the resulting potentially negative production values that can be obtained are discussed in Milchunas and Lauenroth (1992, 2001) and Milchunas (2009) and briefly below in the bias section. The isotope turnover method does not require a uniform labeling. Only the isotope turnover method is discussed below. Other isotope methods (bomb $\Delta^{14}C$ and elevated CO_2 $\delta^{13}C$) are discussed here only in Sect. 16.4.4.

In the isotope turnover method, loss of isotope (root decay) is considered the inverse of root production at steady state, and the time necessary for a complete turnover is estimated by regressing isotope quantity (or percent remaining) over time to the *X*-axis intercept or the time of no remaining isotope (Dahlman and Kucera 1965; Milchunas and Lauenroth 1992, 2001). Average root biomass is then divided by the number of years for complete turnover to estimate annual production, and differences among years (nonsteady state conditions) adjusted for by differences in biomass among years.

It is important to note that root exudation and sloughing is a component of root input of C to soil. Exudation and sloughing is a component of root production that can be estimated using isotope pulse labeling that cannot be estimated by other root production methods. Exudation can represent a substantial component of root production (Biondini et al. 1988; Milchunas et al. 1985; Milchunas and Lauenroth 1992). However, field estimates of exudation based on pulse labeling are subject to a number of problems and biases above those invoked for root production estimates, are probably underestimates, and are not discussed in this chapter. It is also important to note that much information concerning plant allocation and the incorporation of plant carbon into soil can be studied using pulse labeling (van Noordwijk et al. 1994; Swinnen 1994; Swinnen et al. 1995; Stewart and Metherell 1999; Saggar and Hedley 2001), but will also not be covered here.

16.4.1 Isotope Pulse Labeling Methods

The first step in labeling plants for the isotope turnover method is to calculate the amount of isotope that will be necessary to detect in the pool of interest, over a complete turnover time for roots. The "pool of interest" in this context can be just roots or also soil. The soil C pool is very large, so quantities of label necessary will be much greater if soil C dynamics is of interest. Plant respiration losses are also a large percentage of C initially taken up by plants, but is not a component of primary production. These are all system specific values and require a thorough assessment of literature most similar to the one to be studied. An important additional consideration includes the CO_2 compensation point of the plants, since isotope in the atmosphere after that level has been reached will no longer be taken up. Multiple, small "chaser" amounts of $^{12}CO_2$ can be added to the tents to allow for further uptake of $^{14}CO_2$ once compensation points are reached if time for labeling permits. Monitoring of CO_2 levels in the tents as plants draw it down helps obtain good efficiency of label uptake in the shortest time period. Time for labeling can be constrained by air temperatures rising in the tents to critical heat-stress levels if some sort of cooling system is not used. Fans should move air in the tents to reduce leaf-boundary-layer resistance and assure stomatal uptake (Fig. 16.6b).

Some plants will close stomata mid-day to conserve water, as will highly water-stressed plants regardless of time of day, limiting uptake of CO_2. Further, isotope from a pulse label is preferentially allocated to actively growing organs. Labeling in the cool of the morning avoids potential daily stress periods and helps keep tent temperatures low. In order to assure that live roots in all depths of the soil profile are labeled, it is important to add supplemental water if there are dry parts of the soil profile. Otherwise, roots in a dry zone of the profile may be underestimated if that zone remains dry while labile isotope is allocated. On the other hand, dry and wet parts of the growing season may be of interest to study in annual systems, but roots in dry zones not initially labeled in a long-lived perennial system can remain unlabeled for the many years in the future necessary for a complete turnover of all roots, thereby removing that initial dry zone from consideration. Tent edge effects and sample core distances from each other should be based on knowledge of the horizontal distribution of roots from the aboveground edge of individuals in the plant community.

After labeling, samples are obtained by standard coring and root separation from soil methods (see Sect. 16.6). One of the most critical aspects of sampling for root production in the isotope turnover method is determining when labile (or soluble) carbon compounds have all been depleted in the plant and (1) are in structural material, (2) respired or exudated out of the plant, and (3) aboveground–belowground fluxes via translocation have ceased. The potential problems if these conditions have not been achieved prior to calculation of root turnover are discussed in the next section below, as are the limitations for use in some plant systems. Various laboratory fractionation methods are available to determine labile versus structural carbon fractions, the simplest of which is the Van Soest (1967), Van Soest and Wine (1967) neutral detergent fiber (NDF) method with some modifications (see Milchunas and Lauenroth 1992). Methods for laboratory isotope analyses will not be discussed here.

16.4.2 Potential Biases, Problems, and Advantages of Isotope Pulse Labeling

There is some confusion concerning assumptions and potential biases for isotope turnover versus isotope dilution methods of production. The assumption of a uniform label throughout root tissues is not a requirement necessary to avoid bias in the isotope turnover method of calculating turnover from a pulse labeling, but is a requirement in the isotope dilution method. A uniform label means the same proportion of $^{13\text{ or }14}C$:^{12}C throughout all root age classes and throughout all fiber components of different turnover times (hemicellulose, cellulose, and lignins). Since label is laid down preferentially in actively growing locations, some old roots will decompose first, and if these roots do not contain isotope-C that loss will not be accounted for and will result in an underestimation of loss and even negative production values if isotope-C concentrations actually increase after all uptake has ceased (Milchunas and Lauenroth 1992). This is not a requirement or potential bias of the isotope turnover method, because the relatively more heavily labeled young root will proceed through all life stages and decomposition stages, and that longevity is the metric of interest rather than dilution of isotopes. However, the assumption that roots labeled at one time of pulse is representative of life spans and decomposition rates at some other time may or may not be met, and this is the primary potential bias of the isotope turnover method. Root cohort longevities may vary with season of birth (initiation). Differences in fiber component labeling with season or time of year has not been investigated, but may also represent a potential bias. How these can be mitigated to some extent in some systems is described in the next section.

A requirement of all isotope labeling methods for root production is that all labile C (soluble C) has been incorporated into structural tissue. Two problems can occur if this condition has not been met. First, aboveground–belowground allocation of labile compounds can occur. If isotope label is still being translocated belowground, that would be counted as new growth rather than only loss of isotope after the pulse as required, and turnover time estimates would be biased higher resulting in lower production estimates (or conversely lower turnover coefficients and lower production estimates). Second, even if above- and belowground allocation has stabilized, the presence of labile C means that a large part of that can be lost as root respiration. Respiration is not production, but an inefficiency of production, and is not transferable to secondary organisms or soil C. Loss of isotope label through respiration would be accounted for as loss of root production and result in a bias that decreased turnover times or increased turnover estimates (see also earlier section). Therefore, estimates of root production by isotope turnover must wait until all labile/soluble labels have been incorporated into structural tissue or tissue that is not capable of being respired or translocated. Some root loss is possible during this period, and results in a bias in the direction of an underestimate. As with all biases in all methods, different biases can oppose or reinforce each other in terms of the final outcome, and this is discussed in the section on comparisons of different methods below.

A great advantage of isotope labeling methods is that there are no artificial surfaces or growth mediums as in ingrowth or minirhizotron methods, since plants are labeled with no alteration of soils or roots. Isotope turnover is also the only method that can be used to assess crown production in bunch grasses. Crowns are the often ignored, large component of biomass in many grassland ecosystems, although turnover can be low (Milchunas and Lauenroth 1992, 2001). Exudation and sloughing are other transfers of C to soil and soil organisms that can only be estimated through isotope labeling methods, although field estimates of exudation are crude (Milchunas and Lauenroth 1992).

All methods of estimating root production require separation of roots from soil and are subject to biases in that procedure (see Sect. 19.6). However, there are two additional potential problems of separating soil from roots after the initial flotation or had picking. The first is the large amount of ash present in even floated-out, washed roots (Table 16.1) and this is easily dealt with by muffle furnace/ash corrections. There is a second problem with isotope methods. Isotope labeled soil can also adhere to new roots growing long after pulse labeling, and this contamination can result in a bias that produces extended turnover times (Milchunas and Lauenroth 2001). Thus, isotope ash (not just ash) corrections for root biomass samples collected for analyses are also necessary, but how bulk soil and rhizosphere soil may affect corrections has not been investigated. Another reason for an extended tail on isotope content of roots from a pulse labeling is that a few roots may live a very long period of time in perennial systems (Eissenstat and Yanai 1997; Milchunas and Lauenroth 2001), and this has also been observed in bomb $\Delta^{14}C$ studies of root production (Gaudinski et al. 2001; Tierney and Fahey 2002; Matamala et al. 2003; Trumbore and Gaudinski 2003). Separate regressions for turnover of these two pools of roots may be necessary to avoid bias in this case. Milchunas (2009) calculated estimates of root production for a semiarid, long-lived perennial shortgrass steppe grassland of 170 g/m^2/year when not accounting for soil ^{14}C contamination of roots or a second long-lived group of roots (or the long tail of the distribution), 217 g/m^2/year when adjusting for soil ^{14}C contamination of roots, and 185 g/m^2/year when adjusting for both (all estimates ash corrected).

16.4.3 Application of Isotope Pulse Labeling Methods in Different Ecosystems

Problems in applying the isotope turnover method of root production may be greatest at both extremes of root system longevities. At the very slow end, the practicality of the method may be limited simply by the time necessary to obtain time-to-complete-turnover estimates of both large body of root cohorts under the hump of the distribution and the long-lived tail of the bell distribution. The time necessary for a complete turnover of these two respective root pools was estimated at 5.3 and 9.8 years as semiarid shortgrass steppe (see Fig. 1 in Milchunas 2009).

In annual and some perennial systems with relatively short root longevities, a limitation may be at the other end of the sampling period spectrum; the time necessary for isotope in the labile pool to be incorporated into structural tissue and stabilize in terms of above- and belowground allocation, translocation fluxes, and respiration losses of isotope. Unfortunately, there is little evidence for the latter in particular. Swinnen et al. (1994a) observed that allocation in field ^{14}C labeled spring wheat was complete after about three weeks, but had to assume that root respiration of ^{14}C was also restricted to that period. However, winter wheat behaved differently, with allocation of ^{14}C to roots near the end of growth (Swinnen et al. 1995). Milchunas et al. (1985) labeled 45-day-old winter wheat and 95-day-old *Bouteloua gracilis* (the dominant grass in the slow-moving shortgrass steppe) under growth chamber conditions and observed labile ^{14}C 62 days later in senescent wheat plants and 40 days later at the end of the experiment for the grass. While this suggests a limitation for the wheat, the phenology of growth in the chamber was not similar to over-wintering winter wheat in the field. Differences among studies of all labeling wheat illustrate the need to assess labile allocation and labile to structural C dynamics from pulse labeling, and this may limit application of the isotope turnover method of root production in annual systems unless labile pathways are assessed and adjusted for.

16.4.4 Isotope Pulse Labeling Estimate Values Relative to Other Methods

The usual report in the limited literature comparing isotope and minirhizotron estimates is that isotope methods using either bomb $\Delta^{14}C$ or elevated CO_2 $\delta^{13}C$ methods (Gaudinski et al. 2001; Tierney and Fahey 2002; Matamala et al. 2003; Trumbore and Gaudinski 2003) or pulse isotope methods (van Noordwijk et al. 1994; Swinnen et al. 1995; Milchunas et al. 2005b; Milchunas 2009) produce lower estimates of production due to long turnover times associated with isotope methods that include a portion of roots that have long life spans. Although it is sometimes assumed that this is due to a relatively greater weighting of coarser roots in isotope compared with minirhizotron methods in forest systems, this was also observed in grassland where all roots are fine roots (Milchunas 2009). However, Tierney and Fahey (2002) were able to reconcile some of this difference by considering differences in longevities of roots. Similarly, Milchunas (2009) found good agreement between isotope turnover and minirhizotron but only when isotope calculations were both adjusted for label contamination in soil adhering to roots and calculating production using regular and a long lived root pool, *and* when minirhizotron estimates were based on regression of time to equilibrium for new length growth to length loss which bypasses some of the critical biases that can be associated with the minirhizotron method (see Sect. 16.4.3). Thus, root production mean estimate from isotope methods not adjusting for label contamination and using one early regression not accounting for long-lived roots was 170 g/m^2/year compared with a minirhizotron estimate of 414 g/m^2/year based on new length growth/total average length which is subject to some of the critical biases discussed in the minirhizotron section. However, root production estimates converged between the two methods when isotope estimates considered both a two-phase regression for long-lived roots and soil isotope contamination (185 g/m^2/year) and minirhizotron calculations were based on regression of time to equilibrium for new length growth to length loss (160 g/m^2/year). These results are encouraging, although many more comparisons will be necessary to fully establish how these two relatively new methods compare.

16.5 Standing Root Biomass Processing Methods

All methods of root production considered here required root separation from soil, either as a direct part of the method (ingrowth, isotope turnover) or to convert estimates of turnover to biomass (minirhizotron). BNPP is a very basic ecosystem process and the separation of roots from soil is a very basic task, but how this separation is accomplished can have as much effect on estimates as the method of

Fig. 16.7 Separating roots from soil by the flotation method using bucket design from Lauenroth and Whitman (1971) and an outdoor washing bench. The top sieve on the bucket is a coarse mesh and used only to break up soil clumps. Soil settles to bottom of bucket and roots float to top and pass through spout onto 0.5 mm mesh screen for capture

production. This section will only briefly assess a few of the more important issues in root–soil separation.

There are basically two methods of separating roots from soil: flotation/washing-wet sieving (Fig. 16.7) and dry sieving–hand picking. Hand picking smaller roots from soil after dry sieving out larger roots is a tedious labor intensive task, and root recoveries are low. However, some people suggest that washing/floating out roots removes some internal soluble root components, while others suggest that the losses of isotope observed in wash water are from contamination of roots with exudates, mucilages, and or rhizosphere soil embedded in the roots (Swinnen 1994; Milchunas and Lauenroth 2001). It is difficult to test whether the isotope loss during the flotation, washing process is due to loss from root material or loss from soil contamination, but the differences in root biomass estimates between dry and wet processing are great. Pregitzer et al. (2000) found that only about 60% of roots are recovered in very careful hand sorting from soil compared to washing from soil in an automated root elutriator which recovered greater than 99% of fine roots from a forest (size of openings in elutriator containers not given). Milchunas (2009) estimates approximately 64–71% of roots are retained from dry soil hand picking (detailed picking and lifting of fines by static wand) compared to flotation and capture on 0.5 mm sieves of grassland roots (loss from the 0.5 mm sieves was not estimated) (Fig. 16.8). Based on ^{14}C loss, Swinnen (1994) found that from 4.8 to 23.8% of label from 0.3 to 0.6 mm roots was lost when washing from soil, from 9.8 to 41.1% lost after washing hand-picked roots less than 0.3 mm, and from 28.9 to 40.2% lost after washing hand picked soil, with all ranges increasing from young to old spring barley and winter wheat plants using a sieve size of 0.6 mm. Swinnen (1994) concluded that hand picking losses were much greater than washing losses for young plants. However, some of the loss from washing hand picked roots was

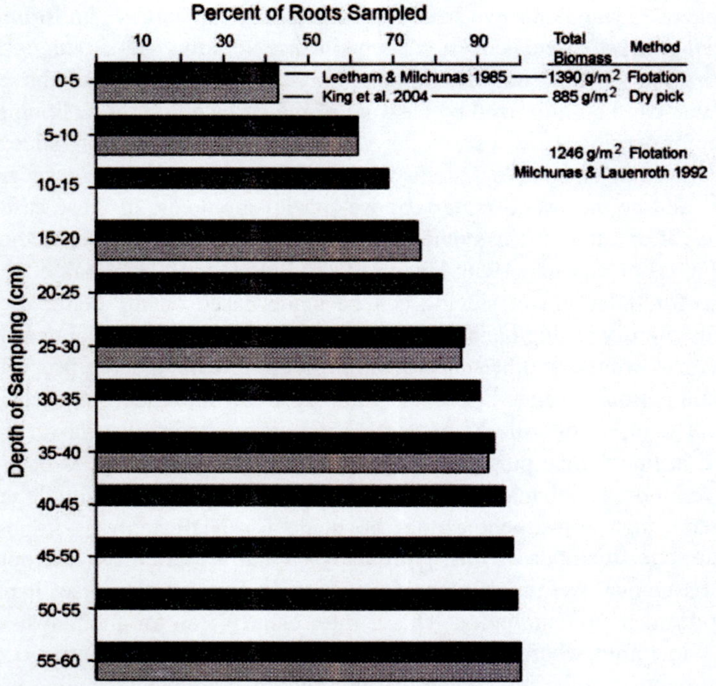

Fig. 16.8 Root biomass depth distribution (%) through the soil profile for shortgrass steppe and the total root biomass for two studies using flotation method and one using careful dry picking by tweezers and lifting of fine roots using a static wand after a coarse dry sieving. Data from Leetham and Milchunas (1985), King et al. (2004), and Milchunas and Lauenroth (1992), and ash corrected to an organic matter basis

most likely from exudates, mucilages, and soil ^{14}C contamination as well. Losses from washing roots from the soil can also be soil organic matter rather than root material. This brings up the important question of when root material becomes soil organic matter and some other issues as well before suggesting possible best options. Note, however, that the above percentage differences among root–soil separation methods are not trivial, but can be as large as differences among methods for estimating turnover, even though relative depth distributions may be similar.

The inability to identify live from dead roots in some plant systems and the questionable identities in others was discussed in the minirhizotron methods section (Sect. 16.3.2). Loss or disappearance of roots was therefore suggested as a standard method, but this determination too is not clear-cut when separating roots from soil. If the disappearance of roots is the criteria of loss from minirhizotron image, then the same criteria of when a root is no longer a root should be used for the root biomass estimate that the minirhizotron turnover estimate is applied to, or bias can be introduced. However, both decisions can be a subjective. The most objective way to do this is to allow sieve size determine when a root becomes soil organic matter, and use a standard sieve size for all studies for comparative uniformity. This works well in most native ecosystems when using flotation/wet-sieving separation and requires only sorting out of minimal amounts of insect fecal and aboveground material that can be minimized by clearing of the surface litter and clipping below crowns before coring. However, crop systems where aboveground residue is tilled into the soil profile and native systems with high organic matter layers that are a mix of above- and belowground material pose special problems. In these cases, hand separations after flotation/wet sieving are necessary, and line-transect methods may help in this task (Newman 1966; Hendrick and Pregitzer 1993; Wenk et al. 2006).

A case for flotation/wet sieving can be made based on the better recoveries compared with dry sifting/hand picking, and because of the great labor savings of the former. However, another consideration is the effects of large proportions of soil contamination imbedded in roots (Table 16.1) on subsequent nutrient or fiber analyses. The effect of soil-^{14}C contamination on estimates of turnover has been illustrated in the isotope pulse labeling section (19.4.2). These types of problems can be even more problematic for other analyses. For example, a simple contamination percentage adjustment cannot be made for NDF analyses important in isotope turnover methods or for lignin analyses that are often run on roots as an index of tissue quality for decomposition. This is because the analyses itself alters the proportions of contamination. This is most clearly seen for an analysis such as sulfuric acid lignin where concentrated acid used in the extraction process can dissolve a fraction of the soil contamination fraction, thereby altering lignin values. Because ash contamination values can be very high (Table 16.1), the best option is a root–soil separation that minimizes soil contamination, which is most likely wash flotation/wet sieving. Even then, soil contamination levels can be so high that running soil samples alone through the sequential NDF, ADF, lignin and adjusting the root values for hemicelluloses, cellulose, and lignin based on soil losses during the NDF, ADF, and lignin process and the sample total ash content can help adjust for the root–soil contamination effect (Milchunas and Lauenroth 1992). Ash correction values can vary widely among plant systems. Roots growing through very hard, compacted, dry soil can have very high ash correction values (Table 16.1). In other cases, Pregitzer et al. (2000) report values for a *Populus tremuloides* forest of less than 5%. Because of this wide range among systems, and often very high values for some, standard practice should be to always report ash corrected root biomass estimates (Böhm 1979; Bledsoe et al. 1999; Lauenroth 2000). The removal of this variable contamination factor (1) allows more accurate cross-site syntheses from studies, (2) lowers variance within a data set for more robust statistical analyses of that particular study (Table 16.1), (3) reduces the possibility of artifact treatment effects if soils/conditions for soil imbedding into roots varies with treatment, and (4) increases accuracy of subsequent nutrient analyses.

Based on the earlier discussions and other reports in the literature, several standard practices for root separation from soil are recommended. A wash flotation/wet sieving method is preferred over dry sieving hand picking. While expensive elutriators are available, inexpensive and efficient flotation work-stations and flotation buckets are easily constructed (Fig. 16.7). A sieve size of 0.5 mm openings has most often been recommended as a standard size for wet sieving collection of floated roots. Ash corrected values for root biomass and production should be standard practice. Cores for root sampling should be at least 50–70 mm inside diameter, because this minimizes fragmentation and loss of root material during separation and better spans spatial heterogeneity of roots (Bledsoe et al. 1999). The diameter of cores must be measured at the inside diameter of the bit tip, not the tube (see Fig. 16.3), because small differences in actual sampling diameter result in large differences in estimates of root production when extrapolated to a meter square basis. Depth of sampling that recovers at least 80% of the total root system is desirable. Authors, reviewers, and editors will help others interpret and use root data by insuring that these important details are included in all published papers, regardless of whether all are followed or not.

References

Austin AT, Vivanco L (2006) Plant litter decomposition in a semi-arid ecosystem controlled by photodegradation. Nature 442:555–558

Arnone JA III, Zaller JG, Spehn EM, Niklaus PA, Wells CE, Körner C (2000) Dynamics of root systems in native grasslands: effects of elevated atmospheric CO_2. New Phytol 147:73–85

Balesdent J, Balabane M (1996) Major contribution of roots to soil carbon storage inferred from maize cultivated soils. Soil Biol Biochem 28:1261–1263

Biondini M, Klein DA, Redente EF (1988) Carbon and nitrogen losses through root exudation by *Agropyron cristatum, A. smithii* and *Bouteloua gracilis*. Soil Biol Biochem 20:477–482

Bledsoe CS, Fahey TJ, Day FP, Ruess RW (1999) Measurement of static root parameters. In: Robertson GP, Bledsoe CS, Coleman D, Sollins P (eds) Standard soil methods for long-term ecological research. Oxford University Press, New York, pp 413–435

Böhm W (1979) Methods of studying root systems. Springer, Berlin

Brandt LA, King JY, Hobbie SE, Milchunas DG, Sinsabaugh R (2010) The role of photodegradation in surface litter decomposition across a grassland ecosystem precipitation gradient. Ecosystems 13:765–781

Brandt LA, King JY, Milchunas DG (2007) Effects of altered ultraviolet radiation, litter chemistry, and precipitation on litter decomposition in shortgrass steppe. Global Change Biol 13:2193–2205

Caldwell MM, Camp LB (1974) Belowground productivity of two cool desert communities. Oecologia 17:123–130

Curtis PS, O'Neill EG, Teeri JA, Zak DR, Pregitzer KS (1994) Belowground responses to rising atmospheric CO_2: Implications for plants, soil biota and ecosystem processes. Plant Soil 165:1–6

Dahlman RC, Kucera CL (1965) Root productivity and turnover in native prairie. Ecology 46:84–89

Eissenstat DM, Yanai RD (1997) The ecology of root lifespan. Adv Ecol Res 27:1–60

Fahey TJ, Bledsoe CS, Day FP, Ruess RW, Smucker AM (1999) Fine root production and demography. In: Robertson GP, Bledsoe CS, Coleman D, Sollins P (eds) Standard soil methods for long-term ecological research. Oxford University Press, New York, pp 437–455

Fahey TJ, Hughes JW (1994) Fine root dynamics in a northern hardwood forest ecosystem, Hubbard Brook Experimental Forest, NH. J Ecol 82:533–548

Fitter AH, Self GK, Wolfenden J, vanVuuren MMI, Brown TK, Williamson L, Graves JD, Robinson D (1996) Root production and mortality under elevated atmospheric carbon dioxide. Plant Soil 187:299–306

Gaudinski JB, Trumbore SE, Davidson EA, Cook AC, Markewitz D, Richter DD (2001) The age of fine-root carbon in three forests of the eastern United States measured by radiocarbon. Oecologia 129:420–429

Geisler D, Ferree DC (1984) Response of plants to root pruning. Hort Rev 6:155–188

Gijsman AJ, Floris J, van Noordwijk M, Brouwer G (1991) An inflatable minirhizotron system for root observation with improved soil-tube contact. Plant Soil 134:261–270

Hendrick RL, Pregitzer KS (1992) The demography of fine roots in a northern hardwood forest. Ecology 73:1094–1104

Hendrick RL, Pregitzer KS (1993) The dynamics of fine root length, biomass and nitrogen content in two northern hardwood ecosystems. Can J For Res 23:2507–2520

Hendricks JJ, Hendrick RL, Wilson CA, Mitchell RJ, Pecot SD, Guo D (2006) Assessing the patterns and controls of fine root dynamics: an empirical test and methodological review. J Ecol 94:40–57

Hendricks JJ, Nadelhoffer KJ, Aber JD (1993) Assessing the role of fine roots in carbon and nitrogen cycling. Trends Ecol Evol 8:174–178

Hendrickson AH, Veihmeyer FJ (1931) Influence of dry soil on root extension. Plant Physiol 6:567–576

Hertel D, Leuschner C (2002) A comparison of four different fine root production estimates with ecosystem carbon balance data in a *Fagus-Quercus* mixed forest. Plant Soil 239:237–251

Higgens PAT, Jackson RB, Des Rosiers JM, Field CB (2002) Root production and demography in a California annual grassland under elevated atmospheric carbon dioxide. Global Change Biol 8:841–850

Huck MG, Hoogenboom G, Peterson CM (1987) Soybean root senescence under drought stress. In: Taylor HM (ed) Minirhizotron observation tubes: methods and applications for measuring rhizosphere dynamics. Am Soc Agron, Madison WI, pp 109–121

Jackson RB, Canadell J, Ehleringer JR, Mooney HA, Sala OE, Schulze ED (1996) A global analysis of root distributions for terrestrial biomes. Oecologia 108:389–411

Johnson MG, Tingey DT, Phillips DL, Storm MJ (2001) Advancing fine root research with minirhizotrons. Environ Exp Bot 45:263–289

Joslin JD, Wolfe MH (1999) Disturbance during minirhizotron installation can affect root observation data. Soil Sci Soc Am J 63:218–221

King JY, Mosier AR, Morgan JA, LeCain DR, Milchunas DG, Parton WJ (2004) Plant nitrogen dynamics in shortgrass steppe under elevated atmospheric carbon dioxide. Ecosystems 7:147–160

Lauenroth WK (2000) Methods of estimating belowground net primary production. In: Sala OE, Jackson RB, Mooney HA, Howarth RW (eds) Methods in ecosystem science. Springer, New York, pp 58–71

Lauenroth WK, Sala OE, Milchunas DG, Lathrop RW (1987) Root dynamics of *Bouteloua gracilis* during short-term recovery from drought. Funct Ecol 1:117–124

Lauenroth WK, Whitman WC (1971) A rapid method for washing roots. J Range Manage 24:308–309

Leetham JW, Milchunas DG (1985) The composition and distribution of soil microarthropods in the shortgrass steppe in relation to the soil water, root biomass, and grazing by cattle. Pedobiologia 28:311–325

Lopez B, Sabate S, Garcia C (1996) An inflatable minirhizotron system for stony soils. Plant Soil 179:255–260

Lucak M, Godbold DL (2010) Fine root biomass and turnover in southern taiga estimated by root inclusion nets. Plant Soil 331:505–513

Lund ZF, Pearson RW, Buchanan GA (1970) An implanted soil mass technique to study herbicide effects on root growth. Weed Sci 18:279–281

Matamala R, Gonzalez-Meler MA, Jastrow JD, Norby RJ, Schlesinger WH (2003) Impact of fine root turnover on forest NPP and soil C sequestration potential. Science 302:1385–1387

Majdi H, Pregitzer K, Morén A, Nylund J, Ågren G (2005) Measuring fine root turnover in forest ecosystems. Plant Soil 276:1–8

McNaughton SJ, Milchunas DG, Frank DA (1996) How can primary productivity be measured in grazing ecosystems? Ecology 77:974–977

Milchunas DG (2009) Estimating root production: comparison of 11 methods in shortgrass steppe and review of biases. Ecosystems 12:1381–1402

Milchunas DG, Lauenroth WK (1992) Carbon dynamics and estimates of primary production by harvest, C^{14} dilution, and C^{14} turnover. Ecology 73:593–607

Milchunas DG, Lauenroth WK (2001) Belowground primary production by carbon isotope decay and long-term root biomass dynamics. Ecosystems 4:139–150

Milchunas DG, Lauenroth WK, Burke IC, Detling JK (2008) Effects of grazing on vegetation in the shortgrass steppe. In: Lauenroth WK, Burke IC (eds) Ecology of the shortgrass steppe: a long-term perspective. Oxford University Press, New York, pp 389–446, Chapter 16

Milchunas DG, Lauenroth WK, Singh JS, Hunt HW, Cole CV (1985) Root turnover and production by ^{14}C dilution: implications of carbon partitioning in plants. Plant Soil 88: 353–365

Milchunas DG, Mosier AR, Morgan JA, LeCain D, King JY, Nelson JA (2005a) Root production and tissue quality in a shortgrass steppe exposed to elevated CO_2: Using a new ingrowth method. Plant Soil 268:111–122

Milchunas DG, Morgan JA, Mosier AR, LeCain D (2005b) Root dynamics and demography in shortgrass steppe under elevated CO_2, and comments on minirhizotron methodology. Global Change Biol 11:1837–1855

Milchunas DG, Mosier AR, Morgan JA, LeCain DL, King JY, Nelson JA (2005c) Elevated CO_2 and defoliation effects on a shortgrass steppe: forage quality versus quantity for ruminants. Agric Ecosystems Environ 111:166–184

Nadelhoffer KJ (2000) The potential effects of N deposition on fine root-production in forest ecosystems. New Phytol 147:131–139

Neill C (1992) Comparison of soil coring and ingrowth methods for measuring belowground production. Ecology 73:1918–1921

Newman EI (1966) A method of estimating total length of root in a sample. J Appl Ecol 3:139–145

Norby RJ (1994) Issues and perspectives for investigating root responses to elevated atmospheric carbon dioxide. Plant Soil 165:9–20

Norby RJ, Jackson RB (2000) Root dynamics and global change: seeking an ecosystem perspective. New Phytol 147:3–12

Persson HA (1978) Root dynamics in a young Scots pine stand in central Sweden. Oikos 30:508–519

Persson HA (1979) Fine-root production, decomposition and mortality in forest ecosystems. Vegetation 41:101–109

Pregitzer KS, Zak DR, Maziasz J, DeForest J, Curtis PS, Lussenhop J (2000) Interactive effects of atmospheric CO_2 and soil-N availability on fine roots of *Populus tremuloides*. Ecol Applic 10:18–33

Pritchard SG, Prior SA, Rogers HH, Davis MA, Runion GB, Popham TW (2006) Effects of elevated atmospheric CO_2 on root dynamics and productivity of sorghum grown under conventional and conservation agricultural management practices. Agric Ecosystems Environ 113: 175–183

Pritchard SG, Rogers HH (2000) Spatial and temporal deployment of crop roots in CO_2-enriched environments. New Phytol 147:55–71

Rees RM, Bingham IJ, Baddeley JA, Watson CA (2005) The role of plants and land management in sequestering soil carbon in temperate arable and grassland ecosystems. Geoderma 128: 130–154

Ruess RW, Hendrick RL, Burton AJ, Pregitzer KS, Sveinbjornssön B, Allen MF, Maurer GE (2003) Coupling fine root dynamics with ecosystem carbon cycling in black spruce forests of interior Alaska. Ecol Monogr 73:643–662

Saggar S, Hedley CB (2001) Estimating seasonal and annual carbon inputs, and root decomposition rates in a temperate pasture following field ^{14}C pulse-labelling. Plant Soil 236:91–103

Sala OE, Austin AT (2000) Methods of estimating aboveground net primary production. In: Sala OE, Jackson RB, Mooney HA, Howarth RW (eds) Methods in ecosystem science. Springer, New York, pp 31–57

Schenk HJ, Jackson RB (2002) Rooting depths, lateral root spreads and below-ground/above-ground allometries of plants in water-limited ecosystems. J Ecol 90:480–494

Sims PL, Singh JS, Lauenroth WK (1978) The structure and function of ten western North American grasslands, I. Abiotic and vegetational characteristics. J Ecol 66:251–285

Singh JS, Lauenroth WK, Hunt HW, Swift DM (1984) Bias and random errors in estimators of net root production: a simulation approach. Ecology 65:1760–1764

Singh JS, Lauenroth WK, Steinhorst RK (1975) Review and assessment of various techniques for estimating net aerial primary production in grasslands from harvest data. Bot Rev 41:181–232

Steen E (1983) Usefulness of the mesh bag method in quantitative root studies. In: Atkinson D (ed) Plant root growth, an ecological perspective. Brit Ecol Soc Special Publ No 10. Blackwell, London, pp 75–86

Steen E (1991) Variation of root growth in a grass ley studied with a mesh bag technique. Swedish J Agric Res 14:93–97

Steingrobe B, Schmid H, Claassen N (2000) The use of the ingrowth core method for measuring root production of arable crops B influence of soil conditions inside the ingrowth core on root growth. J Plant Nutr Soil Sci 163:612–622

Steingrobe B, Schmid H, Claassen N (2001) The use of the ingrowth core method for measuring root production of arable crops B influence of soil and root disturbance during installation of the bags on root ingrowth into the cores. Eur J Agron 15:143–151

Stewart DPC, Metherell AK (1999) Carbon (^{13}C) uptake and allocation in pasture plants following field pulse-labelling. Plant Soil 210:61–73

Swinnen J (1994) Rhizodeposition and turnover of root-derived organic material in barley and wheat under conventional and integrated management. Agric Ecosystems Environ 51:115–128

Swinnen J, Van Veen JA, Merckx R (1994a) ^{14}C pulse-labelling of field-grown spring wheat: an evaluation of its use in rhizosphere carbon budget estimations. Soil Biol Biochem 26:161–170

Swinnen J, Van Veen JA, Merckx R (1994b) Losses of ^{14}C from roots of pulse-labeled wheat and barley during washing from soil. Plant Soil 166:93–99

Swinnen J, Van Veen JA, Merckx R (1995) Root decay and turnover of rhizodeposits in field-grown winter wheat and spring barley estimated by ^{14}C pulse labeling. Soil Biol Biochem 27:211–217

Tierney GL, Fahey TJ (2001) Evaluating minirhizotron estimates of fine root longevity and production in the forest floor of a temperate broadleaf forest. Plant Soil 229:167–176

Tierney GL, Fahey TJ (2002) Fine root turnover in a northern hardwood forest: a direct comparison of the radiocarbon and minirhizotron methods. Can J For Res 32:1692–1697

Tierney GL, Fahey TJ (2007) Estimating belowground primary productivity. In: Fahey TJ, Knapp AK (eds) Principles and standards for measuring primary production. Oxford University Press, Oxford, pp 120–141

Trumbore SE, Gaudinski JB (2003) The secret lives of roots. Science 302:1344–1345

Van Noordwijk M, Brouwer G, Koning H, Meijboom FW, Grzebisz W (1994) Production and decay of structural root material of winter wheat and sugar beet in conventional and integrated cropping systems. Agric Ecosyst Environ 51:99–113

Van Soest PJ (1967) Development of a comprehensive system of feed analysis and its application to forages. J Anim Sci 26:119–128

Van Soest PJ, Wine RH (1967) Use of detergents in the analysis of fibrous feeds. IV. Determination of plant cell-wall constituents. J Assoc Off Agric Chem 50:50–55

Wenk RC, Battles JJ, Jackson RD, Bartolome JW, Allen-Diaz B (2006) An accurate and efficient method for sorting biomass extracted from soil cores using point-intercept sampling. Soil Soc Am J 70:851–855

Withington JM, Elkin AD, Bułlaj B, Olesińnski J, Tracy KN, Bouma TJ, Oleksyn J, Anderson LJ, Modrzyńnski J, Reich PB, Eissenstat DM (2003) The impact of material used for minirhizotron tubes for root research. New Phytol 160:533–544

Chapter 17
Minirhizotrons in Modern Root Studies

Teofilo Vamerali, Marianna Bandiera, and Giuliano Mosca

Abstract In recent years, minirhizotrons have received increasing interest in field studies for characterising several biological processes, such as fine root production, root longevity, mycorrhization and parasitism, as collecting repeated video or digital images allows the fate of individual roots to be followed in time.

This review summarises recent technical improvements in root observation and imaging, comparing results from various construction materials and installation angles. Both glass and transparent acrylic or polycarbonate tubes have shown a relatively low influence on root behaviour; tilted tubes are preferred to vertically oriented ones, in order to avoid artefacts, and to horizontal ones, which involve considerable soil disturbance during positioning in their surroundings in the open field. Precise camera positioning systems and new high-resolution (600 DPI and more) digital sensors are currently replacing external-tracked tubes and analog sensors, enhancing image capturing and quality. This allows for frequent root observation throughout the plant cycle and seasons, required for accurate estimation of root dynamics.

Although manual and semi-automatic image analysis requires timing and tedious work, in the last few years minirhizotron systems have been increasingly used, thanks to fundamental advances in image analysis. The development of new algorithms for root detection and measurement in minirhizotron images has speeded up processing and research. The most interesting results in image analysis derive from the integration of luminance intensity- and geometry-based root vs. non-root classifiers. Discrepancies still remain between minirhizotron data and

T. Vamerali (✉)
Department of Environmental Sciences, University of Parma, Viale G.P. Usberti 11/A, 43100 Parma, Italy
e-mail: teofilo.vamerali@unipd.it

M. Bandiera • G. Mosca
Department of Environmental Agronomy and Crop Sciences, University of Padova, Viale dell'Università 16, 35020 Legnaro, Padova, Italy

S. Mancuso (ed.), *Measuring Roots*,
DOI 10.1007/978-3-642-22067-8_17, © Springer-Verlag Berlin Heidelberg 2012

reference core sampling (root quantification) and carbon isotope methods (root turnover). Underestimation in root length mainly regards the top 0.3 m soil layer, and irreconcilable differences emerge between root life-span and root carbon residence time.

A critical point in minirhizotron studies is exact identification of root activity, and most of the previous work was based on manual or semi-automatic visual evaluation (e.g., root appearance/disappearance, colour, shrinking, blotting, cortex degradation). Natural root fluorescence under UV light is only a reliable feature of root activity in some plant species, but interesting perspectives are expected from visible and near-infrared spectral images. Roots of GFP-engineered (green fluorescent protein) plants can also be distinguished in mixed plant stands by specific camera adaptation (light wavelengths, filters).

17.1 Learning from Minirhizotron Observations

Besides their principal functions of nutrient and water uptake and anchorage, fine roots play an important role in soil carbon recycling (Richter et al. 1999), but insights into root turnover is still hampered by poor accessibility.

Conventional root methods are based on destructive soil sampling, such as pinboards (Schuurman and Goedewaagen 1971), ingrowth mesh bags (Steen 1991) and soil coring (Neill 1992). Repeated non-destructive observations generally involve the use of transparent interfaces, like rhizotrons, minirhizotrons and transparent walls. Minirhizotrons have overcome the high costs of underground large rhizotrons (Böhm 1979), and are extensively used in several plant species to characterise fine root production, branching patterns and phenology, mortality and longevity (Hendrick and Pregitzer 1996; Majdi 1996; Johnson et al. 2001; Guo et al. 2008). Minirhizotrons have also been applied to studying root interactions with a variety of soil organisms such as mycorrhizae (Majdi 2001; Pritchard et al. 2008b), nematodes (De Ruijter et al. 1996; Smit and Vamerali 1998), N-fixing bacteria and root nodules (Pritchard and Rogers 2000) and soil fauna (Steinaker and Wilson 2008). In ecological studies, the use of minirhizotrons has profitably highlighted close relationships between *Collembola* populations, fine root production and mycorrhizal colonisation in forests and grasslands (Steinaker and Wilson 2008). Some consequences of climate change have also been studied in the long term by monthly minirhizotron observations, allowing to detect a reduction of mycorrhizal root tips in deep soil and an increase in shallow layers of a loblolly pine forest subjected to free-air-CO_2 enrichment (Pritchard et al. 2008b). As regards parasites, for instance, the lower root decay and retardation in root growth of nematode-infested potatoes explained the less efficient nutrient uptake and reduced leaf growth and yield (Smit and Vamerali 1998).

The main advantage of minirhizotron systems is that repeated root images can be recorded for extended time periods (Hendrick and Pregitzer 1996; Majdi 1996). Specific root segments can be tracked in time by high-resolution video cameras at desired time intervals from birth to death, without having significant influence on plant rooting (Johnson et al. 2001; Tierney and Fahey 2001).

Minirhizotrons were first described by Bates (1937) as glass tubes inserted into the soil, closed at the bottom and sealed with a metal lid at the top. In that study, roots of *Lolium perenne* L. were observed by means of a mirror and an electric bulb mounted at the end of a metal rod. Minirhizotron systems nowadays typically consist of transparent tubes permanently inserted into the soil, with a miniature video camera which is moved along the tubes at selected positions, and a PC for camera control (focus, light intensity) and image acquisition (Fig. 17.1). Recent digital sensors have allowed images to be stored in digital form directly in the computer memory, without the quality losses which may occur in analog–digital image conversion (Basile et al. 2007; Faget et al. 2010).

Few commercial companies market their minirhizotron systems (e.g., Bartz Technology, http://www.bartztechnology.com; CID Inc., http://www.cid-inc.com) and they are generally quite expensive. The Bartz Technology minirhizotron camera system (Bartz Technology Co., Santa Barbara, CA, USA) has been cited in root studies by several authors (Johnson et al. 2000; Norby et al. 2004; Brown et al. 2009), but many scientists have built their own instrumentation (e.g., Johnson and Meyer 1998). For instance, a cheap home-made equipment was used by Vamerali et al. (1999, 2003a). It consisted of 1.5- or 3-m long 45° angled PVC tubes (50- and 57-mm inner and outer diameters, respectively) sealed by silicone plugs at their ends, and a colour video camera (Pulnix TMC-X, f 12 mm) to explore their central upward sector (1.12 rad) (Figs. 17.2 and 17.3). In root observation, the essential key is repeatability in camera positioning. A suitable camera drive is generally made of a wheel-carriage system and felt pads, together with a guiding handle for correct positioning and image recording.

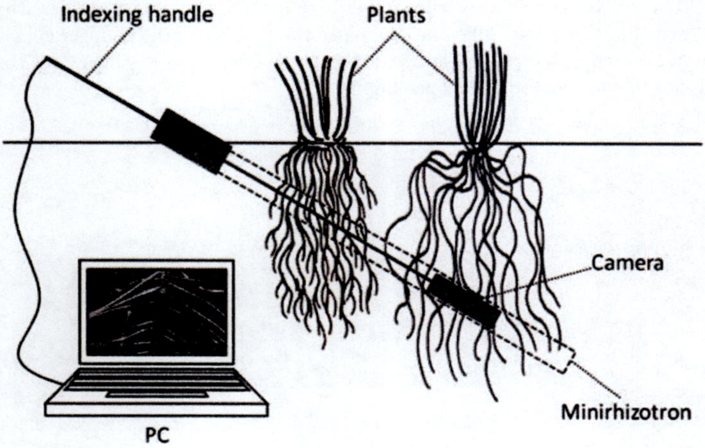

Fig. 17.1 Sketch of a minirhizotron root observation system. Analog cameras need PC connection through a frame grabber for image digitalisation

Fig. 17.2 (**a**) Suspended platform corer for soil drilling in minirhizotron installation; (**b**) installation of 45°-angled minirhizotron tube close to maize rows (at 80 mm); (**c**) upper ends of angled minirhizotrons covered with PVC caps and sealed with rubber plugs in sugar beet; (**d**) home-made mechanical indexing handle for camera positioning

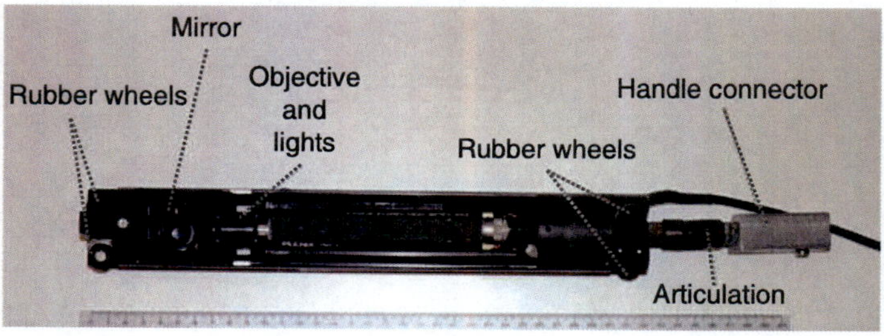

Fig. 17.3 Details of home-made analog video camera used by Vamerali et al. (1999, 2003) for minirhizotron root observations (scale in cm)

Minirhizotron observation systems generally allow a small sector of the soil–tube interface to be viewed, from 18 × 13 mm (Bartz Technology) to 32 × 24 mm (Horizontal × Vertical, Vamerali et al. 1999), thus requiring a large number of recordings. Instead, in the CID-Inc system (Vancouver, Washington, USA), larger images (216 × 196 mm, 1,700 × 1,540 pixels) are captured along the whole circumference of the tube by a rotating scanner head.

17.2 Minirhizotrons: Technical Notes and Installation

Minirhizotrons generally consist of long transparent tubes made of various materials, allowing their external surface to be reached by growing roots. Research was first carried out with tubes made of glass (Böhm et al. 1977), but this was rapidly replaced by more resistant and cheaper synthetic materials such as acrylic (Perspex or Plexiglas, i.e., polymethylmetacrylate) (Coleman et al. 1996; Vamerali et al. 1999), polycarbonate (Joslin and Wolf 1998) and cellulose acetate butyrate (CAB) (Pietola 2005). The tubes usually have a round section, and only rare citations refer to square tubes (van Noordwijk et al. 1985). They may vary greatly in size, length and section, although they are commonly 1.5–2 m long and have wall thicknesses of 3–7 mm. Their diameter should fit the camera size, and frequent inner and outer diameters are 50 and 57 mm, respectively. Although larger diameters improve precision and overcome local soil variability, they may obstruct root growth to some extent.

It has been ascertained that construction material generally has little influence on root growth (Brown and Upchurch 1987), but shorter root survival has been observed near acrylic tubes compared with butyrate ones in conifers, and the opposite in some deciduous broad-leaved species (Withington et al. 2003). The main criteria guiding choice are material costs and market availability, although some specific material drawbacks may influence decisions. For instance, synthetic material-made tubes are less prone to breaking than glass ones (Kosola 1999), but their flexibility may obstruct camera movements when large roots grow nearby (e.g., sugar beet, trees). Among synthetics, polycarbonate is more scratch-resistant than acrylic (van Noordwijk et al. 1985), and both cause some camera light reflectance, which is inconvenient in automatic image analysis, since it requires corrections (Vamerali et al. 2002). According to Taylor and Böhm (1976) and Withington et al. (2003) glass is the least influential material, due to its silica sand composition.

Some success in root observation has been obtained with inflatable minirhizotrons, for which two main systems have been developed in order to ensure good soil–tube contact and reduce altered root growth at the surface. One system consists of a rigid central transparent acrylic tube coated by a pressurised (10–20 kPa) transparent cellulose acetate (Merrill et al. 1987) or polyvinyl (Merrill 1992) film. As a second option, a rubber inner-tube for motorcycle tyres is inflated within a metal grid-cage and removed just before root observation, allowing investigations without interposed walls (Gijsman et al. 1991; López et al. 1996).

The use of minirhizotrons in natural, undisturbed ecosystems requires their installation in intact profiles, because soil compaction round the tubes can alter root growth (McMichael and Taylor 1987). Approaches involving soil excavation or its packing round the tubes can stimulate N release and modify bulk density (Johnson et al. 2001). In these cases, some time must elapse after minirhizotron installation, so that root dynamics can return to being typical of undisturbed areas and data collection can be started. A wait of 6–12 months after installation is suggested (Johnson et al. 2001; Joslin et al. 2006), but longer lapses (up to 17 months) are also cited in the literature (Pritchard et al. 2008b). Reasonable optimisation of the soil–tube contact can often be achieved by making holes with augers or corers, of the same diameter as the tube, preferably before sowing or planting out. For standing plants, suspended platforms can be used to avoid or minimise treading round the minirhizotrons (Phillips et al. 2000).

In sandy, uniform, well-packed soil, the presence of rigid observation tubes does not seem to affect root patterns particularly (Vos and Groenwold 1987; Ephrath et al. 1999), although preferential growth at the soil–tube interface may occur in fine-textured soils. In this case, greater root deepening is expected near the tubes than in the bulk soil (Vos and Groenwold 1987; De Ruijter et al. 1996) due to lower penetration resistance (Taylor and Böhm 1976; Huck and Taylor 1982). This is related to tube inclination, which can affect root patterns and orientation. Some authors have successfully experimented with vertical tubes (Bragg et al. 1983; Hansson et al. 1994; De Ruijter et al. 1996; Majdi and Kangas 1997; Ephrath et al. 1999), but undesired preferential root growth at the soil–tube interface has been observed in various crops, such as spring oat (*Avena sativa* L.) and potato (*Solanum tuberosum* L.) (Bragg et al. 1983; De Ruijter et al. 1996). Horizontal tubes can be installed both in underground corridors (Fig. 17.4) (van de Geijn et al. 1994; Smit et al. 1994a; Smit and Groenwold 2005) and in the open (Dubach and Russelle 1995). This option allows a large number of images to be collected at one specific depth, thus reducing the number of replicate tubes required to satisfy statistical needs, but they have to be placed at various depths. In addition, horizontal tubes must be positioned with greater accuracy, especially in compartments, as they may bend as the soil settles (Ephrath et al. 1999). In the field and in conventional lysimeters, tubes are commonly inserted sideways, at installation angles varying from 21° (Box and Johnson 1987) to 69° (Box Jr. 1993; Ruess et al. 2003) from the vertical; 45° and 30° are the most frequently used inclinations. Angled tubes provide only one observation at any given depth, but they reveal better fine root distribution than vertical tubes (Bragg et al. 1983; Johnson et al. 2001). Angled installation is also easier than the horizontal choice, because extensive soil trenching is not required. Pagès and Bengough (1997) found that lower installation angles (close to the vertical) led to underestimation of rooting depth, and argued that the likelihood of the deepest roots encountering the tubes depended on their density and direction. Accurate tube installation generally provides a good estimation of root deepening, as observed by Vamerali et al. (2003a, b) comparing minirhizotrons and auger sampling in sugar beet.

Fig. 17.4 Video-recording system (Bartz Technology) in horizontal minirhizotron tubes at Wageningen (NL) Rhizolab underground corridor which is now disused (picture: courtesy of A. L. Smit)

In root research, tube positions with respect to plants vary widely in the literature, mainly depending on species, planting system (row crops, small grain cereals, natural ecosystems) and installation angle. Random installation is generally preferable in the case of forest trees or grasses, with tested densities of one tube for every 60 m^2 with 2.4 × 2.4 m spacing in the case of *Pinus taeda* L. (Pritchard et al. 2008a), or one tube for every 180 m^2 for black spruce (*Picea mariana* L.) (Ruess et al. 2003). In some cases, horizontal minirhizotrons have been buried at a shallow depth (35 mm), as experimented in mixed stands of sugar maple (*Acer saccharum* Marsh.), American beech (*Fagus grandifolia* Ehrh.) and yellow birch (*Betula alleghaniensis Brit.*) (Tierney and Fahey 2001). In order to reduce root damage during installation and subsequent root proliferation of existing grasses, angled minirhizotrons can be inserted outside the plots at a distance of at least 0.12 m, towards the plants (van der Krift and Berendse 2002). In shrubs, some research has been done by placing the bottom end of angled tubes under the centre of the plant or inside the edge of the canopy projection (Phillips et al. 2006).

In row crops, angled minirhizotrons are recommended parallel to rows and near the plants (Williams and Weil 2003) or at various distances in inter-rows (Vamerali et al. 2003a) (Fig. 17.2). More rarely, tubes are set perpendicularly to rows, a solution adopted for horizontal minirhizotrons in large wooden boxes cultivated with maize by Faget et al. (2009).

The tubes must be sealed at the ends with rubber or metal plugs in order to prevent water from entering; tube warming and light leaks – which both inhibit root growth – can be avoided by painting the above-ground end first black and then white (Majdi et al. 1992) or covering it with a white cotton-cloth sack.

17.3 Camera Settings and Image Capturing

17.3.1 Camera Positioning

Accuracy in camera positioning is a cardinal step when acquiring root images, especially in root turnover study, since this implies comparisons of the same minirhizotron sector at different dates. During image acquisition, small vertical (sometimes horizontal) camera displacements from one observation to another may occur, producing shifted images. This bias requires alignment procedures, and macros are available in several image analysis programs (e.g., NIH-Image, KS 300) based on reference points common to the time-series image set (Smit and Zuin 1996). Depending on the extent of camera displacement, only a smaller part of the image stack can finally be cropped for processing (Fig. 17.5). In the Bartz Technology system, satisfactory positioning repeatability is obtained by an indexing handle (a small wheel rolling into dents) which is moved within a cap locked at the top end of the tube. Rotational index release of the cap allows viewing round the tube circumference. Johnson and Meyer (1998) developed a mechanical handle for fast, precise camera movement, which allows a great number of frames (up to 20–30 per minute) to be taken. The handle combines a ratchet advance mechanism with another mechanism which locks the handle to the minirhizotron tube, ensuring high precision in camera repositioning (>99%).

The CCD (Charge Coupled Device) sensor of the camera is commonly hosted within a miniature cylindrical metal cage (aluminium or steel) which travels

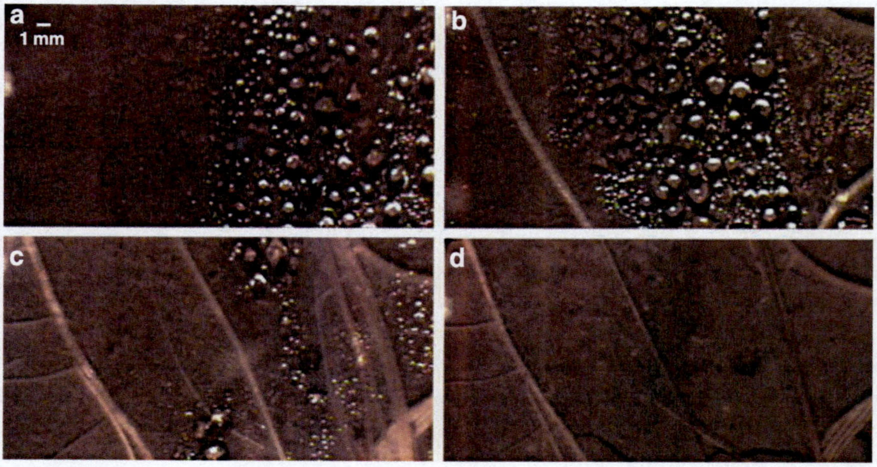

Fig. 17.5 Sugar beet 600-DPI minirhizotron root images collected at constant depth (0.9 m) on four dates: (**a**) May 28 (79 DAS, days after sowing); (**b**) June 10; (**c**) June 26; (**d**) September 2. Images (15.4 × 31 mm) obtained by cropping common area of superimposed original frames (23 × 32 mm)

along the minirhizotron through the handle. Scratches on the inner-tube side and horizontal camera displacements are avoided through a carriage-wheel system and/or external camera foam rings, which should ensure close camera–tube contact. In the camera, angled objectives are needed to view roots growing upwards; when the minirhizotron diameter is not sufficiently large, a 45°-angled mirror is placed in front of the lens (Fig. 17.3).

17.3.2 Image Resolution and Lighting System

Camera resolution must be carefully evaluated when a minirhizotron study is carried out, as it affects the size of observed roots. For a specific physical image size (field view), the number of pixels can be varied considerably, allowing different spatial resolutions. The Bartz Technology camera acquires images with a grid size of 640 × 480 pixels, corresponding to an area of 18 × 13.5 mm, allowing for a theoretical resolution of 28 μm per pixel. As at least two pixels are needed to safely represent the diameter of an elongated object, only roots >~60 μm in diameter can be detected. Lower resolutions were achieved by the analog camera of Vamerali et al. (1999), its images being 32 × 24 mm and 736 × 560 pixels in size (43 μm per pixel), and by the digital sensor of Faget et al. (2010) which acquired frames of 640 × 480 pixels on a 26 × 19.5 mm grid (40.6 μm per pixel). It should be noted that some grass species and crops have very small root diameters, often much lower than 100 μm. In an attempt to apply commercially available webcams to self-made camera systems, Faget et al. (2010) argued that greater accuracy could be reached through 1.3 megapixel sensors (1,280 × 1,024 pixels).

The small field view of the cameras generally used in minirhizotron studies does not produce image distortion but, if detected, this must be taken into account in measuring geometric root features. The camera field view is generally fixed, and in some equipment only it can vary several times (e.g., Bartz Technology, from 18 × 13.5 mm to 3.5 × 2.6 mm). Inside the tubes, roots are observed through a remote-control lighting system, with white and occasionally ultraviolet light (lateral glass-objective mini-neon tubes). White light or LED are preferable to the yellow light produced by low-voltage (usually 6 V) incandescent bulbs, which may change the internal tube temperature. In the BTC Borescope Color Camera System (Bartz Technology), the light is conducted down the length of a borescope via fibre-optic light guides from a light source at the soil surface; this equipment may be hosted in tubes with diameters varying from 6 to 25 mm.

17.3.3 Observation Intervals

When working on root dynamics, one of the basic choices is to define suitable time observation intervals in order to maximise accuracy in estimating fine root production and mortality. It is commonly recognised that increased frequency in root observations yields more reliable data, but there have been few long-term studies evaluating the importance of sampling frequency. The interval between recordings generally results as a compromise between labour intensity (image acquisition and processing) and accuracy (Johnson et al. 2001). Notwithstanding this, rare or irregular observations may lead to underestimation of fine root production and mortality, and awareness of the extent of error deriving from this source of variation is necessary (Tingey et al. 2003).

A wide range of sampling intervals has been tested in the literature, and suitable values change depending on plant species and annual or seasonal turnover intensity. This is closely related to root life-span, as new roots may rapidly grow and die between long observation intervals without being detected.

In forest trees and conifers, frequent observations are necessary in late spring and early summer, when the largest root production and mortality peaks occur. In loblolly pine, regular monthly sampling was adopted by both Pritchard et al. (2008a) and Baker et al. (2001) in a forested floodplain, although the latter author suggested intensifying observations because of irregular root dynamics. Samplings in winter months are generally reduced (Aerts et al. 1992) or eliminated (Kloeppel and Gower 1995; Ruess et al. 1998). In *Pinus palustris* Mill., weekly observations are needed to capture the behaviour of short-lived roots (life-span 10.5 d), which were found to account for 21–37% of total fine root mortality (Stevens et al. 2002). Passing to monthly sampling, these authors underestimated length production by 15% and overestimated median root life-span by 60%. In Douglas fir (*Pseudotsuga menziesii* (Mirb.) Franco) seedlings, underestimation in root production almost doubled (28% vs. 54%) when a 2-week interval was replaced by either 4- or 8-week ones (Johnson et al. 2001). Later studies confirmed that root production and mortality are the parameters mainly affected by observation frequency in both Douglas fir and littleleaf linden (*Tilia cordata* Mill.) (e.g., Tingey et al. 2003).

In annual species, root images are generally recorded with high frequency. Smit and Vamerali (1998) used a 30-day interval in potato, and Vamerali et al. (2003) bi-weekly in sugar beet. In alfalfa (*Medicago sativa* L.), Dubach and Russelle (1995) reported that a 7-day interval was to be preferred to a 4-week one, for correct estimation of root mortality, and Goins and Russelle (1996) suggested intensifying recordings early in the upper 0.2-m depth of soil, as more than 50% of roots were produced within 7 weeks from sowing. Conversely, in *Lotus corniculatus* L., the influence of observation interval on root mortality within a 1–8-week range was negligible (variations <1.3%) (Dubach and Russelle 1995).

17.4 Living Vs. Dead Roots: The Minirhizotron's Point of View

Studies of root dynamics must correctly discriminate dead from living roots. In minirhizotrons, the physiological status of roots is usually assessed by visual evaluation, e.g., disappearance (e.g., Norby et al. 2004), colour darkening

(Hendrick and Pregitzer 1993; Wells and Eissenstat 2001), contour smoothing, and diameter shrinking, which imply loss of continuity with the soil (Passioura 1991; Veen et al. 1992). These criteria are quite subjective if they are not supported by physiological investigations, and may change according to plant species. In many herbaceous and woody species, the youngest active roots are white in colour, and this allows them to be easily classified. In some species (e.g., *Sorghum bicolor* L.), functional roots are both white and brown, so that disappearance and deterioration of structural integrity should be used as criteria (Pritchard et al. 2006). Roots may darken without losing their functionality as a result of secondary growth (dycots) or suberisation, and this complicates their visual evaluation. Under visible light, difficulty in distinguishing roots against organic debris may also arise (Wang et al. 1995).

Some authors have proposed observation under UV light, as root fluorescence has been found to be related to root activity (Fe absorption and elongation rate) (Dyer and Brown 1983). An attempt to estimate living and dead roots under visible and UV light was made by Wang et al. (1995) on a large set of plant species, i.e., *Agrostis borealis* H., *Paspalum dilatatum* L., *Solidago* spp., *Erigeron canadensis* L., *Rubus cuneifolius* Pursh, *Liquidambar styraciflua* L., *Quercus nigra* L. and *Pinus taeda* L. Comparing minirhizotron observations with TTC-positive roots (triphenyltetrazolium chloride test), the above authors found underestimation of living roots under visible light and overestimation under UV. Visible light was found to be more appropriate in grass-dominated communities, since some dead roots apparently still fluoresced and were misidentified as living. Instead, UV light seems to be more suitable for shrubs and woody plants, as suberification and secondary growth darken roots, although most fine roots never develop significant secondary thickening. A different conclusion was drawn by Smit and Zuin (1996), who studied root fluorescence of *Brassica oleracea* L. var. *gemmifera* D.C. (Brussels sprout) and *Allium porrum* L. (leek) under UV light. The two species showed different responses in terms of fluorescence intensity and dynamics. Leek roots fluoresced more strongly than those of Brussels sprout, the roots of which had a weak background contrast. In Brussels sprout, root fluorescence (UV) and luminance (visible) decreased with root ageing, respectively slowly and rapidly. The opposite behaviour was found in leek, as root fluorescence increased over time, suggesting that functionality is not reflected by the fluorescence intensity, which is probably genetically determined. In this regard, screening on some hundreds of soybean accessions revealed that 10% of them lacked root fluorescence under UV light, as the controlling genes were not present in all accessions (Delannay and Palmer 1982). Regardless of its reliable application, Smit and Zuin (1996) believed that a UV light may be interesting in the case of transparent roots, which are hardly visible under white light.

Advances in root status classification (activity/death) are coming from visible/near-infrared (VIS–NIR) spectroscopy, which is increasingly used. The equipment size used by Nakaji et al. (2010) is not compatible with standard minirhizotron tubes, and the large hyperspectral digital camera is positioned outside a clear glass pot. In hybrid poplar cuttings, VIS–NIR (356–976 nm) spectral images allow accurate root classification (number of roots), as reflectance decreases with root ageing, regardless of soil moisture conditions. In roots extracted from core samples of various grass species, such as *Anthoxanthum odoratum* Webb and Brethel., *Dactylis glomerata* L., *Festuca arundinacea* Schreb., *Festuca rubra* L. and *Lolium perenne* L., Picon-Cochard et al. (2009) also achieved a very low classification error (only 15%) using near-infrared reflectance spectroscopy (NIRS).

Roots of GFP-engineered (green fluorescent protein) plants (e.g., maize, Italian ryegrass, soybean) can also be distinguished in mixed plant stands by specific camera adaptation (light wavelengths, filters) (Faget et al. 2009, 2010). The light required to excite the protein can be generated by LEDs that cover the narrow spectrum of 440–460 nm, whilst longpass filters (>515 nm) block the wavelengths outside the GFP fluorescence spectrum.

17.5 Image Processing

Time-consuming manual analysis of minirhizotron images (e.g., counting, measuring root traces) limits the number of frames analysed, but the great complexity of root images still hampers the development of completely automated procedures (Richner et al. 2000). Several authors have proposed approaches for automatic measurement of root length and diameter (Nater et al. 1992; Murphy and Smucker 1995; Dowdy et al. 1998; Zeng et al. 2010), and dedicated commercial and freeware applications are available. NIH-Image is a general public domain software frequently used in root research, in which task-specific macros can be added to the program. Among dedicated software, some research has been carried out with the freeware RMS – Root Measurement System (University of Georgia, Athens) (Ingram and Leers 2001) and Rootfly (Clemson University, SC, USA) (Zeng et al. 2008). RooTracker (Dave Tremmel, Duke University Phytotron, Durham, NC, USA) (Withington et al. 2003) and WinRHIZO Tron (Regent Instruments, Quebec, Canada) are examples of dedicated commercially distributed software, and KS 300 (Carl Zeiss Vision GmbH, Eching bei München, Germany), used by Vamerali et al. (1999), is a general-purpose program.

Automatic procedures generally include algorithms for increasing root-background contrast, which facilitates later image thresholding (binarisation) (Kimura and Yamasaki 2003). The major critical step is root discrimination from the background, which often includes undesired extraneous objects (e.g., light soil particles, water droplets, insects) (see Fig. 17.5). A more complete way of discriminating roots against non-root objects should consider both luminance and geometric features. Vamerali et al. (1999) developed a simple procedure, working on sugar beet colour images (24-bit RGB) acquired in different backgrounds (soil types). In their study, the greatest root-background contrast was detected in the red band, but the method worked better on the blue channel. This allowed correct root classification in lesser contrasting regions, where the red band failed because of inverted luminance hierarchy (background > roots). The method used an exponential algorithm of

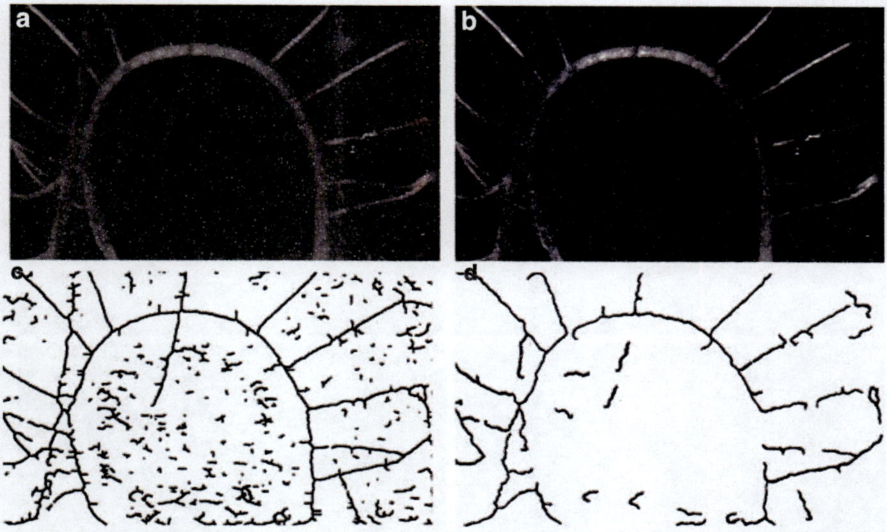

Fig. 17.6 Main steps of image processing, as recommended by Vamerali et al. (1999): (**a**) original image (grey-scale of blue band); (**b**) image after contrast enhancement (exponential algorithm); (**c**) image after skeletonisation; (**d**) root skeletons filtered with minimum root length

contrast stretching in order to make thresholding more effective, and a fixed threshold luminance to produce binary images (Fig. 17.6). A fixed threshold value may not be suitable for the whole image set and for all observation dates, indicating that various thresholding methods should be tested, such as those based on maximisation of luminance variance between foreground and background pixel populations (Vamerali et al. 2009). These authors argued that the most suitable luminance threshold decreases with root ageing, and better correlations between estimated and actual root length were obtained by image fragmentation and local independent analysis (Fig. 17.7).

After thresholding, roots may be distinguished from any bright non-root objects if specific interval values of length, length-to-diameter ratio and elongation index (perimeter^2/area) are adopted (Dowdy et al. 1998; Murphy and Smucker 1995). Root length may be estimated as fibre length with the perimeter and area of root objects (Vamerali et al. 2003b); alternatively, length may be represented by the area of 1-pixel wide root segments yielded by skeletonisation or boundary-erosion algorithms, suitably corrected by the coefficient 1.12 to take random root orientation into account (Smit et al. 1994b). Kimura and Yamasaki (2003) found good correlations between skeleton length and actual root length in both 300- and 600-DPI resolution images, but diameter overestimation occurred, especially at lower resolutions.

Accurate measurement of fine root production and mortality requires individual roots to be tracked across multiple sampling dates. Root tracking requires unambiguous root identification and measurement, and individuals must be identified through a time series despite changes in colour, shape and position. Currently, the

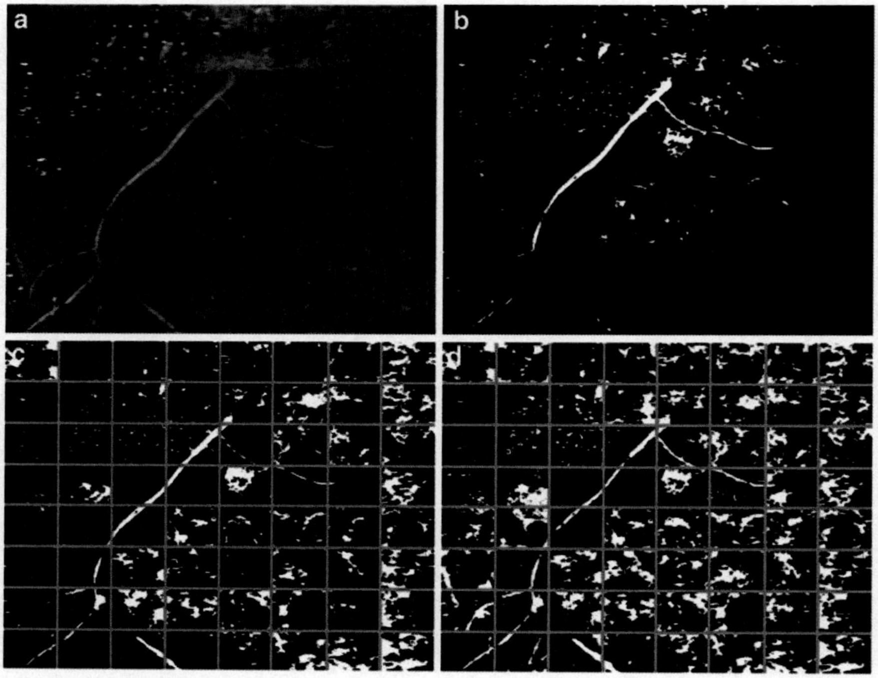

Fig. 17.7 An approach to minirhizotron root image processing based on image subdivision and independent local analysis (Vamerali et al. 2009): (**a**) grey-scale of blue band; (**b**) fixed threshold applied to entire image or (**c**) to 64 sub-areas separately; (**d**) automatic threshold on 64 sub-images with bi-modal approximation (maximum variance between foreground–background pixel populations)

most advanced methods are semi-automatic, with manual recording and tracking of root location, length and diameter by means of the computer mouse or LCD tablets coupled with software such as RooTracker (Majdi and Andersson 2005; Gunderson et al. 2007; Strand et al. 2008) or KS 300 (Vamerali et al. 2003a). In detail, roots must be drawn at their first appearance with a specific colour and their fate followed in subsequent observations. When the tracks are superimposed on the following sampling date, new roots are traced in a different colour, and previous tracks corresponding to dead roots are deleted. An automated procedure is then used to measure the coloured tracks separately.

Zeng et al. (2006) described a method for automatic detection, labelling and measurement of young fine roots as individuals in fruit and ornamental trees. Since young roots are usually light in colour and darken with age, the proposed algorithm detects light-coloured roots by object shape and luminance distribution. The algorithm was improved by boosting a "strong" classifier to discriminate root vs. non-root objects, derived by incorporation of 5 "weak" intensity- and geometry-based classifiers, i.e., grey level histogram distribution, interior intensity edges, eccentricity, approximate line symmetry and boundary parallelism (Zeng et al. 2008). The above authors concluded that subsequent automatic root tracking over time is possible, based on position and geometry. Further improvements in this procedure have recently been made, by modelling the formation of line segments composing roots by a Gibbs point process, and determining the parameters and thresholds of the algorithm by a data-driven approach (Zeng et al. 2010). Root segments are formed by grouping seed points into linear structures, which are then combined and validated by geometric techniques. After root centrelines have been found, root regions are detected by means of a recursively bottom-up region-growing method.

17.6 Can We Believe What Minirhizotrons See?

Correct evaluation of root production and turnover is generally difficult, due to multiple potential biases associated with both old and recent investigation methods. Hendricks et al. (2006) concluded that contrasting theories on allocation of above-ground/below-ground plant resources are partially caused by different root investigation methods, as noted in several studies (Hendricks et al. 1993; Nadelhoffer 2000; Norby and Jackson 2000).

Minirhizotrons allow fast, continuous root observation, but data discrepancies often emerge when comparisons are made with other methods, such as core sampling and the carbon isotope method. In open trials on potato, vertical minirhizotrons have recorded a greater number of roots than core sampling in deeper soil layers, explained by preferential root growth along the tube (De Ruijter et al. 1996). In the field experiment by those authors, among the experimental factors studied, cultivar choice, soil temperature and nematode infestation did not affect the comparability of the two methods but, when soil compaction was examined, reliable results were only achieved by auger sampling.

There is much evidence in the literature of root underestimation by minirhizotrons, although a linear relationship between soil coring (RLD – root length density, root DW) and minirhizotron data (root length, root number density) is often reported (e.g., Box and Ramseur 1993). Correlation coefficients of more than 0.55 have been found in various plant species (e.g., acacia, wheat) in a large soil profile, except for a great underestimation in the top 0.1-m depth (Ephrath et al. 1999). In some experiments, overestimation in root quantification occurred with minirhizotrons, in comparison with core sampling and trench profiles, with better correspondence among methods beneath 0.3-m depth (Parker et al. 1991). The common root underestimation by minirhizotrons in the upper soil layer involves a large variety of plant species, such as maize, barley, faba bean and almond tree (Majdi et al. 1992; Heeraman and Juma 1993; Wiesler and Horst 1994; Franco and Abrisqueta 1997) and seems to be due to poor soil–tube contact and light leaks (Levan et al. 1987). In maize, which can easily reach a maximum depth of 1.5 m, an exponential RLD decrease is generally assumed with soil cores beneath 0.3 m, whereas a linear trend emerges from minirhizotron observations. For this crop, the relationship between minirhizotrons and soil cores in terms of RLD changed with time, even when observations from the soil surface layer were ignored, and better correlation between methods occurred at early growth stages (Samson and Sinclair 1994). Representative data can be retrieved from minirhizotrons, even when root area measurements are compared with soil core root biomass throughout the soil profile of contrasting species, such as black walnut, northern red oak and maize (Jose et al. 2001).

Data discordances are also observed in many fine root longevity studies, comparing minirhizotrons with the few C-isotope derived estimates (Strand et al. 2008). Systematic overestimation of root turnover and the opposite skewing of root age distributions are expected in hardwood forests studied with minirhizotrons in comparison with radiocarbon methods (Tierney and Fahey 2002). Different approaches in calculating turnover may be responsible for this discrepancy, as minirhizotrons use a number-based measure such as median longevity, while the C-isotope method applies a mass-based measure (mean residence time of C) (Guo et al. 2008). In polyannual species, biases may also be ascribed to the heterogeneity of fine root populations (Guo et al. 2004) and to the tendency of minirhizotrons to overemphasise the importance of rapidly cycling roots (first- and second-order), but discounting the effect of less dynamic fine roots (Tierney and Fahey 2002).

References

Aerts R, Bakker C, De Calluwe H (1992) Root turnover as determinant of cycling of C, N, and P in a dry heathland ecosystem. Biogeochemistry 15:175–190

Baker TT III, Conner WH, Lockaby G, Stanturf JA, Burke MK (2001) Fine root productivity and dynamics on a forested floodplain in South Carolina. Soil Sci Soc Am J 65:545–556

Basile B, Bryla DR, Salsman M, Marsal J, Cirillo C, Johnson RS, Dejong TM (2007) Growth patterns and morphology of fine roots of size-controlling and invigorating peach rootstocks. Tree Physiol 27:231–241

Bates GH (1937) A device for the observation of root growth in the soil. Nature 139:966–967

Böhm W (1979) Methods of studying root systems. Springer, Heidelberg

Box JE Jr (1993) Use of the minirhizotron-miniature video camera technique for measuring root dynamics. Geoderma 56:133–141

Box JE Jr, Johnson JW (1987) Minirhizotron rooting comparisons of three wheat cultivars. In: Taylor HM (ed) Minirhizotron observation tubes: methods and applications for measuring rhizosphere dynamics. ASA special publication number 50. American Society of Agronomy, Madison, WI

Box JE Jr, Ramseur EL (1993) Minirhizotron wheat root data: comparisons to soil core root data. Agron J 85:1058–1060

Bragg PL, Govi G, Cannell RQ (1983) A comparison of methods, including angled and vertical minirhizotrons, for studying root growth and distribution in a spring oat crop. Plant Soil 73:435–440

Brown DA, Upchurch DR (1987) Minirhizotrons: a summary of methods and instruments in current use. In: Taylor HM (ed) Minirhizotron observation tubes: methods and applications for measuring rhizosphere dynamics. ASA special publication number 50. American Society of Agronomy, Madison, WI, pp 15–30

Brown ALP, Day FP, Stover DB (2009) Fine root biomass estimates from minirhizotron imagery in a shrub ecosystem exposed to elevated CO_2. Plant Soil 317:45–153

Coleman MD, Dickson RE, Isebrands JG, Karnosky DF (1996) Root growth and physiology of potted and field-grown trembling aspen exposed to tropospheric ozone. Tree Physiol 16:145–152

De Ruijter FJ, Veen BW, Oijen M (1996) A comparison of soil core sampling and minirhizotrons to quantify root development of field-grown potatoes. Plant Soil 182:301–312

Delannay X, Palmer RG (1982) Four genes controlling root fluorescence in soybean. Crop Sci 22:278–281

Dowdy R, Smucker A, Dolan M, Ferguson J (1998) Automated image analyses for separating plant roots from soil debris elutriated from soil cores. Plant Soil 200:91–94

Dubach M, Russelle MP (1995) Reducing the cost of estimating root turnover with horizontally installed minirhizotrons. Agron J 89:258–263

Dyer D, Brown DA (1983) Relationship of fluorescent intensity to ion uptake and elongation rates of soybean roots. Plant Soil 72:127–134

Ephrath JE, Silberbush M, Berliner PR (1999) Calibration of minirhizotron readings against root length density data obtained from soil cores. Plant Soil 209:201–208

Faget M, Herrera JM, Stamp P, Aulinger-Leipner I, Frossard E, Liedgens M (2009) The use of green fluorescent protein as a tool to identify roots in mixed plant stands. Funct Plant Biol 36:930–937

Faget M, Liedgens M, Stamp P, Flütsch P, Herrera JM (2010) A minirhizotron imaging system to identify roots expressing the green fluorescent protein. Comput Electron Agric 74:163–167

Franco JA, Abrisqueta JM (1997) A comparison between minirhizotron and soil coring methods of estimating root distribution in young almond trees under trickle irrigation. J Hortic Sci 72:797–805

Gijsman AJ, Floris J, Van Noordwijk M, Brouwer G (1991) An inflatable minirhizotron system for root observation with improved soil/tube contact. Plant Soil 134:261–269

Goins GD, Russelle MP (1996) Fine root demography in alfalfa (*Medicago sativa* L.). Plant Soil 185:281–291

Gunderson JJ, Knight JD, Van Rees KCJ (2007) Impact of ectomycorrhizal colonization of hybrid poplar on the remediation of diesel-contaminated soil. J Environ Qual 36:927–934

Guo D, Mitchell RJ, Hendricks JJ (2004) Fine root branch orders respond differently to carbon source–sink manipulations in a longleaf pine forest. Oecologia 140:450–457

Guo D, Li H, Mitchell RJ, Han W, Hendricks JJ, Fahey TJ, Hendrick RL (2008) Fine root heterogeneity by branch order: exploring the discrepancy in root turnover estimates between minirhizotron and carbon isotopic methods. New Phytol 177:443–456

Hansson AC, Zhao AF, Andrén O (1994) Fine-root growth dynamics of two shrubs in semiarid rangeland in inner Mongolia, China. Ambio 23:225–228

Heeraman DA, Juma NG (1993) A comparison of minirhizotron, core and monolith methods for quantifying barley (*Hordeum vulgare* L.) and fababean (*Vicia faba* L.) root distribution. Plant Soil 148:29–41

Hendrick RL, Pregitzer KS (1993) The dynamics of fine root length, biomass, and nitrogen content in two northern hardwood ecosystems. Can J For Res 23:2507–2520

Hendrick RL, Pregitzer KS (1996) Application of minirhizotrons to understand root function in forests and other natural ecosystems. Plant Soil 185:293–304

Hendricks JJ, Nadelhoffer KJ, Aber JD (1993) Assessing the role of fine roots in carbon and nitrogen cycling trends. Ecol Evol 8:174–178

Hendricks JJ, Hendrick RL, Wilson CA, Mitchell RJ, Pecot SD, Guo D (2006) Assessing the patterns and controls of fine root dynamics: an empirical test and methodological review. J Ecol 94:40–57

Huck MG, Taylor HM (1982) The rhizotron as a tool for root research. Adv Agron 35:1–35

Ingram KT, Leers GA (2001) Software for measuring root characters from digital images. Agron J 93:918–922

Johnson MG, Meyer PF (1998) Mechanical advancing handle that simplifies minirhizotron camera registration and image collection. J Environ Qual 27:710–714

Johnson MG, Phillips DL, Tingey DT, Storm MJ (2000) Effects of elevated CO_2, N-fertilization, and season on survival of ponderosa pine fine roots. Can J For Res 30:220–228

Johnson MG, Tingey DT, Phillips DL, Storm MJ (2001) Advancing fine root research with minirhizotrons. Environ Exp Bot 45:263–289

Jose S, Gillespie AR, Seifert JR, Pope PE (2001) Comparison of minirhizotron and soil core methods for quantifying root biomass in a temperate alley cropping system. Agroforest Syst 50:161–168

Joslin JD, Wolf MH (1998) Impacts of water input manipulations on fine root production and mortality in a mature hardwood forest. Plant Soil 204:165–174

Joslin JD, Gaudinski JB, Torn MS, Riley WJ, Hanson PJ (2006) Fine-root turnover patterns and their relationship to root diameter and soil depth in a ^{14}C-labeled hardwood forest. New Phytol 172:523–535

Kimura K, Yamasaki S (2003) Accurate root length and diameter measurement using NIH Image: use of Pythagorean distance for diameter estimation. Plant Soil 254:305–315

Kloeppel BD, Gower ST (1995) Construction and installation of acrylic minirhizotron tubes in forest ecosystems. Soil Sci Soc Am J 59:241–243

Kosola KR (1999) Laparoscopic sampling of roots of known age from an expandable-wall minirhizotron system. Agron J 91:876–879

Levan MA, Ycas JW, Hammel JW (1987) Light leak effects on near-surface soybean rooting. In: Taylor HM (ed) Minirhizotron observation tubes: methods and applications for measuring rhizosphere dynamics. ASA special publication number 50. American Society of Agronomy, Madison, WI, pp 89–98

López B, Sabate S, Gracia C (1996) An inflatable minirhizotron system for stony soils. Plant Soil 179:255–260

Majdi H (1996) Root sampling methods – applications and limitations of the minirhizotron technique. Plant Soil 185:255–258

Majdi H (2001) Changes in fine root production and longevity in relation to water and nutrient availability in a Norway spruce stand in northern Sweden. Tree Physiol 21:1057–1061

Majdi H, Andersson P (2005) Fine root production and turnover in a Norway spruce stand in northern Sweden: effects of nitrogen and water manipulation. Ecosystems 8:191–199

Majdi H, Kangas P (1997) Demography of fine roots in response to nutrient applications in a Norway spruce stand in southwestern Sweden. Ecoscience 4:199–205

Majdi H, Smucker AJM, Persson H (1992) A comparison between minirhizotron and monolith sampling methods for measuring root growth of maize (*Zea mays* L.). Plant Soil 147:127–134

McMichael BL, Taylor HM (1987) Applications and limitations of rhizotrons and minirhizotrons. In: Taylor HM (ed) Minirhizotron observation tubes: methods and applications for measuring rhizosphere dynamics. ASA special publication number 50. American Society of Agronomy, Madison, WI, pp 1–13

Merrill SD (1992) Pressurized-wall minirhizotron for field observation of root growth dynamics. Agron J 84:755–758

Merrill SD, Doering EJ, Reichman GA (1987) Application of a minirhizotron with flexible, pressurized walls to a study of corn root growth. In: Taylor HM (ed) Minirhizotron observation tubes: methods and applications for measuring rhizosphere dynamics. ASA special publication number 50. American Society of Agronomy, Madison, WI, pp 131–143

Murphy SL, Smucker AJM (1995) Evaluation of video image analysis and line-intercept methods for measuring root systems of alfalfa and ryegrass. Agron J 87:865–868

Nadelhoffer KJ (2000) The potential effects of N deposition on fine root-production in forest ecosystems. New Phytol 147:131139

Nakaji T, Noguchi K, Oguma H (2010) Classification of rhizosphere components using visible-near infrared spectral images. Plant Soil 310:245–261

Nater EA, Nater KD, Baker JM (1992) Application of artificial neural system algorithms to image-analysis of roots in soil. 1. Initial results. Geoderma 53:237–253

Neill C (1992) Comparison of soil coring and ingrowth methods for measuring belowground production. Ecology 73:1918–1921

Norby RJ, Jackson RB (2000) Root dynamics and global change: seeking an ecosystem perspective. New Phytol 147:3–12

Norby RJ, Ledford J, Reilly CD, Miller NE, O'Neill EG (2004) Fine-root production dominates response of a deciduous forest to atmospheric CO_2 enrichment. Proc Natl Acad Sci USA 101:9689–9693

Pagès L, Bengough AG (1997) Modelling minirhizotron observations to test experimental procedures. Plant Soil 189:81–89

Parker CJ, Carra MKV, Jarvisa NJ, Puplampua BO LVH (1991) An evaluation of the minirhizotron technique for estimating root distribution in potatoes. J Agric Sci 116:341–350

Passioura JB (1991) Soil structure and plant growth. Aust J Soil Res 29:717–728

Phillips DL, Johnson MG, Tingey DT, Biggart C, Nowak RS, Newsom JC (2000) Minirhizotron installation in sandy, rocky soils with minimal soil disturbance. Soil Sci Soc Am J 61:761–764

Phillips DL, Johnson MG, Tingey DT, Catricala CE, Hoyman TL, Nowak RS (2006) Effects of elevated CO_2 on fine root dynamics in a Mojave Desert community: a FACE study. Glob Change Biol 12:61–73

Picon-Cochard C, Pilon R, Revaillot S, Jestin M, Dawson L (2009) Use of near-infrared reflectance spectroscopy to predict the percentage of dead versus living grass roots. Plant Soil 317:309–320

Pietola LM (2005) Root growth dynamics of spring cereals with discontinuation of mouldboard ploughing. Soil Till Res 80:103–114

Pritchard SG, Rogers HH (2000) Spatial and temporal deployment of crop roots in CO_2-enriched environments. New Phytol 147:55–71

Pritchard SG, Prior SA, Rogers HH, Davis MA, Runion GB, Popham TW (2006) Effects of elevated atmospheric CO_2 on root dynamics and productivity of sorghum grown under conventional and conservation agricultural management practices. Agric Ecosyst Environ 113:175–183

Pritchard SG, Strand AE, McCormack ML, Davis MA, Finzi AC, Jackson RB, Matamala R, Rogers HH, Oren R (2008a) Fine root dynamics in a loblolly pine forest are influenced by free-air-CO_2-enrichment: a six-year-minirhizotron study. Glob Change Biol 14:1–15

Pritchard SG, Strand AE, McCormack ML, Davis MA, Oren R (2008b) Mycorrhizal and rhizomorph dynamics in a loblolly pine forest during 5 years of free-air-CO_2-enrichment. Glob Change Biol 14:1–13

Richner W, Liedgens M, Biirgi H, Soldati A, Stamp P (2000) Root Image analysis and interpretation. In: Smit AL, Bengough AG, Engels C, van Noordwijk M, Pellerin S, van de Geijn SC (eds) Root methods – a handbook. Springer, Heidelberg, pp 304–341

Richter DD, Markewitz D, Trumbore SE, Wells CG (1999) Rapid accumulation and turnover of soil carbon in a re-establishing forest. Nature 400:56–58

Ruess RW, Hendrick RL, Bryant JP (1998) Regulation of fine root dynamics by mammalian browsers in early successional Alaskan taiga forests. Ecology 79:2706–2720

Ruess RW, Hendrick RL, Burton AJ, Pregitzer KS, Sveinbjornssön B, Allen MF, Maurer GE (2003) Coupling fine root dynamics with ecosystem carbon cycling in black spruce forests of interior Alaska. Ecol Monogr 73:643–662

Samson BK, Sinclair TR (1994) Soil core and minirhizotron comparison for the determination of root length density. Plant Soil 161:225–232

Schuurman JJ, Goedewaagen MAJ (1971) Methods for the examination of root systems and roots: methods in use at the Institute for Soil Fertility for eco-morphological root investigations, 2nd edn. Centre for Agricultural Publishing and Documentation in Wageningen Pudoc, Wageningen

Smit AL, Groenwold J (2005) Root characteristics of selected field crops: data from the Wageningen Rhizolab (1990–2002). Plant Soil 272:365–384

Smit AL, Vamerali T (1998) The influence of potato cyst nematodes (*Globodera pallida*) and drought on rooting dynamics of potato (*Solanum tuberosum* L.). Eur J Agron 9:137–146

Smit AL, Zuin A (1996) Root growth dynamics of Brussels sprouts (*Brassica oleracea* var *gemmifera*) and leeks (*Allium porrum* L) as reflected by root length, root colour and UV fluorescence. Plant Soil 185:271–280

Smit AL, Groenwold J, Vos J (1994a) The Wageningen rhizolab – a facility to study soil-root-shoot-atmosphere interactions in crops. 2. Methods of root observations. Plant Soil 161:289–298

Smit AL, Sprangers JFCM, Sablik PW, Groenwold J (1994b) Automated measurement of root length with a three-dimensional high-resolution scanner and image analysis. Plant Soil 158:145–149

Steen E (1991) Usefulness of the mesh bag method in quantitative root studies. In: Atkinson D (ed) Plant root growth. Blackwell, Oxford, pp 75–86

Steinaker DF, Wilson SD (2008) Scale and density dependent relationships among roots, mycorrhizal fungi and collembola in grassland and forest. Oikos 117:703–710

Stevens GN, Jones RH, Mitchell RJ (2002) Rapid fine root disappearance in a pine woodland: a substantial carbon flux. Can J For Res 32:2225–2230

Strand AE, Pritchard SG, McCormack ML, Davis MA, Oren R (2008) Irreconcilable differences: fine-root life spans and soil carbon persistence. Science 319:456–458

Taylor HM, Böhm W (1976) Use of acrylic plastic as rhizotron windows. Agron J 68:693–694

Tierney GL, Fahey TJ (2001) Evaluation of minirhizotron estimates of fine root longevity in the forest floor of a temperate broadleaf forest. Plant Soil 229:167–176

Tierney GL, Fahey TJ (2002) Fine root turnover in a northern hardwood forest: a direct comparison of the radiocarbon and minirhizotron methods. Can J For Res 32:1692–1697

Tingey DT, Phillips DL, Johnson MG (2003) Optimizing minirhizotron sample frequency for an evergreen and deciduous tree species. New Phytol 157:155–161

Vamerali T, Ganis A, Bona S, Mosca G (1999) An approach to minirhizotron root image analysis. Plant Soil 217:183–193

Vamerali T, Ganis A, Bona S Mosca G (2002) Advances in minirhizotron root image analysis. In: Proceedings of the VII ESA Congress, 15–18 July 2002, Córdoba, Spain, pp 419–420

Vamerali T, Ganis A, Bona S, Mosca G (2003a) Fibrous root turnover and growth in sugar beet (*Beta vulgaris* var *saccharifera*) as affected by nitrogen shortage. Plant Soil 255:169–177

Vamerali T, Guarise M, Ganis A, Bona S, Mosca G (2003b) Analysis of root images from auger sampling with a fast procedure: a case of application to sugar beet. Plant Soil 255:387–397

Vamerali T, Ganis A, Mosca G (2009) Methods for thresholding minirhizotron root images In: Proceedings of the seventh international symposium "Root Research and Applications" RootRAP, 2–4 September 2009, Boku, Vienna, Austria, pp 1–2 (CD-ROM)

van de Geijn SC, Vos J, Groenwold J, Goudriaan J, Leffelaar PA (1994) The Wageningen rhizolab – a facility to study soil-root-shoot-atmosphere interactions in crops. 1. Description of main functions. Plant Soil 161:275–287

van der Krift TAJ, Berendse F (2002) Root life spans of four grass species from habitats differing in nutrient availability. Funct Ecol 16:198–203

van Noordwijk M, De Jager A, Floris J (1985) A new dimension to observations in minirhizotrons: a stereoscopic view on root photographs. Plant Soil 86:447–453

Veen BW, Van Noordwijk M, De Willigen P, Boone FR, Kooistra MJ (1992) Root-soil contact of maize, as measured by a thin-section technique. III. Effects on shoot growth, nitrate and water uptake efficiency. Plant Soil 139:131–138

Vos J, Groenwold J (1987) The relation between root growth along observation tubes and in bulk soil. In: Taylor HM (ed) Minirhizotron observation tubes: methods and applications for measuring rhizosphere dynamics. ASA special publication number 50. American Society of Agronomy, Madison, WI, pp 39–50

Wang Z, Burch WH, Mou P, Jones RH, Mitchell RJ (1995) Accuracy of visible and ultraviolet light for estimating live root proportions with minirhizotrons. Ecology 76:2330–2334

Wells CE, Eissenstat DM (2001) Marked differences in survivorship among apple roots of different diameter. Ecology 82:882–892

Wiesler F, Horst WF (1994) Root growth of maize cultivars under field conditions as studied by the core and minirhizotron method and relationships to shoot growth. J Plant Nutr Soil Sci 157:351–358

Williams SM, Weil RR (2003) Crop cover root channels may alleviate soil compaction effects on soybean crop. Soil Sci Soc Am J 68:1403–1409

Withington JM, Elkin AD, Bułaj B, Olesiński J, Tracy KN, Bouma TJ, Oleksyn J, Anderson LJ, Modrzyński J, Reich PB, Eissenstat DM (2003) The impact of material used for minirhizotron tubes for root research. New Phytol 160:533–544

Zeng G, Birchfield ST, Wells CE (2006) Detecting and measuring fine roots in minirhizotron images using matched filtering and local entropy thresholding. Mach Vision Appl 17:265–278

Zeng G, Birchfield ST, Wells CE (2008) Automatic discrimination of fine roots in minirhizotron images. New Phytol 177:549–557

Zeng G, Birchfield ST, Wells CE (2010) Rapid automated detection of roots in minirhizotron images. Mach Vision Appl 21:309–317

Chapter 18
Fine Root Turnover

Martin Lukac

Abstract Fine roots constitute an interface between plants and soils and thus play a crucial role in forest carbon, nutrient and water cycles. Their continuous growth and dieback, often termed turnover of fine roots, may constitute a major carbon input to soils and significantly contribute to belowground carbon cycle. For this reason, it is of importance to accurately estimate not only the standing biomass of fine roots, but also its rate of turnover. To date, no direct and reliable method of measuring fine root turnover exists. The main reason for this is that the two component processes of root turnover, namely growth and dieback of fine roots, nearly always happen in the same place and at the same time. Further, the estimation of fine root turnover is complicated by the inaccessibility of tree root systems, its labour intensiveness and is often compounded by artefacts created by soil disturbance. Despite the fact that the elucidation of the patterns and controls of forest fine root turnover is of utmost importance for the development of realistic carbon cycle models, our knowledge of the contribution of fine root turnover to carbon and nutrient cycles in forests remains uncertain. This chapter will detail all major methods currently used for estimating fine root turnover and highlight their advantages as well as drawbacks.

18.1 Why Measure Root Turnover?

Fine roots are responsible for the bulk of water and nutrient acquisition and form the most active and short-lived part of the root system. A lively debate is ongoing in the root research community as to what is a fine root. This is largely fuelled by the fact

M. Lukac (✉)
Department of Agriculture, Development and Policy, University of Reading, Whiteknights
RG6 6AR, UK
e-mail: m.lukac@reading.ac.uk

S. Mancuso (ed.), *Measuring Roots*,
DOI 10.1007/978-3-642-22067-8_18,

that fine roots are defined on a functional basis (resource uptake), but have to be measured according to a morphological one (root diameter). For simplicity, and to promote comparability of results, fine roots are often assumed to be less than 1 or 2 mm in diameter. Coarse roots, on the other hand, are thought to live for significantly longer periods and fulfil roles related to resource transportation and tree stability. Their biomass will eventually contribute to soil biogeochemical cycles, but this usually happens after many years or decades and their overall contribution to carbon and nutrient cycling is significantly smaller. About 33% of global net primary production is assumed to be used in fine root production and functioning (Jackson et al. 1997). As a result, fine root systems contribute substantially to global terrestrial carbon (C) cycle and are a major reservoir of C (Vogt and Persson 1991). Gower et al. (1997) suggested that the proportion of assimilated C allocated to fine roots is greater in coniferous forests than deciduous forest, and fine root production is thought to constitute a greater proportion of total biomass production in boreal forests characterised by low soil fertility and temperatures (Ruess et al. 1996). However, in an analysis of over 140 sites in Europe, Finer et al. (2007) showed that standing fine root biomass is greater in beech (*Fagus sylvatica* L.) than Norway spruce (*Picea abies* L. Karst) and Scots pine (*Pinus sylvestris* L.), but the amount of fine root biomass is strongly linked to above ground biomass. Fine roots have a much shorter lifespan than coarse roots, as a consequence, their biomass varies both seasonally and due to changing environmental conditions. Because of their rapid production, senescence and decomposition, fine roots contribute significantly to forest soil C flux, making accurate measurement of their biomass and rate of production indispensable for closing the boreal forest C budget (Brunner and Godbold 2007).

In order to understand ecosystem functioning and nutrient cycling, but also to address current requirement for accurate models of forest C cycle, we need to quantify fine root production, mortality and associated turnover rates, as well as root necromass turnover rates. Since roots are the "hidden half" of most terrestrial ecosystems (Waisel et al. 1991), we are severely limited in our ability to directly observe growth, senescence, death and eventual decomposition of entire root systems. As researchers, we are handicapped by the imposed necessity of having to design experimental and sampling protocols without prior knowledge of fine root diameter distribution patterns, morphological characteristics, productivity and mortality, phenology and seasonality (Vogt et al. 1998). Root turnover rate is usually defined as the number of times fine root biomass is replaced each year and constitutes the inverse of root longevity. For example, a turnover rate of 2 yr^{-1} implies that the entire fine root system dies and grows back twice a year. Having said that, no widely accepted definition of fine root turnover exists as the actual calculation of the rate depends on the root system parameters being measured (biomass, necromass, root length, root growth and disappearance, etc.). Most commonly, the rate of turnover is established by comparing an estimate of annual fine root production to an estimate of standing fine root biomass. There are several methods for measuring fine root production in situ, all of which have drawbacks which influence the reliability of the observation, mostly related to the fact that a certain degree of soil environment disturbance is necessary before commencing the measurement. The challenge in estimating fine root turnover is obvious: one has to measure a process which is not clearly defined, is composed of two concurrent biomass fluxes (growth and dieback), while using methods which change environmental parameters and may affect the turnover rate as a result! The following sections outline the methods most commonly used to estimate root turnover, point out their advantages and drawbacks and provide some illustrations of their use.

18.2 Sequential Coring

Sequential soil coring was among the first methods used to estimate changes in fine root biomass and necromass over time (Nadelhoffer and Raich 1992). The method relies on the assumptions that live roots (biomass) enter a measurable dead root pool (necromass) and that the changes of both biomass and necromass pools are indicative of the rate of turnover. This method makes use of an auger corer to make destructive sampling of root systems. An intact soil core, usually around 5–10 cm in diameter, is extracted from the soil profile, all soil is washed away while both live (at the time of sampling) and dead roots are kept and weighed. This type of coring is very well suited for establishing standing root biomass, but understandably suffers from reliance on several assumptions when used to calculate root turnover. As the name suggests, root sampling has to be carried out repeatedly in the same location to establish annual and inter-annual variation in the amount of root biomass and necromass present in the soil. The more frequent the sampling, the more reliable root turnover estimates are thought to be, since shorter intervals between samplings lessen the importance of underlying assumptions inherent to this method. Fine root production and turnover are then obtained by several calculation methods which make use of sequential coring data, the most common being the minimum–maximum calculation. This equates the difference between the minimum and the maximum standing fine root biomass to the annual fine root production (Edwards and Harris 1977). Root turnover is then calculated by relating annual production to standing biomass.

As already mentioned, this method relies on several assumptions, the (in-) validity of which make this method subject to criticism (Majdi et al. 2007). As an indirect method of measuring fine root turnover, it assumes that fine root production and mortality occur at different times (e.g., roots are produced in the spring–summer and shed in autumn–winter in Northern forests). Further, it assumes that no unobserved peaks or troughs in standing root biomass occur between sampling dates as the existence of these would cause an underestimate of root production. It has been suggested that sequential coring is also sensitive to accumulation of sampling errors as the frequency of sampling increases (Sala et al. 1988) and that the principal problem affecting the accuracy of root turnover predictions is obtaining reliable estimates of in situ root decomposition. The reason

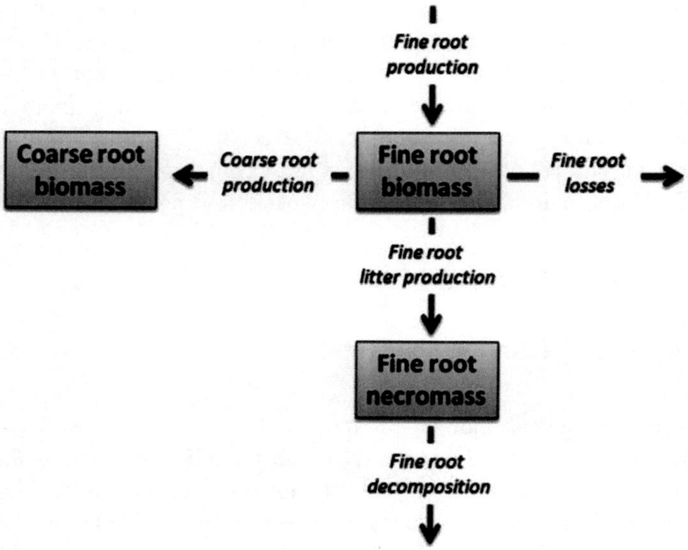

Fig. 18.1 A schematic compartment model describing biomass pools (*bold*) and fluxes (*italics*) associated with fine root turnover calculations

for the latter is the fact that most calculation methods used to obtain turnover rates from sequential coring data rely on necromass estimates. It is notoriously difficult to extract dead root biomass from the soil, mainly because an arbitrary threshold has to be imposed to distinguish root necromass from decomposed plant material. This artificially divides an otherwise continuous process of organic matter transformation from freshly abscised fine roots to soil humus.

A simple compartment model describing the pools and fluxes of fine root biomass in the soil is usually applied to analyse sequential coring data. An example of such a model can be seen in Fig. 18.1, the actual analysis is then carried out by applying a decision matrix which specifies the equations used to calculate fine root production between two sampling dates according to the magnitude and the direction of the changes in biomass and necromass pools.

18.3 Ingrowth Cores and Meshes

Ingrowth methods measure fine root production directly by introducing a "time zero" threshold within the soil profile which defines the initiation of root growth. They have been successfully used to measure root production in a variety of ecosystems and can be used to estimate root turnover in conjunction with root coring. This method is especially suitable for systems with fast root growth and for assessing effects of experimental manipulations on root production. The greatest advantage of ingrowth methods is their low cost and relative simplicity. Ingrowth cores, or ingrowth bags, define a 3-dimensional root-free zone within the soil profile, inside which root growth is observed for a defined period of time. The most common approach is to make use of holes and soil material created by soil coring. All root material is extracted from the cores; the remaining soil is then repacked and inserted as a root free cylinder or bag of soil into the soil profile separated from it by a mesh or a net. The manipulation of soil during root extraction and subsequent re-packing constitutes the biggest drawback of the ingrowth core method – the chemical and physical structure of the soil is severely altered and differs from the surrounding undisturbed soil. When using ingrowth cores to measure root production in, for example, arable crops this is not of major importance as the soil profile has been already thoroughly mixed (Steingrobe et al. 2000). However, when measuring root production in an undisturbed natural ecosystem, ingrowth coring is fraught with difficulty and is very tedious as each ingrowth core must be made to resemble the original soil profile. This is especially difficult in soils with multiple and well-defined horizons, as each of these have to be re-created inside the core. A further complication arises from the fact that fine roots colonising the soil inside an ingrowth core are initially growing in an absence of competition from other roots and may exhibit faster growth as a result. This may lead to an overestimation of fine root production, together with excessive fine root growth related to the injuries and cuts affected by coring. Once colonised, ingrowth cores are extracted from the soil. All roots found within the cores represent root production during the period of time for which the cores were left in the soil (Fig. 18.2).

To limit soil disturbance and the need to re-construct soil horizons, a modification of the ingrowth core method was developed by using a two-dimensional net inserted into the soil. Fahey and Hughes (1994) used the net method in a study of a Northern hardwood dominated by *Acer saccharum*. In this investigation, Fahey and Hughes (1994) used 2.5 mm mesh screens set at an angle of 45° in the soil, and used the number of root intersection per screen to gain a relative measure of fine

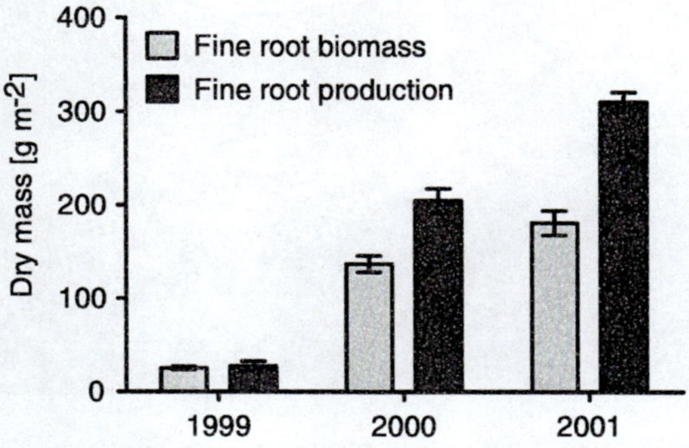

Fig. 18.2 Fine root biomass measured by sequential coring and fine root production measured by ingrowth cores in three poplar species grown from hardwood cuttings fir 3 years. Rate of turnover is calculated by dividing production by biomass

root production, but also of fine root disappearance. More recently, Lukac and Godbold (2010) proposed the use of nylon nets inserted vertically into the soil profile. Net insertion does not alter the soil profile properties as the only effect of inserting the net may be soil compaction in the immediate vicinity of the net. All soil horizons are left undisturbed, and it is possible to install numerous replicate nets within a relatively short time. Once a sufficient time has passed, the nets are cut out from the soil profile together with 2–3 cm of soil either side of the net plane. The width of the net, together with the thickness of the soil cut-out, then make the basis for a volumetric estimate of root production. The ingrowth net method does not disturb the soil and offers a way of checking for injury stimulated growth of fine roots. Since the net is extracted together with the surrounding soil, all roots which were severed during net installation and which sprouted numerous new roots are visible because they would be attached to the net (Fig. 18.3).

A critical point which has to be considered when using either ingrowth cores or nets is the period of time for which they are left in the soil. This is largely dependent on the expected rate of root growth, but also – perhaps more importantly – on the likely rate of root abscission and decomposition. If left in the soil for too long, the ingrown fine roots may start to turn over, resulting in an underestimate of root production. The period of ingrowth core incubation varies greatly with the type of vegetation which is being observed, but also with prevailing environmental conditions. Ecosystems where fast root growth and root decomposition is likely may need several cohorts of ingrowth cores or nets to capture all fine root production (Lukac et al. 2003). On the other hand, ingrowth cores may be left in

Fig. 18.3 Ingrowth net with fine roots of Norway spruce in Russian taiga. The net was left in the soil from September 2002 until April 2004, the protrusion of the net by growing fine roots is clearly visible

the soil for up to 2 years in places where plant growth and organic matter decomposition is limited by environmental conditions (Ostonen et al. 2005).

18.4 Minirhizotrons

Similarly to ingrowth method, the minirhizotron technique introduces a time threshold from which fine root growth is observed. The time frame for root observation using minirhizotrons is defined by their installation, settling-in period and their useful life which is dependent on the relative activity of the ecosystem where they are installed. As explained in Chap. 16, minirhizotrons are transparent tubes which can be inserted in the soil either vertically, horizontally or at an angle – 45° being the most commonly used inclination of minirhizotron tubes (Majdi and Andersson 2005). Due to the non-destructive nature of minirhizotron observations, the same root can be observed repeatedly in selected intervals, which can range from days to years. Roots are frequently observed from their initiation until their death and eventual disappearance. Still or video photography is usually used to record root activity in conjunction with software created to analyse resulting images. Minirhizotrons are usually used to estimate root production, longevity and mortality of fine root fractions smaller than 1 mm. By comparing images taken from the same minirhizotron on different dates, it is possible to identify and measure growth of individual roots over their lifespan. Image analysis allows for several quantitative parameters to be obtained, these data often include root length production, root length mortality, longevity, root length density and root diameter (Hendrick and Pregitzer 1996). Alongside quantitative data, minirhizotrons have the potential to generate qualitative information relating to root branching, colour or mycorrhizal colonisation.

Once installed in the soil, minirhizotrons cannot immediately be used to obtain reliable root production data. A suitable lag is advised before root observations commence to allow for root density equilibration and for root processes at the soil–minirhizotron interface to become representative of the wider soil. The reason for this lag is the soil disturbance caused by minirhizotron installation, which often involves soil drilling and compaction. Similarly to ingrowth cores, fine root growth (and possibly fine root longevity) in the immediate vicinity of the minirhizotron tube may be stimulated by increased nutrient availability caused by soil disturbance and by an absence of root competition (Joslin and Wolfe 1999). Minirhizotron observations suffer from two additional drawbacks; the difficulty of ascertaining root death and the problematic conversion of minirhizotron estimates of root growth to unit area basis. The first limitation is related to the fact that if roots are only classified as dead when they disappear, the overall root longevity may be overestimated (and root turnover underestimated). Roots may stop any physiological activity and even die without any apparent change to their appearance, persisting in minirhizotron images for prolonged periods of time. This limitation is notable in ecosystems where root decomposition is limited by temperature or lack

Fig. 18.4 A scanned image of a minirhizotron tube installed under *Alnus glutinosa* in Bangor, North Wales. The scan was taken in late April and shows older roots from previous year (*darker roots*), as well as recent fine root growth (*white roots*)

of oxygen. Conversely, many an example have been made of roots classified as dead, only to spring back to life again and start producing new root tips. The difficulty inherent in converting minirhizotron estimates to root production per unit area arises from the fact that all minirhizotron observations are carried out over a 2-dimensional surface of the tube, with the "depth" of the observation only assumed (Fig. 18.4).

Several statistical approaches were developed and applied to the analysis of time-series data originating from minirhizotron observations. Kaplan–Meier (Kaplan and Meier 1958) and Cox proportional hazard (Cox 1972) regression approaches are probably the most prominent examples of such methods. Kaplan–Meier regression can be used to estimate median survivorship and longevity of a single or multiple cohorts of fine roots. During the observation, fine roots are followed until the pre-determined end of the study period and classified as alive (uncensored) or dead (censored) at each observation point. The median longevity is then calculated from the data, the rationale of using this parameter is based on the fact that some roots are still alive at the end of the study and their longevity is therefore underestimated. The Kaplan–Meier analysis is complicated by physiologically induced gaps in root activity, such as the dormant season in temperate and boreal climates. Cox proportional hazard analysis fits a regression model to minirhizotron data, making use of and taking into account all available covariates. A "death hazard", which can be understood as an immediate probability of root mortality, is then established for each root cohort at any given time. An advantage of this method is the fact that it can be used to investigate the effects of treatments or physiological and environmental variables on root mortality (Green et al. 2005).

18.5 Radiocarbon Measurements

Radiocarbon measurements of root longevity make use of the "bomb" ^{14}C signal originating from the peak of atmospheric nuclear weapon testing in the 1960s. The premise of this method is that age of any organic matter containing carbon assimilated from the atmosphere can be ascertained by measuring its ^{14}C content. This is dependent on the background $^{14}CO_2$ concentration in the atmosphere at the time of fixation by photosynthesis and on the rate of ^{14}C decay. If no mixing of C pools held in organic matter occurs once these were built into organic tissues, an analysis of ^{14}C content can resolve dynamic changes that operate on annual to decadal time scales (Gaudinski et al. 2001). The analysis of ^{14}C data is, however, far from trivial, as the results are partially dependent on the particular mixing model used for data interpretation (Trumbore and Druffel 1995).

In a landmark study, Gaudinski et al. (2001) used ^{14}C analysis to estimate the mean age of fine root C in temperate forests by comparing ^{14}C content of fine roots with isotopic composition of tissue samples of known age. Mean age values obtained for biomass, necromass and mixed fine root C pools from deciduous and coniferous forests indicated that between 3 and 18 years have elapsed since C used to build these roots wax assimilated from the atmosphere. Such values do not correspond with root turnover rates obtained in temperate forests by other methods, since mean fine root turnover in temperate forests, measured by methods other than radiometric, is frequently reported to exceed 1. This implies mean root age shorter than 1 year. There are several alternative explanations as to why fine roots are rich in ^{14}C (a) fine roots, on average, do not have fast turnover rate and the 14C signature obtained by radiometric analyses is indicative of their true age; (b) fine roots are at least partially built from older organic material richer in ^{14}C which is recycled within the plant; or (c) some of the building materials used to construct fine roots is taken up from the soil and therefore is older and more enriched (Trumbore et al. 2006). Tierney and Fahey (2002) tested the second hypothesis by measuring ^{14}C content of roots which they observed growing in a minirhizotron tube. They reported that the ^{14}C signature of such newly grown roots was correspondent to the ^{14}C content of freshly assimilated photosynthate and therefore reflected the true age of these fine roots (Fig. 18.5).

The implications of greater fine root longevity (or slower fine root turnover) are great, especially for estimates of C allocation to belowground structures. In terms of biomass pool size, fine root do not normally represent a significant proportion of the total forest biomass. However, if this biomass pool is characterised by fast turnover rate, there must exists a sizeable C transfer to continuously re-build this pool. If the rate of turnover is slower than thought, as indicated by radiometric studies, this would greatly decrease the C requirement of fine root biomass. The issue is not settled at the present and is subject to considerable debate. Tierney and Fahey (2002) showed that using a normal age distribution underestimated mean root ages in minirhizotron applications and overestimated ages in isotopic applications. Even accepting that fine root age distribution is positively skewed towards the

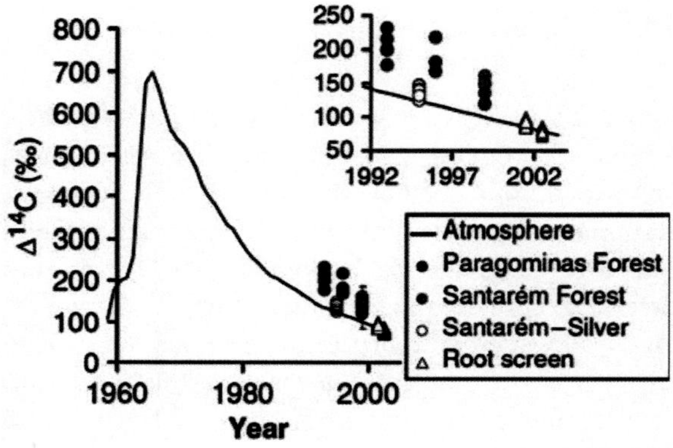

Fig. 18.5 Radiocarbon values in tropical tree fine root material sampled from the soil and from root screens. Roots collected from the soil represent a pool of unknown age, while roots collected from the screen are known to be younger than the screen installation. With permission from Trumbore et al. (2006)

younger roots – implying that there is a lot of short-lived roots and a few long-lived ones – the discrepancy between available estimation of fine root longevity is too great to be ignored and need to be studied further.

18.6 Conclusion

There are several methods for measuring fine root turnover or its inverse, fine root longevity. None of the currently available methods is considered fool-proof, chiefly because no method is able to measure fine root turnover directly. It can be argued that all fine root turnover measurement methods suffer from drawbacks which affect the accuracy of the observations. For this reason, it is advisable not to rely on a single method, but – resources permitting – to employ a combination of techniques to provide verification of turnover estimates. Given its importance for forest ecosystem functioning and significant contribution to the C cycle, there is a need to address the methodological difficulties and to improve the reliability of fine root turnover estimates.

References

Brunner I, Godbold DL (2007) Tree roots in a changing world. J For Res 12:78–82

Cox DR (1972) Regression models and life-tables. J Roy Stat Soc B-Stat Methodol 34:187–220

Edwards NT, Harris WF (1977) Carbon cycling in a mixed deciduous forest floor. Ecology 58:431–437

Fahey TJ, Hughes JW (1994) Fine-root dynamics in a Northern Hardwood Forest Ecosystem, Hubbard Brook Experimental Forest, NH. J Ecol 82:533–548

Finer L, Helmisaari HS, Lohmus K et al (2007) Variation in fine root biomass of three European tree species: Beech (*Fagus sylvatica* L.), Norway spruce (*Picea abies* L. Karst.), and Scots pine (*Pinus sylvestris* L.). Plant Biosyst 141:394–405

Gaudinski JB, Trumbore SE, Davidson A et al (2001) The age of fine-root carbon in three forests of the eastern United States measured by radiocarbon. Oecologia 129:420–429

Gower ST, Vogel JG, Norman JM et al (1997) Carbon distribution and aboveground net primary production in aspen, jack pine, and black spruce stands in Saskatchewan and Manitoba, Canada. J Geophys Res-Atmos 102:29029–29041

Green IJ, Dawson LA, Proctor J et al (2005) Fine root dynamics in a tropical rain forest is influenced by rainfall. Plant Soil 276:23–32

Hendrick RL, Pregitzer KS (1996) Temporal and depth-related patterns of fine root dynamics in northern hardwood forests. J Ecol 84:167–176

Jackson RB, Mooney HA, Schulze ED (1997) A global budget for fine root biomass, surface area, and nutrient contents. Proc Natl Acad Sci USA 94:7362–7366

Joslin JD, Wolfe MH (1999) Disturbances during minirhizotron installation can affect root observation data. Soil Sci Soc Am J 63:218–221

Kaplan EL, Meier P (1958) Nonparametric estimation from incomplete observations. J Am Stat Assoc 53:457–481

Lukac M, Calfapietra C, Godbold DL (2003) Production, turnover and mycorrhizal colonization of root systems of three Populus species grown under elevated CO_2 (POPFACE). Glob Change Biol 9:838–848

Lukac M, Godbold DL (2010) Fine root biomass and turnover in southern taiga estimated by root inclusion nets. Plant Soil 331:505–513

Majdi H, Andersson P (2005) Fine root production and turnover in a Norway spruce stand in northern Sweden: effects of nitrogen and water manipulation. Ecosystems 8:191–199

Majdi H, Nylund JE, Agren GI (2007) Root respiration data and minirhizotron observations conflict with root turnover estimates from sequential soil coring. Scand J Forest Res 22:299–303

Nadelhoffer KJ, Raich JW (1992) Fine root production estimates and belowground carbon allocation in forest ecosystems. Ecology 73:1139–1147

Ostonen I, Lohmus K, Pajuste K (2005) Fine root biomass, production and its proportion of NPP in a fertile middle-aged Norway spruce forest: comparison of soil core and ingrowth core methods. For Ecol Manage 212:264–277

Ruess RW, VanCleve K, Yarie J et al (1996) Contributions of fine root production and turnover to the carbon and nitrogen cycling in taiga forests of the Alaskan interior. Can J Forest Res 26:1326–1336

Sala OE, Biondini ME, Lauenroth WK (1988) Bias in estimates of primary production - an analytical solution. Ecol Model 44:43–55

Steingrobe B, Schmid H, Claassen N (2000) The use of the ingrowth core method for measuring root production of arable crops – influence of soil conditions inside the ingrowth core on root growth. J Plant Nutr Soil Sci 163:617–622

Tierney GL, Fahey TJ (2002) Fine root turnover in a northern hardwood forest: a direct comparison of the radiocarbon and minirhizotron methods. Can J Forest Res 32:1692–1697

Trumbore SE, Druffel ERM (1995) Carbon isotopes for characterizing sources and turnover of nonliving organic matter. In: Zepp RG, Sonntag CK (eds) Role of nonliving organic matter in the Earth's Carbon Cycle. Wiley, Chichester

Trumbore SE, Da Costa ES, Nepstad DC et al (2006) Dynamics of fine root carbon in Amazonian tropical ecosystems and the contribution of roots to soil respiration. Glob Change Biol 12:217–229

Vogt K, Vogt DJ, Bloomfield J (1998) Analysis of some direct and indirect methods for estimating root biomass and production of forests at an ecosystem level. Plant Soil 200:71–89

Vogt KA, Persson H (1991) Measuring growth and development of roots. In: Lassoie JP, Hinckley TM (eds) Techniques and approaches in forest tree ecophysiology. CRC, Boca Raton, FL

Waisel Y, Eshel A, Kalkafi U (1991) Plant roots: the hidden half. Marcel Decker, New York

Index

S. Mancuso (ed.), *Measuring Roots*,
DOI 10.1007/978-3-642-22067-8, © Springer-Verlag Berlin Heidelberg 2012

S

T